Recent Advances In Numerical Methods In Fluids

VOLUME 1

Recent Advances In Numerical Methods In Fluids

VOLUME 1

Edited by

C. TAYLOR

and

K. MORGAN

University College of Swansea, Swansea, U.K.

1980

PINERIDGE PRESS LIMITED

SWANSEA, U.K.

First Published in 1980 by

PINERIDGE PRESS LIMITED

91, West Cross Lane, West Cross, Swansea, U.K.

Recent advances in numerical methods in fluids.
Vol.1
1. Fluid dynamics - Mathematics
2. Numerical calculations
I. Taylor,C II. Morgan,K, b.1945
532'. 051'015117 QA911

ISBN 0=906674-07-7

Printed in Great Britain by

Redwood Burn Limited, Trowbridge and Esher

CONTRIBUTING AUTHORS

F.T. ADAMS, University of Washington, Seattle, Washington, U.S.A.

A.J. BAKER, University of Tennessee, Knoxville, Tennessee, U.S.A.

C.A. BREBBIA, University of Southampton, Southampton, U.K.

P. CHANG, University of Washington, Seattle, Washington, U.S.A.

D.J. EVANS, University of Technology, Loughborough, U.K.

B.A. FINLAYSON, University of Washington, Seattle, Washington, U.S.A.

A.D. GOSMAN, Imperial College of Science and Technology, London, U.K.

G. GLOTZ, Universität Karlsruhe, Karlsruhe, W. Germany.

P.M. GRESHO, Lawrence Livermore Laboratory, Livermore, California, U.S.A.

W.G. HABASHI, Concordia University, Montreal, Canada.

M. HANIN, Technion, Israel Institute of Technology, Haifa, Israel.

T.G. HUGHES, W.S. Atkins and Partners, Swansea, U.K.

R.L. LEE, Lawrence Livermore Laboratory, Livermore, California, U.S.A.

K. MORGAN, University College of Swansea, Swansea, U.K.

T.W. PATTEN, University of Washington, Seattle, Washington, U.S.A.

K. RAITH, Universität Karlsruhe, Karlsruhe, W. Germany.

C.W. RAPLEY, Sunderland Polytechnic, Sunderland, U.K.

V. REITMAN, Technion, Israel Institute of Technology, Haifa, Israel.

R.L. SANI, CIRES/NOAA, University of Colorado, Colorado, U.S.A.

W. SCHÖNAUER, Universität Karlsruhe, Karlsruhe, W. Germany.

D.B. SPALDING, Imperial College of Science and Technology, London, U.K.

C. TAYLOR, University College of Swansea, Swansea, U.K.

W. WOLFSHTEIN, Technion, Israel Institute of Technology, Haifa, Israel.

L.C. WROBEL, University of Southampton, Southampton, U.K.

Preface

The considerable pace at which numerical methods are being developed and utilised in the analysis of Flow Problems prompted the editors to initiate a series of texts devoted to the presentation of Recent Advances in such methods. The present volume is the first of a series in which each chapter deals with a specific aspect of the utilisation of numerical techniques to solve a particular flow problem. The numerical models range from those depicting potential flow to those where two equation models of turbulent flow are considered to be valid.

In Chapter 1 C.A. Brebbia and L.C. Wrobel illustrate how the boundary element method can be used to solve for the steady spatial distribution of potential within a bounded domain or a domain where a free surface exists. It is demonstrated that there is a considerable reduction in the number of equations to be solved and data required for input.

The second chapter is devoted to reporting the current status and suggestions for future developments in the solution of the incompressible Navier Stokes equations in two and three dimensions. The primitive variable approach is used and the temporal problem of vortex shedding behind a cylinder is analysed. An extension of such techniques to a three dimensional spatial domain is discussed.

The contribution by W. Schonauer, G. Glotz and K. Raith on the solution of laminar boundary layer flow is presented in Chapter 3. A self adaptive difference method for solving parabolic equations is presented and used to solve the incompressible laminar and hydraulic boundary layer equations including chemical reactions. 3-D boundary layer problems are also analysed.

Both Newtonian and Visco-elastic fluids are discussed in Chapter 4. The Galerkin and collocation methods are used and compared for efficiency. A non-linear convective Maxwell model is used to depict the visco-elastic properties of the fluid. D.B. Spalding reports on the I.P.S.A. (≡ inter-phase-slip-algorithm) numerical procedure for solving the equations of heat, mass and momentum transfer governing multi-phase flow phenomena. A range of problems can be analysed, using a staggered topological cartesian grid, which include steady and transient processes.

The interaction of an oblique shock wave with both laminar and turbulent boundary layers is the subject of Chapter 6. The basic features of the presentation centre on an A.D.I. method for the solution of relevant equations. Computations are performed for Reynolds numbers 3.0×10^5 and $\sim 1.2 \times 10^7$ at freestream Mach numbers of 2-4 and are compared with experimental results.

New computational methods for the solution of large sparse linear systems derived from the discretisation of parabolic and elliptic partial differential equations. By exploiting the sparseness of the resulting system of equations semi-direct implicit methods of solution are evolved which have specific advantages over direct methods. New matrix factorisation procedures are also investigated aimed at integration with future technological advancements in computer technology.

Numerical Methods in Turbomachinery is reviewed in depth by W.G. Habashi in Chapter 8. Traditional techniques as well as more recent developments are included. The author emphasises that it is imperative to specify tests and criteria by which various methods of solution can be judged. It is envisaged that this would enable designers to rationalise on a method of approach for a particular problem.

Flow in both laminar and turbulent boundary layers is investigated utilising the finite element technique in Chapter 9. The specified intent is to assess the accuracy and convergence of a finite element based implicit solution algorithm and to compare the linear finite element algorithm to the Crank Nicholson finite difference approach. The versatility of the advocated approach is demonstrated via several examples.

In Chapter 10, the one and two equation models of turbulence are incorporated into a finite element based approach to solve the mass flow and momentum equations which can be used to depict turbulent flow. Recent literature where such techniques are employed are included and the approach is demonstrated by solving well defined examples.

It is envisaged that the present volume should provide an insight into the recent developments in a variety of flow problems. In this context it is hoped that the text would prove useful to both practitioners and researchers in the specified subject areas.

C. TAYLOR
K. MORGAN

Note: The views expressed and conclusions drawn in the text are those of the contributing authors.

CONTENTS

CHAPTER 1

STEADY AND UNSTEADY POTENTIAL PROBLEMS USING THE BOUNDARY ELEMENT METHOD

C. A. Brebbia and L. C. Wrobel

Southampton University - U.K.

1. INTRODUCTION

Boundary solutions in the past have been used almost exclusively in solid mechanics [1]. However, they can be of considerable interest for many fluid mechanics problems [2]. This paper attempts to review the fundamentals of a boundary solution technique known as the boundary element method and indicates how it can be applied in practical cases.

In the boundary element method the external surface of the domain is divided into a series of elements over which the functions under consideration are assumed to vary in much the same way as in finite elements. This produces a series of nodal unknowns on the surface of the body only. These unknowns are related through the same influence functions used in the boundary integral equation method. The capability of using different types of surface elements is important as in the past, integral equation formulations were generally restricted to constant sources assumed to be concentrated at a series of points on the boundary. Furthermore, the technique is now being re-examined using weighted residual type statements, which allows for the method to be more easily extended to non-linear, time dependent and other complex problems.

This paper starts explaining how boundary elements can be applied to solve time independent potential problems. The methodology is presented in matrix form for the Poisson's equation considering boundary conditions including free surface. Time dependent problems are discussed in section 3, where a time dependent potential problem is presented. For this case the starting weighted residual statement has to include the initial in addition to the boundary conditions. This type of relation implies a new type of fundamental solution, which depends on time as well as the spatial coordinates. The approach indicated in this paper is of considerable interest as it eliminates the need of integrating step by step on time using finite difference or similar discretizations.

2. TIME INDEPENDENT POTENTIAL PROBLEMS

Let us first consider a potential function u over a domain Ω, where the following governing equation is to be satisfied,

$$\nabla^2 u - p = 0 \qquad \text{in } \Omega \tag{1}$$

The boundary conditions corresponding to this problem are of two types: essential conditions, such as $u = \bar{u}$ on Γ_1 and natural conditions such as $q = \frac{\partial u}{\partial n} = \bar{q}$ on Γ_2. The total boundary is $\Gamma = \Gamma_1 + \Gamma_2$.

For our numerical solution u will be approximated and we can minimize the error thus introduced by weighting the governing equation by a new function u^*. This gives,

$$\int_\Omega (\nabla^2 u - p)\, u^*\, d\Omega = \int_{\Gamma_2} (q - \bar{q})\, u^*\, d\Gamma - \int_{\Gamma_1} (u - \bar{u})\, q^* d\Gamma \tag{2}$$

where $q^* = \frac{\partial u^*}{\partial n}$.

After integrating by parts twice the terms in the Laplacian equation (2) becomes,

$$\int_\Omega (\nabla^2 u^*)\, u d\Omega - \int_\Omega p u^* d\Omega = - \int_{\Gamma_2} \bar{q} u^* d\Gamma - \int_{\Gamma_1} q u^* d\Gamma + \int_{\Gamma_2} u q^* d\Gamma + \int_{\Gamma_1} \bar{u} q^* d\Gamma \tag{3}$$

The function u^* is now assumed to be the fundamental solution of the equation, representing a concentrated potential at a point i, i.e. the solution of

$$\nabla^2 u^* + \Delta_i = 0 \tag{4}$$

where Δ_i is the Dirac delta function. Hence equation (3) can be written as,

$$u_i + \int_{\Gamma_2} u q^* d\Gamma + \int_{\Gamma_1} \bar{u} q^* d\Gamma + \int_\Omega p u^* d\Omega =$$

$$= \int_{\Gamma_2} \bar{q}u^* d\Gamma + \int_{\Gamma_1} qu^* d\Gamma \qquad (5)$$

It can be noted that for an isotropic three dimensional medium the fundamental solution of equation (4) is

$$u^* = \frac{1}{4\Pi r} \qquad (6)$$

where r is the distance from the point of application of the concentrated potential to the point under consideration. For two dimensions the solution is,

$$u^* = \frac{1}{2\Pi} \ \ell n \left(\frac{1}{r}\right) \qquad (7)$$

Other fundamental solutions for potential problems are given in reference [3].

Equation (5) is valid for any point inside the domain, but in order to formulate the problem as a boundary technique one needs to take this point to the boundary. For this case equation (5) can be written as,

$$c_i \, u_i + \int_{\Gamma} uq^* d\Gamma + \int_{\Omega} pu^* d\Omega = \int_{\Gamma} qu^* d\Gamma \qquad (8)$$

Note that we have written the integrals for the whole $\Gamma = \Gamma_1 + \Gamma_2$ boundary but that, depending on which part of the boundary we consider, the $\bar{u}$ or $\bar{q}$ values will be known. We can introduce these boundary conditions at a later stage.

The value of c_i is:

$c_i = 1$ for an internal point.

$c_i = 0$ for an external point.

$c_i = \frac{1}{2}$ for a boundary point on a smooth boundary.

Boundary elements

Equation (8) can now be applied on the boundary of the domain under consideration. This boundary can be divided into n elements. The points where the unknown values are considered are called 'nodes' and are similar to those of finite elements. The main difference is that now elements and nodes are defined only on the Γ boundary. The functions u and q over each boundary element are given by,

$$u = \underline{\Phi}^T \, \underline{u}^n$$
$$q = \underline{\Psi}^T \, \underline{q}^n \qquad (9)$$

and equation (8) is discretized as follows,

$$c_i\, u_i + \sum_{j=1}^{n} \int_{\Gamma_j} u q^* d\Gamma + \sum_{k=1}^{m} \int_{\Omega_k} p u^* d\Omega = \sum_{j=1}^{n} \int_{\Gamma_j} q u^* d\Gamma \qquad (10)$$

Note that m internal elements or cells need to be defined to compute the integrals in Ω but these elements do not introduce any further unknown and hence the problem is still a boundary problem.

We can substitute the u and q values given by (9) into (10) and carry out the integrations (usually numerically). This gives for each node, after assembling, the following equation,

$$c_i\, u_i + \sum_{j=1}^{n} \hat{H}_{ij}\, u_j + B_i = \sum_{j=1}^{n} G_{ij}\, q_i \qquad (11)$$

or

$$\sum_{j=1}^{n} H_{ij}\, u_j + B_i = \sum_{j=1}^{n} G_{ij}\, q_j \qquad (12)$$

where the B_i term is the result of having integrated the domain term and

$H_{ij} = \hat{H}_{ij}$ for $i \neq j$ and $H_{ij} = \hat{H}_{ij} + c_i$ for $i = j$

The whole set can be written in matrix form as follows,

$$\underset{\sim}{H}\, \underset{\sim}{U} + \underset{\sim}{B} = \underset{\sim}{G}\, \underset{\sim}{Q} \qquad (13)$$

Note that n_1 values of u and n_2 values of q are known on the boundary, hence equation (13) can be reordered in such a way that all the unknowns are on the left hand side, i.e.,

$$\underset{\sim}{A}\, \underset{\sim}{X} = \underset{\sim}{F} \qquad (14)$$

where $\underset{\sim}{X}$ is the vector of unknowns u's and q's.

When the surface is not smooth at the point 'i' the $c_i = \frac{1}{2}$ value is no longer valid. However, we can always calculate the diagonal terms of $\underset{\sim}{H}$ by the fact that when a uniform potential is applied over the whole boundary, the normal derivatives must be zero. Hence equation (13) in the absence of body forces becomes,

$$\underset{\sim}{H}\, \underset{\sim}{U} = \underline{0} \qquad (15)$$

where $\underset{\sim}{U}$ is a uniform potential. Thus the sum of all the elements of $\underset{\sim}{H}$ in any row ought to be zero and the value of the coefficient on the diagonal can be easily calculated once the off-diagonal coefficients are all known, i.e.

$$H_{ii} = - \sum_{\substack{j=1 \\ (i \neq j)}}^{n} H_{ij} \qquad (16)$$

Once the values of the boundary unknowns are found, the values of u and q anywhere can be calculated using equation (8). The internal fluxes are obtained by computing the derivatives of (8).

Free surface condition

Boundary conditions such as free surfaces can, in many cases, be treated simply by approximating the location of the surface, which is not known a priori, and iterating until the free surface condition is fulfilled.

In cases of seepage the condition is that at any point on the free surface the potential head u equals the elevation head from a fixed reference plane. This condition is also valid for cases such as flow over a spillway when the velocity head can be neglected, i.e., when the height of water over the nappe is small by comparison with the spillway height [4].

The problem is governed by a Laplace's equation for the potential u. For the solution, an initial guess is assumed for the free surface and the problem is solved for u applying the condition $q = 0$ (there are no flux through the free surface). The calculated potential at every nodal point on the free surface is then compared against its elevation; if the difference between these two values is greater than a maximum acceptable error, this difference is algebraically added on the elevation of the nodal point, and a new iteration is carried out.

The elements on the G and $\underset{\sim}{H}$ matrices - equation (13) - corresponding to the influence of fixed boundary nodes on other fixed boundary nodes will remain constant during the analysis, hence they can be computed once and stored. The potentials at internal points are calculated just after the correct position of the free surface has been found. With finite elements instead the internal potentials need to be computed during all iterations.

Applications

i) Heat conduction problem The surface temperature distribution of an infinitely long circular cylinder of radius R is specified as shown in figure (1). We wish to find the steady temperature of the cylinder. The formulation of the problem is,

$$\frac{\partial^2 T}{\partial r^2} + \frac{1}{r}\frac{\partial T}{\partial r} + \frac{1}{r^2}\frac{\partial^2 T}{\partial \theta^2} = 0 \qquad (17)$$

which is the Laplace equation in polar coordinates.

Note that there is a discontinuity on the temperature for the point (R, 0) and (R, Π). An analytical solution of the problem can be obtained as a Fourier series of the type [5],

$$\frac{T(r,\theta)}{T_o} = \frac{1}{2} + 2 \sum_{n=1}^{\infty} \frac{1}{n\Pi} \left(\frac{r}{R}\right)^n \sin n\theta, \; n = 1, 3, 5, \ldots \qquad (18)$$

Results for some internal points (figure 2) are compared in table 1 for two different discretizations using constant elements, assuming $T_o = R = 1$. The agreement is very good even for the coarse discretization. As the analytical solution is expanded as a sine series, it is not capable of predicting the distribution of fluxes. The normal fluxes will present a singularity as shown in figure (3) for the fluxes on the boundary. The numerical results, for both discretizations, tend to represent this singularity as shown in figure (4).

Point	B.E.M. (24 elements)	B.E.M. (48 elements)	Analytical
1	0.500	0.500	0.500
2	0.799	0.796	0.795
3	0.776	0.774	0.773
4	0.689	0.688	0.687
5	0.500	0.500	0.500
6	0.311	0.312	0.313
7	0.224	0.226	0.227
8	0.201	0.204	0.205

Table 1. Temperature at internal points

ii) <u>Lake circulation</u> Flow in lakes can be approximated to provide an initial estimate of the circulation, which can then be checked against the full shallow water equations. The flow is governed by the following linearised equations, obtained by neglecting the inertia terms in the momentum equations [6],

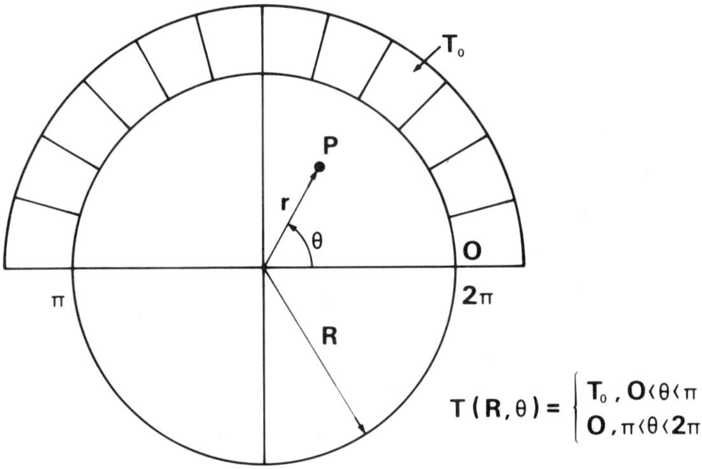

Fig 1. Surface temperature of an infinitely long circular cylinder.

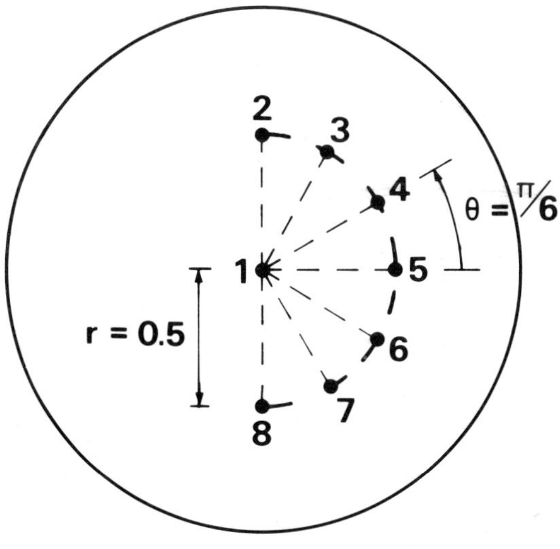

Fig 2. Location of internal points.

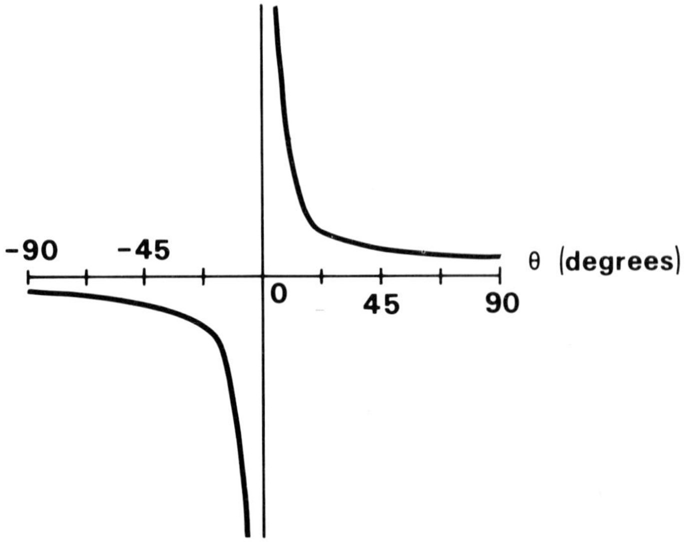

Fig 3. Singularity of the normal fluxes.

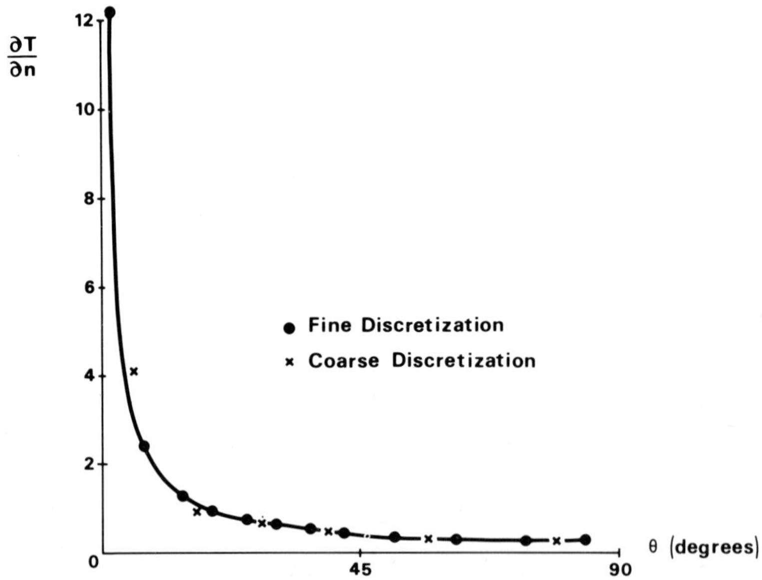

Fig 4. Results for the normal fluxes.

$$- f\, q_2 + \rho\, g\, H \frac{\partial \eta}{\partial x_1} + (\tau_1{}^s - \tau_1{}^b) = 0 \tag{19}$$

$$f\, q_1 + \rho g\, H \frac{\partial \eta}{\partial x_2} + (\tau_2{}^s - \tau_2{}^b) = 0$$

and the continuity formula,

$$\frac{\partial q_1}{\partial x_1} + \frac{\partial q_2}{\partial x_2} = 0 \tag{20}$$

where:
f = Coriolis parameter

q_1, q_2 = vertically integrated velocity components in x_1, x_2 directions

ρ = mass density

g = acceleration of gravity

$H = h + \eta$ = total depth of water

h = depth with relation to the mean water level

η = elevation of the free surface

τ^s = wind stresses

τ^b = bottom friction stresses

If the η values are much smaller than the h we can write $H \tilde{-} h$, hence,

$$- f\, q_2 + \rho gh \frac{\partial \eta}{\partial x_1} + (\tau_1{}^s - \tau_1{}^b) = 0$$

$$f\, q_1 + \rho gh \frac{\partial \eta}{\partial x_2} + (\tau_2{}^s - \tau_2{}^b) = 0 \tag{21}$$

Assuming the τ^b terms to be linearly proportional to the mean momentum components,

$$\tau_1{}^b = \gamma q_1,\ \tau_2{}^b = \gamma q_2 \tag{22}$$

we can cross-differentiate equations (21) and afterwards subtract both equations. Assuming that the derivatives of h are negligible (i.e. the bottom slope is small) this gives, taking continuity into consideration, the following equation,

$$\frac{\partial \tau_1{}^s}{\partial x_2} - \frac{\partial \tau_2{}^s}{\partial x_1} = \gamma \left(\frac{\partial q_1}{\partial x_2} - \frac{\partial q_2}{\partial x_1}\right) \tag{23}$$

One can propose a stream function Ψ such as,

$$q_1 = \frac{\partial \Psi}{\partial x_2}, \quad q_2 = -\frac{\partial \Psi}{\partial x_1} \tag{24}$$

and formula (23) becomes,

$$\nabla^2 \Psi = \frac{1}{\gamma} W \tag{25}$$

where

$$W = \frac{\partial \tau_1{}^s}{\partial x_2} - \frac{\partial \tau_2{}^s}{\partial x_1} \tag{26}$$

Note that we have included the Coriolis parameter but assumed it constant for all the lake, i.e. the lake is small enough to allow the neglect of local variations in the Coriolis forces.

We analysed, using the above formulation, the wind circulation in Lagoa dos Patos, Brazil (figure 5), discretizing the lake into 96 constant elements. As a first numerical example, we calculate the stream lines for the flow in and out of the lake without wind effects, taking $\Psi = 0$ for the west shore of the lake and $\Psi = 1$ for the east shore. Results are shown in figure (6). For this case, the governing equation becomes a Laplace equation.

If we consider the right hand side of equation (25) equal to 1, X_1 and X_2, this allows for a superposition of three different sets of results in order to obtain any solution of the type,

$$\nabla^2 \Psi = A + BX_1 + CX_2 \tag{27}$$

where the right hand side represents a quadratic wind stress distribution. This term is included in the analysis by dividing the domain into cells and integrating numerically over all the cells. Results for A = 1, B = −3, C = 0 are shown in figure (7).

iii) Free surface seepage through a soil block Figure (8) shows the boundary element discretization for a free surface flow through a block of porous medium. The upstream and downstream levels are maintained at 19.7 in and 1.3 in, respectively, above the horizontal impervious base. The boundary conditions of the problem are: (1) u = 19.7 on the upstream face; (2) $\partial u/\partial n = 0$ on the bottom (impervious) surface; (3) u = 1.3 on the downstream face; (4) $\partial u/\partial n = 0$ on the free surface, including node 11. Note that the initial guess of the free surface was arbitrarily assumed to be a straight line. The final position of the free surface was obtained by iteration as previously discussed.

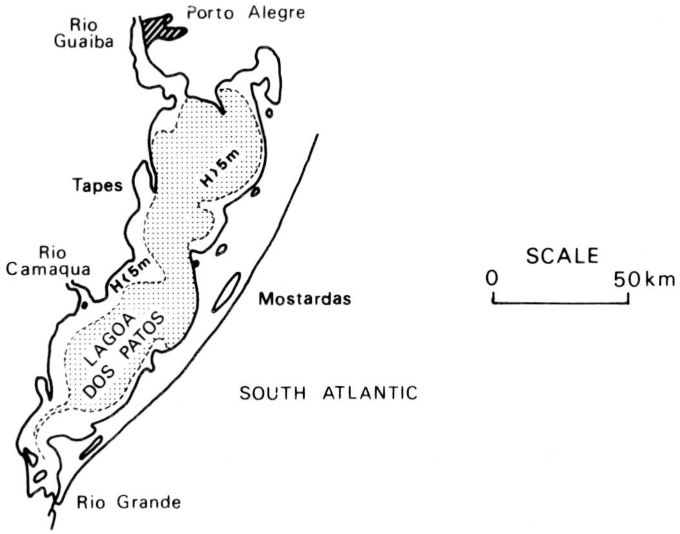

Fig 5. Geometry of Lagoa dos Patos.

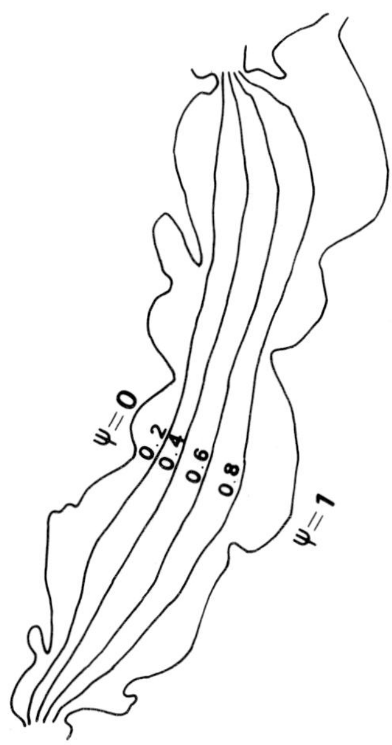

Fig 6. Flow pattern for potential flow.

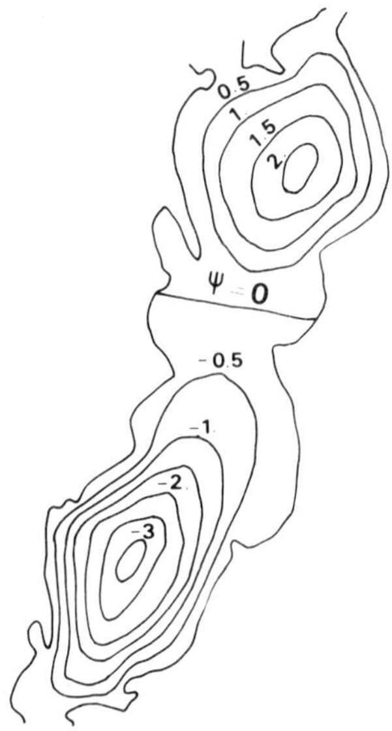

Fig 7. Two – celled mean circulation driven by a quadratic wind stress distribution.

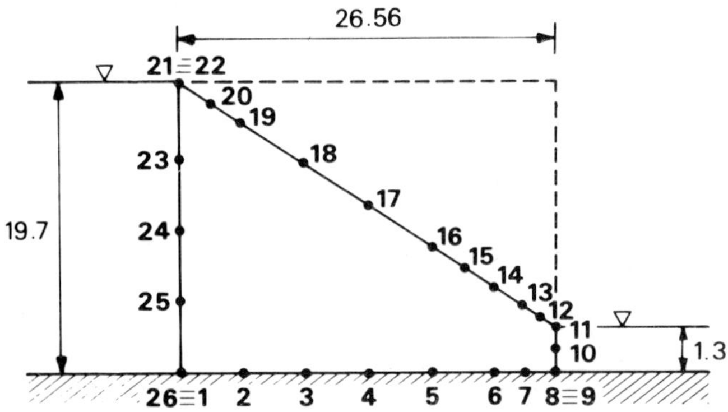

Fig 8. Boundary elements discretization of soil block with a free surface.

Results are shown in figure (9), compared against a finite element solution and an experimental solution obtained from an analogue model [7]. After the 7th iteration the difference between the computed potential head and the elevation of each nodal point along the free surface was less than 0.1%.

iv) Seepage through an earth dam Seepage of water through dams is a common problem of considerable practical importance. For the dam shown in figure (10), the water level on the upstream side is maintained at 6 ft below the crest and the downstream level was taken to be zero. The initial guess of the free surface was arbitrarily assumed and its choice has no significant effects on the final results. The final position of the free surface is shown in figure (10) and compared against finite element results [7].

3. TIME DEPENDENT PROBLEMS

In this section we will consider the case of having a time dependent potential problem governed by the following equation,

$$\nabla^2 u = \frac{\partial u}{\partial t} \tag{28}$$

with boundary conditions of the same type as previously seen and some initial conditions. As the problem is now time dependent, the equation will also be integrated with relation to time. Weighting expression (28) we have,

$$\int_0^t \int_\Omega (\nabla^2 u - \frac{\partial u}{\partial t})\, u^* \, d\Omega d\tau = \int_0^t \int_{\Gamma_2} (q - \bar{q})\, u^* d\Gamma d\tau - \int_0^t \int_{\Gamma_1} (u - \bar{u})\, q^* \, d\Gamma d\tau \tag{29}$$

where $0 \leqslant \tau \leqslant t$.

Integrating by parts twice we find,

$$\int_0^t \int_\Omega (\nabla^2 u^* + \frac{\partial u^*}{\partial t})\, u d\Omega d\tau - \left[\int_\Omega u u^* d\Omega \right]_{\tau = 0}^{\tau = t} + \int_0^t \int_\Gamma q u^* d\Gamma d\tau = \int_0^t \int_\Gamma u q^* d\Gamma d\tau \tag{30}$$

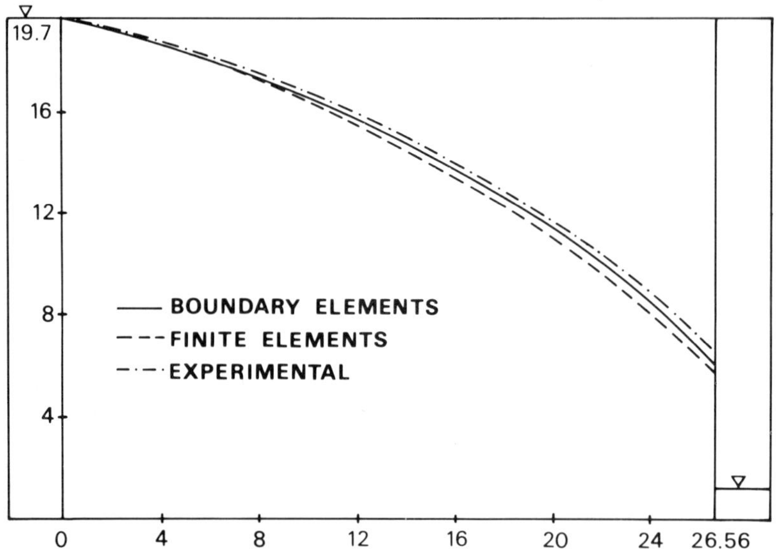

Fig 9. Converged solution for the free surface.

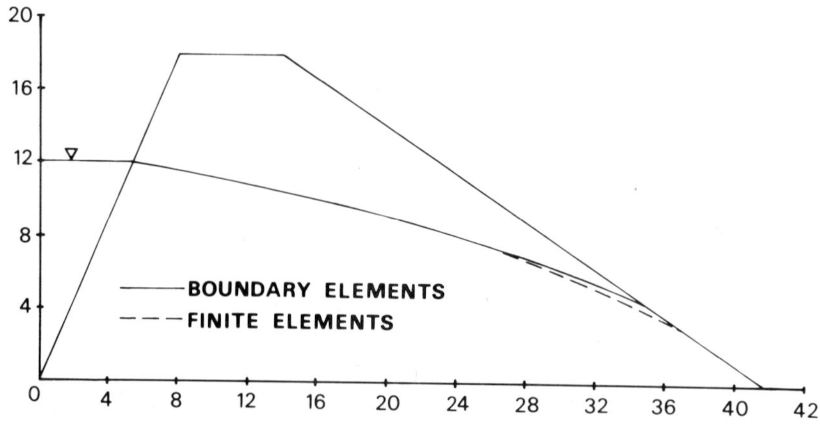

Fig 10. Seepage through an earth dam.

where the $\frac{\partial u^*}{\partial t}$ term was obtained integrating by parts with respect to time.

The corresponding fundamental solution for this equation is [8],

$$u^* = \frac{1}{(4\Pi\xi)^{d/2}} \exp\left(\frac{-r^2}{4\xi}\right) \tag{31}$$

where $\xi = t - \tau$ and d is the number of spatial dimensionality, i.e., d = 3 for three dimensional problems, etc. The fundamental solution possesses the properties,

$$\nabla^2 u^* + \frac{\partial u^*}{\partial t} = 0 \text{ in } \Omega \text{ for all } \tau \tag{32}$$

and for $\tau = t$:

$$\int_\Omega u^* d\Omega = \begin{cases} 0 \text{ for } r \neq 0 \\ 1 \text{ for } r = 0 \end{cases} \tag{33}$$

Substituting this solution into equation (30), one obtains for a point 'i',

$$c_i\, u_i + \int_o^t \int_\Gamma u q^* d\Gamma d\tau = \int_o^t \int_\Gamma q u^* d\Gamma d\tau + \left[\int_\Omega u u^* d\Omega \right]_{\tau = 0} \tag{34}$$

The last term in the above formula corresponds to the initial conditions at $\tau = 0$. Since the fundamental solution itself is time dependent, one does not need to propose an iterative scheme to solve time dependent problems as it is usually done in finite elements or finite differences.

The values of c_i are the same given in section 2. When the surface is not smooth at the point 'i', one can still calculate the diagonal terms of $\underset{\sim}{H}$ by the application of a uniform potential over the whole domain, which will give zero normal fluxes at the boundaries. Notice, however, that now equation (34) can be written as,

$$\underset{\sim}{H}\,\underset{\sim}{U} = \underset{\sim}{G}\,\underset{\sim}{Q} + \underset{\sim}{P} \tag{35}$$

where the vector $\underset{\sim}{P}$ depends on the potentials. Thus the coefficients on the diagonal of $\underset{\sim}{H}$ will be given by,

$$H_{ii} = - \sum_{\substack{j=1 \\ (i \neq j)}}^{n} H_{ij} + P_i \tag{36}$$

The numerical procedure necessary to develop a computer program for time dependent potential problems will now be shown, with reference to the two dimensional unsteady heat condution equation. In the absence of heat generators, this equation can be written as,

$$k\nabla^2 u = \frac{\partial u}{\partial t} \tag{37}$$

where u is temperature, t is time, $k = K/\rho c$, K is the thermal conductivity, ρ the mass density and c the specific heat. The fundamental solution of this equation is [8],

$$u^* = \frac{1}{4\Pi k\ (t-\tau)} \exp\left(-\frac{r^2}{4k\ (t-\tau)}\right) \tag{38}$$

One can divide the boundary of the domain under consideration into elements, as shown in section 2. One also needs now to assume a variation on time for the functions u and q. As these functions very much slower than u^* and q^*, it is a reasonable approximation to assume that they are constant over small intervals of time and perform the time integrations stepwise. Then, equation (34) becomes,

$$c_i\ u_i + \int_\Gamma u \int_{t_1}^{t_2} q^* d\tau d\Gamma = \int_\Gamma q \int_{t_1}^{t_2} u^* d\tau d\Gamma +$$

$$+ \left[\int_\Omega u u^* d\Omega\right]_{\tau=t_1} \tag{39}$$

Deriving equation (38) one finds,

$$q^* = \frac{\partial u^*}{\partial n} = -\frac{d}{8\Pi k^2 (t-\tau)^2} \exp\left(-\frac{r^2}{4k(t-\tau)}\right) \tag{40}$$

where d is the distance from the point 'i' under consideration to the boundary elements. So the inner integral on the l.h.s. of (39) can be carried out as follows,

$$\int_{t_1}^{t_2} q^* d\tau = \frac{d}{2\Pi k} \int_{t_1}^{t_2} -\frac{1}{4k(t_2-\tau)^2} \exp\left(-\frac{r^2}{4k(t_2-\tau)}\right) d\tau =$$

$$= -\frac{d}{2\Pi k r^2} \exp\left(-\frac{r^2}{4k(t_2-t_1)}\right) \tag{41}$$

To evaluate the time integral on the r.h.s. of (39) one needs first to make an appropriate change of variables. Calling,

$$x = \frac{r^2}{4k(t_2-\tau)} \tag{42}$$

the integral becomes,

$$\int_{t_1}^{t_2} u^* d\tau = \frac{1}{\Pi r^2} \int_{t_1}^{t_2} \frac{r^2}{4k(t_2-\tau)} \exp\left(-\frac{r^2}{4k(t_2-\tau)}\right) d\tau =$$

$$= \frac{1}{4\Pi k} \int_a^{\infty} \frac{e^{-x}}{x} dx = \frac{1}{4\Pi k} \quad Ei[a] \tag{43}$$

where

$$a = \frac{r^2}{4k(t_2-t_1)} \tag{44}$$

and $Ei[a]$ is the exponential - integral function, which can be evaluated by series, i.e.

$$Ei[a] = -C - \ln|a| + \sum_{n=1}^{\infty} (-1)^{n-1} \frac{a^n}{n.n!} \tag{45}$$

C being the Euler's constant, C = 0.57721566.....

This series, although convergent for all values of a, is not suitable for computation when $|a|$ is large as it will require a great number of terms in the calculations. To overcome this difficulty, one can use the following approximation [9],

$$Ei[a] \simeq U[a] = e^{-a} \sum_{n=1}^{\infty} (-1)^{n-1} \frac{(n-1)!}{a^n} \tag{46}$$

For large $|a|$, the terms of this series at first decrease very rapidly. Thus, up to the degree of accuracy required in engineering problems, this series is very suitable for computation. Table 2 shows the number of terms required for the original series (equation (45)) and the approximate series (equation (46)) to achieve the same order of accuracy. They are compared against values taken from reference [10], and the table demonstrates the advantage of using the approximate equation. However, care should be taken not to take too many terms for the approximation as the series is divergent.

Values of a	Reference [10]	Equation (45)	Number of terms	Equation (46)	Number of terms
3	0.1305×10^{-1}	0.1304×10^{-1}	18	0.1475×10^{-1}	3
4	0.3779×10^{-2}	0.3773×10^{-2}	21	0.3577×10^{-2}	4
5	0.1148×10^{-2}	0.1143×10^{-2}	24	0.1173×10^{-2}	5
6	0.3601×10^{-3}	0.3514×10^{-3}	27	0.3570×10^{-3}	6
7	0.1155×10^{-3}	0.1111×10^{-3}	30	0.1159×10^{-3}	7
8	0.3767×10^{-4}	0.4054×10^{-4}	32	0.3762×10^{-4}	8
9	0.1245×10^{-4}	0.2334×10^{-4}	35	0.1245×10^{-4}	9

Table 2. Values of Ei[a] obtained with single precision in an IBM 360/195

Equation (39) now becomes,

$$c_i u_i - \frac{1}{2\Pi k} \int_\Gamma \frac{u\ d}{r^2} \exp\left(-\frac{r^2}{4k(t_2-t_1)}\right) d\Gamma =$$

$$= \frac{1}{4\Pi k} \int_\Gamma q\ \mathrm{Ei}[a]\, d\Gamma + \frac{1}{4\Pi k(t_2-t_1)} \int_\Omega u \exp\left(-\frac{r^2}{4k(t_2-t_1)}\right) d\Omega \tag{47}$$

The integrations on space are carried out numerically. In this paper, constant elements were used and the boundary integrals calculated applying a 6 points Gauss quadrature formula. The integration over the domain was performed dividing it into triangular cells and using Hammer's quintic quadrature scheme [6]. The temperature on each integration point on the cells was evaluated at the end of each time step in order to determine the initial conditions for the next time step. It is worth noting that we had no numerical problems in doing so even for points located very near the boundary.

Applications

An infinitely long rectangular prism at zero initial temperature was analysed under different boundary conditions. The discretization employed for all cases was 36 equal elements and comparison of the results are made against available analytical solutions. The numerical values assumed are $\ell = 4$, $L = 5$ for the cross-section and for convenience the parameter k was taken to be unity.

First example

The first case studied was when the temperature over all the boundary was changed suddenly from zero to unity. The analytical solution of this problem is known [8], i.e.,

$$u = 1 - \frac{16}{\Pi^2} \sum_{n=o}^{\infty} \frac{(-1)^n}{2n+1} \exp\left(\frac{-k(2n+1)^2\Pi^2 t}{4L^2}\right) \cos\left(\frac{(2n+1)\Pi x}{2L}\right)$$

$$\sum_{n=o}^{\infty} \frac{(-1)^n}{2n+1} \exp\left(\frac{-k(2n+1)^2\Pi^2 t}{4\ell^2}\right) \cos\left(\frac{(2n+1)\Pi y}{2\ell}\right) \tag{48}$$

Temperatures at some internal points obtained using boundary elements are compared in figures (11) and (12) with the results obtained with the above analytical solution. The results are in very close agreement.

Second example

A case of mixed boundary conditions, i.e. two insulated opposing sides ($y = \pm\ell$) and the two others subject to a uniform unity temperature was also analysed. This

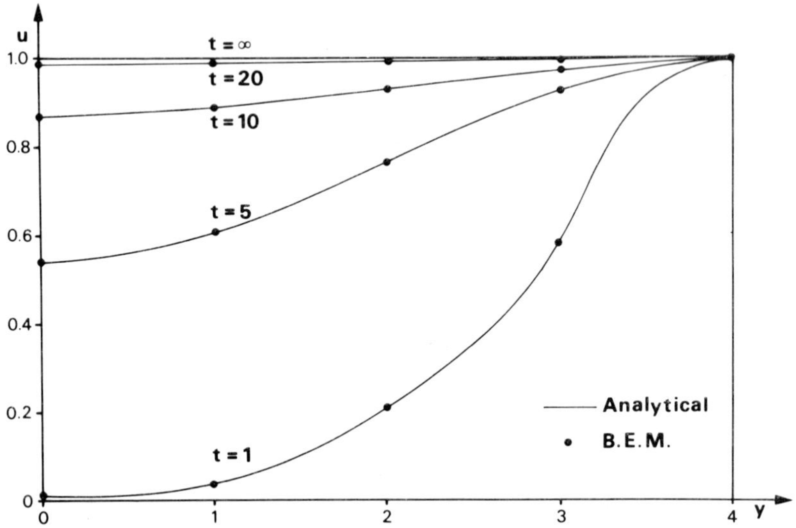

Fig 11. Temperatures along semidiagonal A-B.

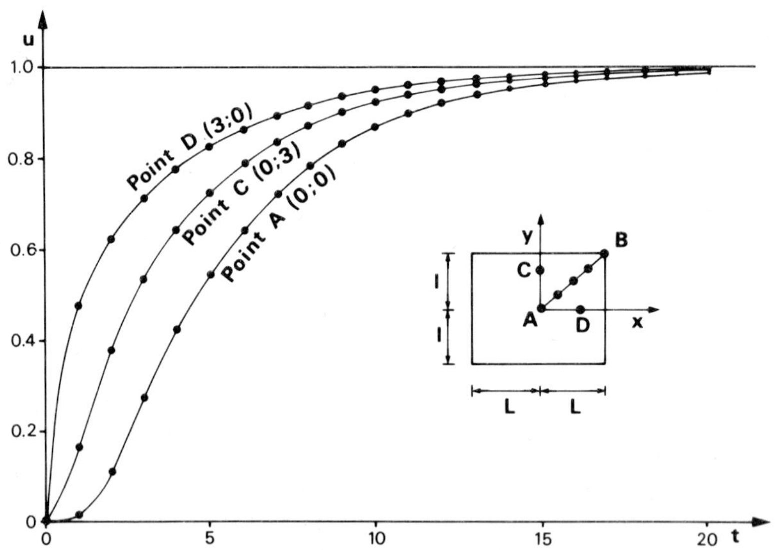

Fig 12. Temperatures at some internal points.

problem is equivalent to that of an infinite slab over both surfaces of which a uniform temperature is prescribed. The analytical solution, which is then independent of y, is [8],

$$u = 1 - \frac{4}{\Pi} \sum_{n=o}^{\infty} \frac{(-1)^n}{2n+1} \exp\left(-\frac{k(2n+1)^2\Pi^2 t}{4L^2}\right) \cos\left(\frac{(2n+1)\Pi x}{2L}\right) \tag{49}$$

Temperatures along the x-axis are shown in figure (13) and normal fluxes at the faces $x = \pm L$ in figure (14). The agreement for temperature and flux results is very close even in the proximity of $t = 0$ when there is a singularity for the fluxes.

Third example

The rectangular prism was finally analysed subject to a discontinuous temperature over the boundary. The temperature on the face $x = 0$ was suddenly changed to unity, all the other boundaries remaining at zero temperature. The analytical solution of this problem is [5],

$$u = 2 \sum_{n=0}^{\infty} \frac{(-1)^n}{\lambda_n \ell} \frac{\sinh \lambda_n (2L-x)}{\sinh 2\lambda_n L} \cos \lambda_n y \; +$$

$$+ \; 4 \sum_{k=0}^{\infty} \sum_{m=1}^{\infty} (-1)^{k+m} \frac{2\mu_m L}{\nu_k \ell} \frac{1}{4aL^2} \exp(-at) \sin\mu_m (2L-x)$$

$$\cos \nu_k y \tag{50}$$

where,

$$\lambda_n = \frac{(2n+1)\Pi}{2\ell}, \quad \nu_k = \frac{(2k+1)\Pi}{2\ell}$$

$$\mu_m = \frac{m\Pi}{2L}, \quad a = \nu_k^2 + \mu_m^2 \tag{51}$$

Although there is a discontinuity of temperature at the points $x = 0$, $y = \pm\ell$, which means that the flux at these points is infinity, the results once again are remarkably accurate, as can be seen in figures (15) and (16).

4. CONCLUSIONS

The main conclusion of this paper is that the boundary element method can be applied to solve many fluid mechanics problems and that it presents a series of advantages over more classical domain type techniques such as finite differences. One of the most interesting features of the

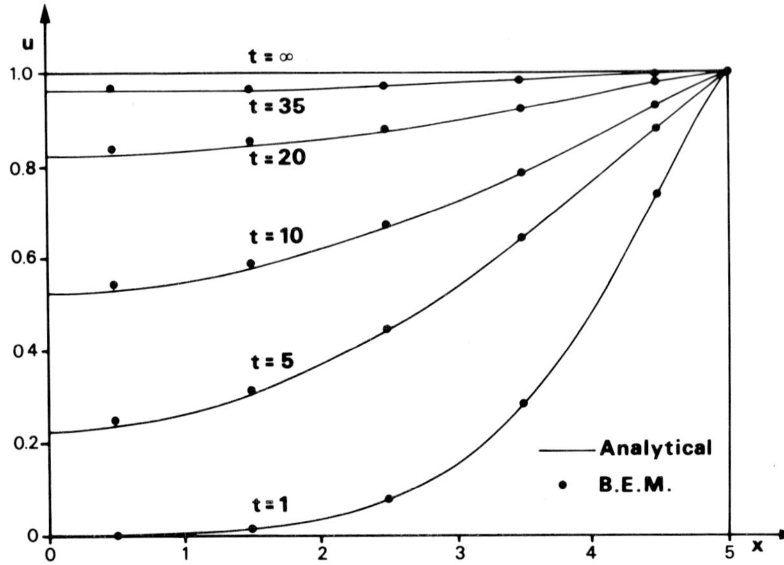

Fig 13. Temperatures along the x-axis.

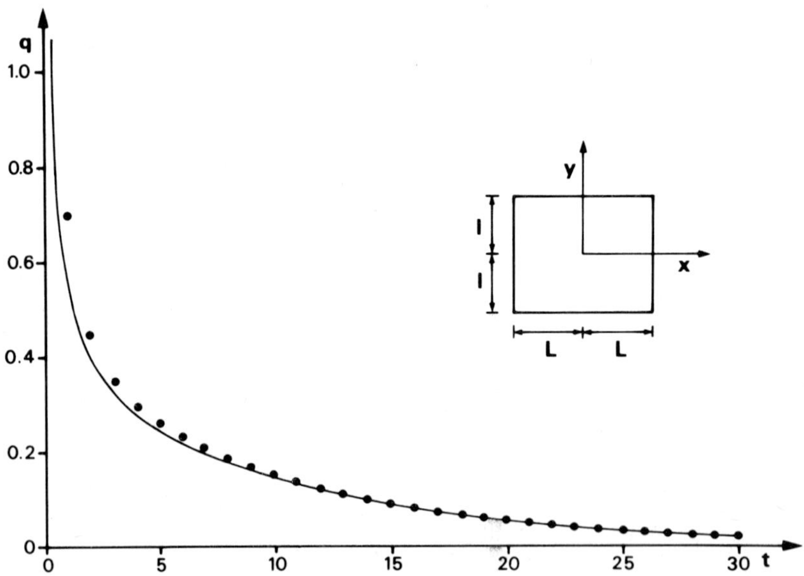

Fig 14. Fluxes at faces $x = \pm L$.

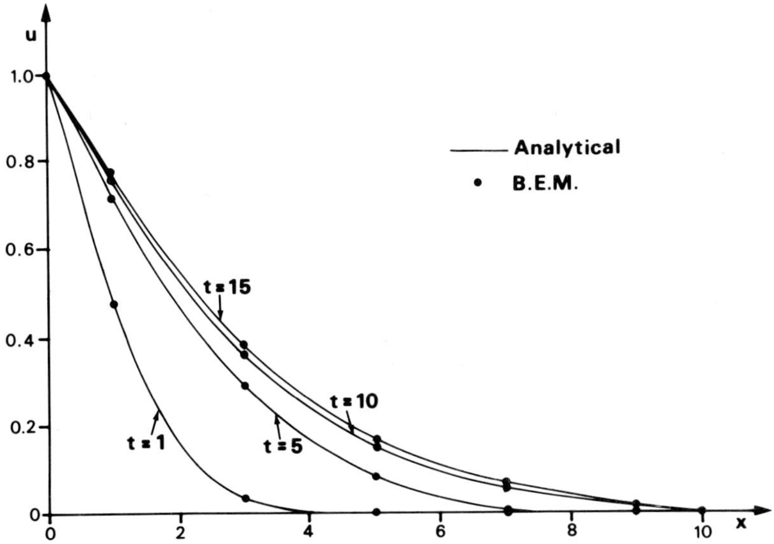

Fig. 15 Temperatures along x-axis

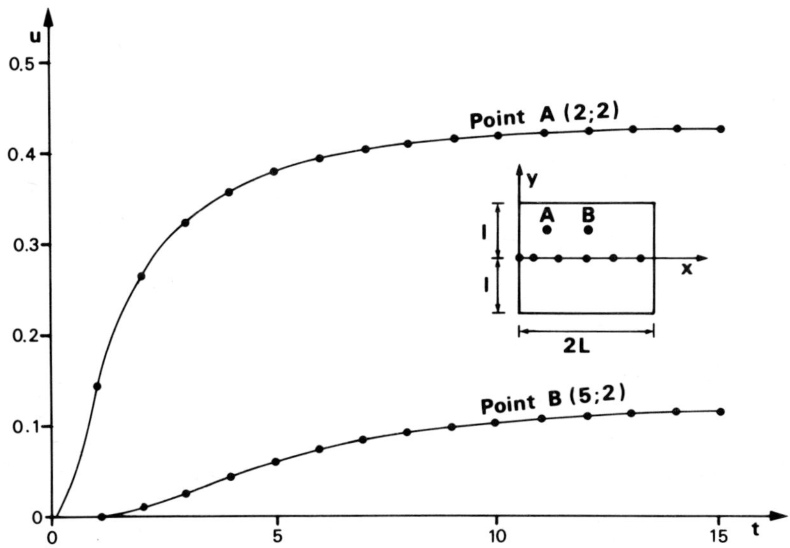

Fig 16 Temperatures at some internal points

technique is that a much smaller resulting system of equations and a considerable reduction in the data required to solve a problem can be achieved. These advantages are more marked in three dimensional problems. In addition, the numerical accuracy of boundary elements can be greater than that of techniques such as finite elements.

Boundary elements are also well suited for solving problems with infinite domains such as those frequently occurring in fluid mechanics and for which the classical domain methods are obviously unsuitable.

In this paper the method has been extended to solve unsteady potential problems. This was done by using a fundamental solution which is function of time as well as space. This new approach is of considerable interest as it eliminates the need of integrating step by step on time using finite differences or similar discretizations. In practical applications, however, time intervals are recommended, but as the examples showed, they can be very large without affecting the accuracy of the solution.

REFERENCES

1. Brebbia, C. A. (editor), "Recent Advances in Boundary Element Methods", Southampton, 1978, Pentech Press.

2. Brebbia C. A. and Wrobel, L. C., "Applications of Boundary Elements in Fluid Flow", Second Int. Conf. on Finite Elements in Water Resources, Imperial College, London, 1978, Pentech Press.

3. Brebbia, C. A., "The Boundary Element Method for Engineers", Pentech Press, London, 1978.

4. Chow, V. T., "Open-Channel Hydraulics", McGraw-Hill, U.S.A., 1959.

5. Arpaci, V. S., "Conduction Heat Transfer", Addison-Wesley Publishing Co., U.S.A., 1966.

6. Connor, J. J. and Brebbia, C. A., "Finite Element Techniques for Fluid Flow", Newnes-Butterworths, London, 1976.

7. France, P. W., Parekh, C. J., Peters, J. C. and Taylor, C. "Numerical Analysis of Free Surface Seepage Problems", Journal of the Irrigation and Drainage Division, A.S.C.E, Vol. 97, IR1, March 1971.

8. Carslaw, H. S. and Jaeger, J. C. "Conduction of Heat in Solids", 2nd ed., Clarendon Press, Oxford, 1959.

9. Bromwich, T. J. I., "An Introduction to the Theory of Infinite Series", 2nd ed., MacMillan and Co., London, 1931.

10. Spiegel, M. R. "Mathematical Handbook of Formulas and Tables", Schaum's Outline Series, McGraw-Hill, 1968.

CHAPTER 2

ON THE TIME-DEPENDENT SOLUTION OF THE INCOMPRESSIBLE NAVIER-STOKES EQUATIONS IN TWO AND THREE DIMENSIONS

Philip M. Gresho and Robert L. Lee
Lawrence Livermore Laboratory, Livermore, CA 94550, USA

and

Robert L. Sani
CIRES/NOAA, University of Colorado, Boulder, CO 80302, USA

INTRODUCTION

In this paper we report on our current status and reflect on future prospects regarding the numerical solution of the viscous, incompressible Navier-Stokes (NS) equations using the finite element method (FEM). Since we also plan to solve these equations in three-dimensions (3-D) (and for other well-known reasons), we have considered only the primitive variable ($\underline{u}$,P) formulation. We will present our two-dimensional (2-D) solution method in some detail, emphasizing the novel (but rigorous, and useful), mildly sophisticated technique developed for integrating the semi-discretized system of ordinary differential equations (ODE's). After demonstrating the capabilities of the 2-D code via a numerical example of Karman vortex shedding behind a circular cylinder, we venture somewhat from reality toward speculation as we discuss issues and options in 3-D, and summarize our current opinions and future plans.

NAVIER-STOKES EQUATIONS AND SPATIAL DISCRETIZATION

The equations of motion and continuity for a constant property, incompressible Newtonian fluid are the NS equations, written here in stress-divergence form,

$$\rho\left(\frac{\partial \underline{u}}{\partial t} + \underline{u} \cdot \nabla \underline{u}\right) = \nabla \cdot \underline{\underline{\tau}} \ . \tag{1a}$$

$$\nabla \cdot \underline{u} = 0, \tag{1b}$$

where the velocity $\underline{u}$ = (u,v) in two-dimensions (2-D) and $\underline{u}$ = (u,v,w) in three-dimensions (3-D), ρ is the fluid density, and

$$\tau_{ij} = - P\delta_{ij} + \mu\left(\frac{\partial u_i}{\partial x_j} + \frac{\partial u_j}{\partial x_i}\right) \tag{1c}$$

is the symmetric stress tensor, P is the pressure, μ is the fluid viscosity, and δ_{ij} is the Kronecker delta. Equation (1) can be used, given appropriate initial and boundary conditions, to obtain the velocity and pressure as functions of space and time.

In an incompressible fluid the pressure is an intrinsic and independent variable of the motion (Aris, 1962) and is not related to any thermodynamic equation of state; it is an implicit variable which instantaneously 'adjusts itself' in such a way that the incompressibility constraint, Equation (1b), remains satisfied. This is one characteristic of an incompressible flow that invariably makes the problem difficult to solve numerically; another is the nonlinear advection terms, $\underline{u} \cdot \nabla \underline{u}$.

The finite element spatial discretization of these equations is performed via the Galerkin method (i.e., on the weak form, see Appendix 1), wherein the velocity and pressure are approximated by

$$\underline{u}^h = \sum_{j=1}^{N} \underline{u}_j(t)\ \phi_j(\underline{x}) \tag{2a}$$

and

$$P^h = \sum_{j=1}^{M} P_j(t)\ \psi_j(\underline{x}) \quad , \tag{2b}$$

where there are N velocity nodes and M pressure "nodes" in the discretized domain and the superscript h indicates a finite dimensional approximation. The basis functions for velocity, $\phi_i(\underline{x})$, are continuous (C^0) piecewise polynomials which are one degree higher than those for the pressure approximation, ψ_i, (these can be either continuous or piecewise discontinuous (C^{-1}) polynomials) for reasons enumerated previously (e.g. Olson 1977; see also Sani et al. 1980).

We have experimented with a number of different quadrilateral 'elements' (basis functions) in 2-D, and found some to be more useful (robust) than others, although we are not aware of a single element without some 'problems'. Most experiments have been performed with steady-state codes and the following elements:

(1) 8-node quadratic velocity (serendipity), 4-node bilinear (C^0) pressure.

This element usually generates good velocities, but the pressures are often quite poor; hence, we have essentially abandoned it (see Huyakorn et al., 1978).

(2) 9-node biquadratic (Lagrange) velocity, 4-node bilinear (C^0) pressure.

This element appears to give somewhat better velocities, much better pressures (ibid.), and has been one of our favorite elements. It too, however, is often far from 'perfect'. For one thing, when streamlines are computed directly from the velocity field via the element boundary integral technique (Gartling, 1974), the lack of exact element mass balances (which also exists, albeit usually less seriously, for the serendipity element) causes an accumulation of error and poor streamline plots. We have 'solved' this minor problem, however, by employing a least squares velocity adjustment process which generates 'mass consistent' nodal velocities (see Sani et al., 1978) prior to computing streamlines. Another, more serious, problem has recently plagued this element, however, which appears to be related to the fact that it is relatively 'unconstrained' (there are fewer continuity constraints than the 'optimum', which we loosely define as "one continuity constraint equation per pair (2-D) of velocity equations - on the average"). The effects of this deficiency often show up in the form of poor velocities on 'coarse' grids or 'tough problems' (flows with detailed, fine-scale eddy structures) and is especially pronounced when buoyancy-coupled flows are attempted, using the Boussinesq equations. For further discussion, and examples of these problems, see Gresho et al. (1980) and Leone (1980).

(3) 4-node bilinear velocity, piecewise constant (C^{-1}) pressure.

This lowest order element has the advantages of simplicity, element-level mass balances, and an optimum constraint ratio, but suffers from one rather subtle problem: the spurious "checkerboard pressure mode." This pressure mode is manifest under certain boundary conditions and is a "zero energy mode" which totally pollutes the pressure field (the velocities are basically unaffected). Fortunately, we have learned how to filter this mode to recover useful pressure fields (see Lee et al., 1979; and Sani et al., 1980).

(4) 9-node biquadratic velocity, discontinuous (C^{-1}) bilinear pressure.

While we have only limited experience with this element, in which the pressure nodes are at the 2 x 2 Gauss points, it has been most favorable; this appears to be a very powerful element. It also enjoys the optimum constraint ratio, generates element-level mass balances, and is, of course, of higher-order accuracy than the previous one. It too, unfortunately, is afflicted by a spurious zero energy pressure mode; we have also devised a successful filter for this element (Sani et al., 1980).

(5) 9-node biquadratic velocity, 5-node (bilinear + constant) pressure.

This new element, which we designed (see Gresho et al., 1980) to relieve some of the shortcomings of "element 2" above, uses a 'mixed' pressure approximation consisting of a sum of the C^0 bilinear pressures of element 2 and the piecewise constant pressure of element 3; the overall pressure approximation is C^{-1} and it simultaneously guarantees element mass balances and improves the constraint ratio. This element performs better than element 2, but not as well as element 4, apparently because it still has too few constraint equations.

(6) and (7) "Penalty" elements.

If the penalty method is applied to the Navier-Stokes equations (see Marshall et al., 1978; Hughes et al., 1979; Bercovier and Engelman, 1979), it is possible to generate solutions which are very close to those of the "optimum" linear and quadratic elements, elements 3 and 4 above. This approach has the decided advantage of cost-effectiveness, since the pressure variable is eliminated a priori and the system of final equations contains only the velocities (the pressure is recovered, when desired, by an inexpensive post-processing scheme). These elements differ from elements 3 and 4 in that (a) the pressure modes are present in a different, somewhat disguised way (they are also smaller), and (b) the incompressibility contraint is not satisfied exactly. They appear to be capable of generating the most cost-effective solutions to the Navier-Stokes equations in 2-D (we will discuss 3-D applications later), at least in those cases (probably the majority) where the penalty approach is effective (as pointed out by Marshall et al. (1978), difficulties can arise if a physical parameter in the problem, such as the Rayleigh number, is very large; e.g. 10^8: in these cases, penalty methods may give poor results from computers which cannot perform accurate arithmetic with both small (O(1)) numbers and very large ($O(10^8-10^{12})$ numbers appearing simultaneously, since the penalty parameter should be much larger than the Rayleigh number. In these cases, the full mixed-interpolation elements, elements 3 and 4, are to be preferred).

In our time-dependent codes, we have thus far employed only elements 1, 2, 3, and 5, and have the most experience with elements 2 and 3.

Returning now to the equation formulation, the expansions in Equation (2) are inserted into the weak (Galerkin) form of Equation (1) (the form which permits C^0 velocity and C^{-1} pressure approximations; see Appendix 1). The result is a coupled system of nonlinear, first-order ODE's in time — the Galerkin FEM equations, written here in a compact matrix form:

$$M\dot{u} + [K + N(u)]u + CP = f \quad , \tag{3a}$$

$$C^T u = 0 \quad , \tag{3b}$$

where now u is a global vector of length DN (D is the dimension of the physical space, 2 or 3) containing all nodal velocity components, P is a global M-vector of nodal pressures, and f is a global vector (length DN) which incorporates all imposed traction boundary conditions (discussed below) on velocity. (Specified nodal values of velocity are imposed directly on the assembled system, typically by deleting the equation in question and transposing all coupling terms to the right hand side (RHS) or by replacing the equation by one with all zeros except for unity 'on the diagonal' and the specified value placed on the RHS.) M is the DN x DN mass matrix, K is the DN x DN viscous stress matrix, N(u) is the DN x DN advection matrix, C is the

DN x M pressure gradient matrix and its transpose, C^T, is the M x 2N divergence matrix. The detailed matrix definitions are also presented in Appendix 1.

It is noteworthy that the Galerkin method leads to discretized approximations of the advection terms which are predominantly of a centered-difference nature, but are actually 'indefinite' with respect to the introduction of artificial dissipation (numerical diffusion); i.e., the quadratic form, $u^T N(u)u$, can be positive or negative. It is, however, easy to generate advection approximations which are absolutely nondissipative (see Lee et al., 1980); if the advective term $\underline{u} \cdot \nabla \underline{u}$ in Equation (1a) is replaced by $1/2\,[\underline{u} \cdot \nabla \underline{u} + \nabla \cdot (\underline{u}\,\underline{u})]$, where $\underline{u}\,\underline{u}$ is the dyadic product, which is equivalent -- in the continuum -- by virtue of Equation (1b), the Galerkin method will generate an 'advection' matrix which guarantees that $u^T N(u)u = 0$, since $N(u)$ is then skew-symmetric. We have recently experimented with this conservation form and compared it to the simpler advective form (ibid.); the results, while not yet conclusive, seem to indicate that the simpler form generally yields results of equivalent accuracy, usually displays little dissipation, but can lead to a nonlinear instability in long-time inviscid calculations. On the other hand, it is also possible to generate advection approximations which are very dissipative; these so-called "upwind" methods (most of which are ad hoc) have received much attention recently (see the ASME Symposium volume referred to in the Gresho and Lee (1979) reference), and are usually claimed to generate more accurate solutions because the results are always smooth. We do not subscribe to the 'smooth is beautiful' or 'smooth implies accurate' notions and do not believe in upwinding, or any other 'wiggle suppressants' (see Gresho and Lee (1979) and Leone and Gresho (1979) for further discussion of these issues).

Boundary conditions

Before discussing time integration methods and initial conditions for Equation (3), it seems relevant to clarify the role of boundary conditions (BC's), which are "built-in" to these spatially discretized equations; in particular, we wish to address appropriate and inappropriate BC's.

Velocity The computational domain is bounded by a piecewise smooth boundary, Γ, along which only the following BC's are permissible: specify u_n or f_n and u_τ or f_τ, where u_n, u_τ are the normal and tangential components of $\underline{u}$ on Γ, and f_n, f_τ are the respective components of the surface traction force and correspond to the natural boundary conditions associated with the weak form of Equation (1) (see Appendix 1); i.e.,

$$f_n = -P + 2\mu \frac{\partial u_n}{\partial n} \tag{4a}$$

is the normal stress and

$$f_\tau = \mu \left(\frac{\partial u_n}{\partial \tau} + \frac{\partial u_\tau}{\partial n} \right) \tag{4b}$$

is the shear stress, where $\underline{n}$ is the outward pointing unit normal and τ

is the unit tangent vector. We have found that traction-free ($f_n = f_\tau = 0$) BC's are quite useful at 'artificial' outflow boundaries, especially when the inertial terms dominate (i.e., at 'high' Reynolds number, Re; traction-free outflow BC's for low Re (or Stokes) flow, however, are not as useful and can, in fact, generate surprising results). If the outflow wants to be unidirectional at the outlet (a situation in which the inertial terms are zero), such as in a fully developed channel, these traction-free BC's are not so useful since they are not compatible with unidirectional flow; here one could use $u_\tau = f_n = 0$, but this BC can cause oscillations (see Leone and Gresho, 1979; also Hutton and Smith, 1979). If the flow field has known symmetry, the application of $u_n = f_\tau = 0$ on the symmetry boundary is useful and appropriate.

Pressure In general, the application of pressure BC's (i.e., the specification of pressure along a portion of Γ) is inconsistent with Equation (1) and thus 'illegal'. It is, unfortunately, too often assumed in the literature that an applied pressure BC (e.g. at an inflow or outflow boundary) is legitimate. In an incompressible flow, the pressure is an implicit variable whose value (pointwise) is determined by the requirement that $\nabla \cdot \underline{u} = 0$. If the pressure is presumed to be known in any portion of the flow field (including Γ), it follows that it is also presumed known that $\nabla \cdot \underline{u} = 0$ there, and one can (must, in fact) discard the continuity equation at any point where P is presumed known. This argument carries over directly to the discretized equations; if the pressure at a boundary node is to be specified, the continuity equation (Equation 3b) must be discarded at that node. The implication of this approach should be clear; whereas there is no algebraic inconsistency associated with such a procedure, there is clearly a physical inconsistency. Since the mass balance constraint must be disregarded at all "specified pressure" nodes, it follows that the required mass balance will (in general) not be obtained at these nodes, nor (of course) globally.

An alternative argument can be presented based on the Poisson equation for pressure, obtained by forming the divergence of Equation (1a), using (1b) and (1c); viz,

$$\nabla^2 P = -\rho[\nabla \cdot (\underline{u} \cdot \nabla \underline{u})] \quad . \qquad (5)$$

While it is well known that any linear combination of Dirichlet (specified pressure) and Neumann (specified normal derivative of pressure) BC's are applicable to the Laplacian operator, it must be recognized that this is a higher-order, derived equation, whose BC's must also be derived from the original equations. Since these are all Neumann-type boundary conditions (from Equation (1)), it follows that specified pressure boundary conditions are generally not consistent with the fundamental governing equations.

If one has an urge to specify pressures, he must suppress it and, instead, be satisfied with approximating the pressure by specifying the normal traction, Equation (4a). In this case the problem is still well-posed and, importantly, all of the continuity equations are retained, as required. The resulting pressures will generally be reasonably close to those which were desired to be specified (i.e., the natural BC causes $2\mu\partial u_n/\partial n$ to be small; at least for Re >> 1, since, in dimensionless form, $f_n = -P + 2Re^{-1}\ \partial u_n/\partial n$), and the correct mass balances will be obtained, both locally and globally.

In Figure 1 we demonstrate the type of solution which obtains by specifying an inlet pressure for the case of steady flow around a cylinder at Re = 20; in addition to the rather suspicious looking streamline patterns, the integrated exit flow is only 88% of the inlet flow. Figure 2 shows the solution using an appropriate inflow BC, and displays the proper global mass balance.

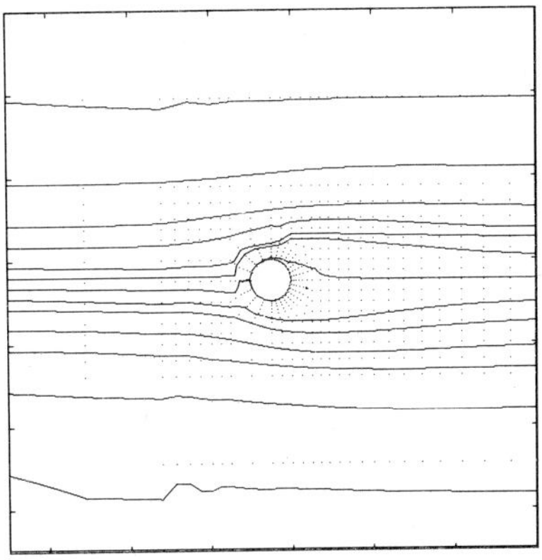

Figure 1. Steady flow around a cylinder at Re=20; the relevant BC's are: u=1, v=0, P=0 at inlet, $f_n=f_\tau=0$ at outlet.

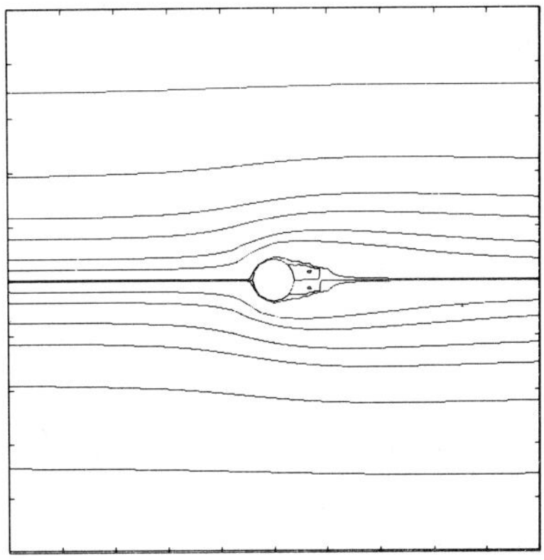

Figure 2. Same as Fig. 1 except for the inlet BC which is: u=1, v=0. The streamlines very close to the cylinder are $\psi = \pm.005$ (the separation streamline is $\psi=0$).

Finally, for the special, but common case of a simulation in which u_n is specified on all of Γ (e.g. contained flow), a pressure datum is required since the pressure is then only obtainable up to an arbitrary additive constant. In this case, one nodal pressure can be specified and the corresponding continuity equation omitted, since it is consistently redundant (assuming, of course, that the specified normal velocities satisfy $\int u_n\, d\Gamma = 0$, a consistency requirement from Equation (1b)).

A SOLUTION METHOD FOR TWO-DIMENSIONAL FLOWS

The role of the pressure and the algebraic constraint equations, (3b), can perhaps be more clearly appreciated by expressing Equation (3) in the equivalent and so-called time-singular form, using partitioned matrices;

$$\begin{bmatrix} M & 0 \\ 0 & 0 \end{bmatrix} \begin{Bmatrix} \dot{u} \\ \dot{P} \end{Bmatrix} + \begin{bmatrix} K+N(u) & C \\ C^T & 0 \end{bmatrix} \begin{Bmatrix} u \\ P \end{Bmatrix} = \begin{Bmatrix} f \\ 0 \end{Bmatrix} , \tag{6}$$

in which the matrix multiplying $(\dot{u}\ \ \dot{P})^T$ is obviously singular. Hence, it is clearly not possible to solve Equation (3) by any purely explicit (time-marching) procedure. This fact, combined with the mass matrix which couples the nodal accelerations, suggests that implicit time integration techniques are appropriate. While they are mandatory for the treatment of the pressure and continuity equation, the remaining terms in the momentum equation (Ku and N(u)u, the viscous and advective terms in Equation (3a)), could be treated explicitly or semi-implicitly if desired. There are of course other reasons to consider implicit methods, most of them focusing on their greater stability — which can often be used to advantage when an explicit method would impose unrealistically (or unaffordably) small time steps, a disadvantage which could seriously hinder the deployment of isoparametric elements to generate efficient, graded meshes since time step limits are often functions of the smallest mesh size. Unfortunately, this greater stability exacts a significant computational price; viz, implicit techniques applied to nonlinear ODE's generate nonlinear algebraic systems whose solution must be obtained once per time step.

Faced with the prospect of solving large, nonlinear algebraic systems and the fact that, as a minimum, the pressure must be solved implicitly, one could (and should) consider compromise techniques. The compromises (from a fully implicit method) generally result in exchanging stability advantages for computational simplicity. We will discuss some of these compromises later, in the context of 3-D flows, after we present and demonstrate what we believe to be a robust, no-compromise implicit method in 2-D.

The following sections will describe in some detail and demonstrate these time integration techniques; here we provide an overview and point out that the scheme is definitely not ad hoc — it is firmly based on the well-developed theory of ODE's from which we extracted a small, but useful portion. By an appropriate combination of two common, second-order-accurate integration techniques, the implicit trapezoid rule (TR) and an explicit Adams-Bashforth (AB) formula, we have developed a stable time integration scheme in which we cheaply and automatically vary the size of the time step, based solely on temporal accuracy requirements, by obtaining a good estimate of the local (single step) time truncation error. This also provides us with useful insight into the prevailing 'physics' of the flow by the associated monitoring of its time scale. For example, the physics may require a small time step to follow the formation of a boundary layer or the shedding of vortices. Or a large time step may be sufficient to follow the slow growth of an attached eddy or a flow which is approaching steady state. In either case, the algorithm will (usually) automatically select the appropriate time step, thus providing a cost-effective method in that the step size is increased whenever possible and decreased only when necessary. The nonlinear systems engendered by the TR algorithm are solved using a "one-step" Newton method, which requires the construction and solution of just one linear system per time step; this unsymmetric linear system is solved via Gaussian elimination without pivoting. The overall time integration scheme can be conveniently described in four stages, as discussed below.

Initialization

The appropriate initial condition for Equation (1) is: any solenoidal velocity field, $\underline{u}_o$ (one satisfying $\nabla \cdot \underline{u}_o = 0$). No initial conditions are required for the pressure since there will be a unique pressure field accommodating any solenoidal velocity field. In the continuum, this pressure field is given by Equation (5) and the appropriate (Neumann !) BC's. In the FEM discretized system, the analogous initial condition is: any velocity field which satisfies the discrete analog of Equation (1b); viz Equation (3b). In fact, if the initial discretized velocity, u_o, does not satisfy $C^T u_o = 0$, the ODE system given by Equation (3) is ill-posed. If $C^T u_o \neq 0$, it is easy to show that the following results will obtain (using TR -- an analogous result will be obtained for any other implicit ODE scheme, such as backward Euler -- recall that explicit ODE methods are not applicable) at the end of the first time step as $\Delta t \to 0$:

$$u_1 \to u_o - M^{-1}(C^T M^{-1} C)^{-1} (C^T u_o) \tag{7a}$$

and

$$(P_1 + P_0)/2 \simeq (C^T M^{-1} C)^{-1} (C^T u_o)/\Delta t \quad ; \tag{7b}$$

i.e., there is a finite jump in velocity, proportional to $C^T u_o$, and the pressure becomes arbitrarily large (the system is ill-posed; a consistent ODE system will give $u_1 \to u_o$, $P_1 \to P_0$ as $\Delta t \to 0$). We have also confirmed these theoretical results numerically: the non-dissipative TR caused essentially undiminished ringing of u and P throughout the integration period; whereas the dissipative backward Euler scheme caused the spurious short time results to be rapidly damped (the system recovers and can even approach steady-state, although the transient 'solution' is devoid of physical significance.)

In order to obtain the initial compatible (with $C^T u_o = 0$) pressure field, we mimic the solution of Equation (5) by solving Equation (3a) and a time-differentiated form of Equation (3b) (which is valid since $C^T u = 0$ for all time); i.e., we must solve

$$\begin{bmatrix} M & C \\ C^T & 0 \end{bmatrix} \begin{Bmatrix} \dot{u}_o \\ P_0 \end{Bmatrix} = \begin{Bmatrix} f - [K+N(u_o)]\, u_o \\ 0 \end{Bmatrix} \tag{8}$$

simultaneously for P_0 and the initial acceleration, $\dot{u}_o$. The extra cost associated with this initialization procedure is amortized over the rest of the computation and also provides additional necessary information because our time step control strategy requires the acceleration vector; the amortization results because Equation (8) needs to be solved only once per problem, the subsequent acceleration vectors being obtained from a simple recursion formula, which will be demonstrated in due course.

Predictor equation for velocity

The variable-step, second-order AB formula applied to $\dot{y} = f$ is (Shampine and Gordon, 1975)

$$y^p_{n+1} = y_n + \frac{\Delta t_n}{2}\left[\left(2 + \frac{\Delta t_n}{\Delta t_{n-1}}\right)\dot{y}_n - \frac{\Delta t_n}{\Delta t_{n-1}}\dot{y}_{n-1}\right] \quad . \tag{9}$$

Since this is an explicit formula, it can only be applied to the velocity in Equation (3); i.e., to (3a). Note that two history vectors of acceleration are required; these are obtained simply and recursively from the TR (corrector) results to be described below. This predictor step is very cheap (relative to the TR step) and, since it performs two important functions, quite cost-effective (it provides a very good predictor for TR -- which contributes substantially to the success of our one-step Newton method, described below — and it permits control of local error). Finally, since $\dot{y}_{n-1}$ is required, the AB formula cannot be applied until the second step (n = 1), where $\dot{u}_0$ is available from Equation (8). Error estimation therefore begins at the completion of the second step.

Corrector step and the solution for pressure

The corrector step uses TR, which is non-dissipative, completely stable (A-stable to be more precise), and also second-order accurate; applied to $\dot{y} = f$ it is

$$y_{n+1} = y_n + \frac{\Delta t_n}{2}(f_n + f_{n+1}) \quad . \tag{10}$$

Being implicit, the TR is applied to the full set of equations, (3a) and (3b), the solution of which yields the final (reported) values of velocity and pressure. Application of the TR to Equation (3) gives

$$\left[\begin{array}{c:c} \frac{2}{\Delta t_n} M+K+N(u_{n+1}) & C \\ \hdashline C^T & 0 \end{array}\right] \left\{\begin{array}{c} u_{n+1} \\ \hdashline P_{n+1} \end{array}\right\} = \left\{\begin{array}{c} M\left(\frac{2}{\Delta t_n} u_n + \dot{u}_n\right) + f_{n+1} \\ \hdashline 0 \end{array}\right\} , \tag{11}$$

where we have utilized our knowledge of the acceleration (see Equation (12) below) to employ the computationally efficient form of the right hand side (RHS) vector. In fact, it is just this acceleration term which distinguishes TR from the first-order accurate, backward Euler scheme (which can be obtained from Equation (11) by omitting $\dot{u}_n$ and replacing $2/\Delta t_n$ by $1/\Delta t_n$). Although we considered the Euler schemes (using forward Euler as a predictor), we rejected them because they are numerically dissipative and the more accurate AB/TR results are available for little extra work.

We recognize Equation (11) as a nonlinear algebraic system A(x)x=b which, when solved for u_{n+1}, P_{n+1} (the details of which are deferred), permits an update of the acceleration vector in preparation for the next time step by 'inverting' the TR in the form

$$\dot{u}_{n+1} = \frac{2}{\Delta t_n}(u_{n+1} - u_n) - \dot{u}_n \quad , \tag{12}$$

where $\dot{u}_n$ is available from the previous application of the same equation. It is important to note that the acceleration vectors are simple to obtain even using the consistent mass matrix (mass lumping saves no money). While it is true that the mass matrix must actually be 'inverted' (ala Equation (8)) to obtain $\dot{u}_o$ (and P_o), this is required only once per problem and provides sufficient benefits to easily justify this incremental cost.

Local time truncation error and time step selection

The estimation of the local time truncation error begins with a Taylor series analysis of both predictor (AB) and corrector (TR) schemes, Equations (9) and (10). Denoting by $y(t_n)$ the exact solution at time t_n and assuming that $y_n = y(t_n)$ (i.e., local error estimates are based on the assumption that the exact solution is known at the beginning of a time step), we have the following error estimates:

$$\text{AB:} \quad y^p_{y+1} - y(t_{n+1}) = -\frac{1}{12}\left(2 + 3\,\frac{\Delta t_{n-1}}{\Delta t_n}\right)\Delta t_n^3\,\dddot{y}_n + O(\Delta t_n^4) \tag{13}$$

$$\text{TR:} \quad y_{n+1} - y(t_{n+1}) \equiv d(y_{n+1}) = \frac{\Delta t_n^3}{12}\dddot{y}_n + O(\Delta t_n^4) \quad , \tag{14}$$

where $d(y_{n+1})$ is the estimate of the local time truncation error of the actual (TR) solution. Since y^p_{n+1} and y_{n+1} are available, Equations (13) and (14) may be used to eliminate the two unknowns, $y(t_{n+1})$ and $\dddot{y}_n$ in favor of the desired result, which is

$$d(y_{n+1}) = \frac{y_{n+1} - y^p_{n+1}}{3\left(1 + \dfrac{\Delta t_{n-1}}{\Delta t_n}\right)} + O(\Delta t_n^4) \quad . \tag{15}$$

This result can be used to estimate the next step size based on the requirement that a (relative) norm of the error for the next step should equal a pre-set input value, ε. From Equation (14) we have

$$\frac{|d(y_{n+2})|}{|d(y_{n+1})|} \cong \left(\frac{\Delta t_{n+1}}{\Delta t_n}\right)^3 \cdot \frac{|\dddot{y}_{n+1}|}{|\dddot{y}_n|} \quad , \tag{16}$$

where $d(y_{n+1})$ is available. Since $\dddot{y}_{n+1} = \dddot{y}_n + O(\Delta t_n)$, Equation (16) permits the solution for Δt_{n+1}, after setting $d(y_{n+2}) = \varepsilon$ and neglecting higher-order terms; viz.,

$$\Delta t_{n+1} = \Delta t_n \left(\varepsilon / |d(y_{n+1})|\right)^{1/3} \tag{17}$$

is used to compute the (potential) next step size.

Applying these techniques to the discretized NS equations requires that error estimates be based solely on velocity as there is no predictor equation for pressure (and none is needed). In our work, we have been using the following combined, weighted RMS norm:

$$|d(y_{n+1})|^2 = \frac{1}{N_u+N_v}\left[\frac{1}{U_{max}^2}\sum_{i=1}^{N_u} d_i^2(u_{n+1}) + \frac{1}{V_{max}^2}\sum_{i=1}^{N_v} d_i^2(v_{n+1})\right] , \qquad (18)$$

where N_u, N_v are the number of nodes for u and v and U_{max}, V_{max} represent the largest values in the u and v velocity vectors at t_{n+1}.

The only user-specified parameter is ε, which can obviously have a significant effect on cost and accuracy. Too large an ε can cause one or more of the following: (1) the theory behind the error estimate to be poor, (2) the well-known oscillatory behavior when large time steps are used with the TR, and (3) a more difficult nonlinear system (Equation (11); y_{n+1} is farther from y_n). Too small an ε, however, will cause excessively expensive (albeit accurate) computations. It has been our experience that $\varepsilon \cong .001$ (.1% relative error per step) is nearly optimum in that $\varepsilon = 10^{-4}$ (which approximately doubles the total number of time steps) does not give significantly different results and $\varepsilon = .01$ will often cause visibly oscillatory solutions. (The numerical oscillations generated by TR, whose amplitudes are theoretically $O(\varepsilon)$ in this method, are only slowly damped, owing to physical diffusion, as $t \to \infty$.) It would of course be desirable to be able to estimate and control the global time integration error but, unfortunately, there is little or no theory to guide us here (the theory does give the result that a local truncation error of $O(\Delta t^3)$ gives a global error which goes like Δt^2; Gear, 1971). It is also noteworthy that this scheme, by virtue of providing an estimate of $\dddot{y}_n$ from Equations (13) and (14), could ostensibly permit a third-order-accurate result by simply subtracting the local truncation error, $\Delta t_n^3\, \dddot{y}_n/12$ from the TR result. This procedure, while perhaps useful at output times, is generally not advisable for the computations since it reduces the stability of the scheme (it is then no longer a TR scheme).

Final details regarding time step selection are deferred until we discuss the solution of Equation (11), since certain additional economic factors will arise.

SOLUTION OF THE NONLINEAR SYSTEM

To solve Equation (11), we have employed several variants of Newton's (or Newton-Raphson's) method, which is quite powerful for time dependent problems because a good first guess is always available (the solution usually changes little between time steps). It has proven especially so when coupled with a predictor equation (which provides an even better first guess) and has eventually led us to our current method: 'one-step Newton'.

Newton's method and chord approximations

The chord approximation to Newton's method for solving $A(x)x=b$ leads to the following iterative sequence of linear systems:

$$J_k\, \delta X_{m+1} = b - A(X_m)X_m \quad ; \quad m = 1,2,\dots \quad , \qquad (19)$$

where $\delta X_{m+1} = X_{m+1} - X_m$ is the change in X between iterations and

$J_k = \partial[A(X_k)X_k]/\partial X_k$ is the Jacobian matrix. If the Jacobian were to be fully updated for each iteration, we would use $k=m$ and be performing 'full-Newton' iterations. Although this is the only way to realize the (asymptotically) quadratic convergence rate associated with Newton's method, it is expensive since J_m must be formed and factored (via an LU decomposition, assuming that direct methods are employed) at each iteration within a single time step. The chord approximation is a potentially cost-effective modification and is based on using an out-of-date Jacobian and accepting the concomitantly slower convergence rate (asymptotically linear). Such an approximation, which must result in fewer costly factorizations, is justifiable since (1) we have a good starting guess from the AB predictor, (2) we can afford more iterations since they are ostensibly much cheaper (involving only the forward-reduction and back-substitution steps on a previously factored matrix) and (3) we will obtain the same result in the converged limit (as $m \to \infty$, $\delta X_{m+1} \to 0$ and we have a solution to $A(x)x = b$ for any convergent method). In fact, not only can we avoid full Jacobian updates within a time step, but we can even permit J to lag several time steps behind and only update it when the chord iteration sequence is converging too slowly or not at all.

We originally developed an algorithm which could use either the chord method or full Newton. Our initial results regarding the cost-effectiveness of the chord method were somewhat discouraging, apparently due to the following factors in combination: (1) Based on the general convergence criterion that the solution of the nonlinear system should be sufficiently accurate that the result does not contaminate the estimation of the local time truncation error, we required an appropriate norm of the change in velocity between successive iterates to be some small fraction of ε. For full Newton, a criterion of 0.1ε was found to work quite well, whereas the chord method gave unreliable time-step results unless the criterion was more like $.01\varepsilon$, which increased the number of iterations. (2) In our codes, which employ a disk-based unsymmetric frontal method (modified from the version published in Hood (1976)), it turned out that the cost of constructing the RHS of Equation (19) was significant enough that the extra iterations associated with the chord method (typically 4-6 vs. 2 for full Newton) often made the total cost comparable to that of full Newton.

Admittedly, our experience with the chord method was brief and it may still be a useful technique in some situations (perhaps in conjunction with a different linear equation solver). However, we have been so favorably impressed by the one-step Newton scheme, described below, that we now use it exclusively.

One-step Newton method

This method is as simple as it sounds; rather than solving the sequence of linear systems of Equation (19), we construct and solve only one linear system per time step, using

$$J_p(X-X_p) = b - A(X_p)X_p \quad , \tag{20}$$

where X_p is the predictor value of X, J_p is the Jacobian based on this predictor, and X is assumed to be the correct solution. Application of Equation (20) to Equation (11), with $X = (u_{n+1}\ P_{n+1})^T$ and $X_p = (u_{n+1}\ 0)^T$ gives (there is no predictor equation for pressure -- and none is needed since it appears linearly);

$$J_p = \left[\begin{array}{c|c} \frac{2}{\Delta t_n}M+K+N(u^p_{n+1})+N'(u^p_{n+1}) & C \\ \hline C^T & 0 \end{array}\right] \quad . \qquad (21)$$

Also, $A(X_p)$ is the matrix on the left side of Equation (11) in which u^p_{n+1} is used to evaluate $N(u_{n+1})$ and b is the RHS vector in Equation (11). Whereas N(u) corresponds to the advection operator, $\underline{u} \cdot \nabla$, $N'(u)$ corresponds to terms like $\partial u/\partial x$, etc.; refer to Appendix 1 for details. All contributions to $A(X_p)$ and J_p are computed analytically from the basis functions, once per problem, with the resulting arrays (doubly and triply-subscripted) stored element-by-element on disk.

When we compared one-step Newton against full Newton on several test problems, with $\varepsilon = .001$, and using $.1\varepsilon$ as the Newton convergence criterion (which virtually always resulted in 2 iterations per step), the results were essentially indistinguishable and we therefore have sufficient confidence in the former (at least for $\varepsilon = .001$) to use it exclusively.

With the solution method of the 'nonlinear' system of the TR step described, we are now in a position to summarize the entire algorithm for advancing the solution by one time step:

1. The predictor for velocity is computed using Equation (9).

2. The Jacobian matrix (Equation (21)) and the RHS of Equation (20) are assembled (by looping through the elements).

3. The final velocity and pressure are obtained by solving the linear system (20).

4. The acceleration vector is updated via Equation (12).

5. The next potential time step, Δt_{n+1}, is computed from Equations (15), (17), and (18).

The final details of time step modification, which are mostly based on qualitative cost-effectiveness principles and are similar to those used in certain standard ODE software packages (Hindmarsh, 1972; Byrne et al., 1977) are these:

1. If $\Delta t_{n+1} \geq \Delta t_n$, we always accept the increase.

2. If $\gamma\Delta t_n \leq \Delta t_{n+1} < \Delta t_n$, where γ is typically 0.8, we accept the solution but do not change the time step.

3. If $\Delta t_{n+1} < \gamma\Delta t_n$, we reduce Δt and repeat the time step (i.e., we reject the solution at t_{n+1}).

Considering that the assembly of the Jacobian matrix is actually done simultaneously with that of $A(X_p)$ and requires only several additional submatrices, it would appear that the one-step Newton procedure described above has significant advantages over approaches based on either functional iteration or those which involve the formation and solution of one linear system per time step.

Although we have used the AB/TR integration scheme exclusively, we should point out an attractive feature of a similar, and somewhat simpler algorithm; viz, one based on the first-order accurate Euler integration schemes referred to earlier. If the (explicit) forward Euler (FE) scheme is used as a predictor and the (implicit) backward Euler (BE) method as the corrector, an algorithm would result which would be very similar to our second-order scheme; i.e., one could use the FE predictor to solve the BE equations using one-step Newton and employ automatic time-step selection based on an estimate of the local time truncation error. Such a scheme could exhibit a significant advantage over the AB/TR scheme when seeking steady state solutions; viz, the numerically dissipative properties of BE could be used to advantage in that larger (hence, fewer) time steps could be taken (via a larger ε) without causing the "ringing" (oscillations) associated with the non-dissipative TR. We believe that under many conditions this scheme would be quite competitive with iterative methods applied directly to the steady state NS equations. For better transient simulations or for truly time-dependent flows, however, the more accurate and non-dissipative second-order scheme is still to be preferred (it is more cost-effective). This FE/BE algorithm is summarized in Appendix 2.

NUMERICAL RESULTS: DEVELOPMENT OF A KARMAN VORTEX STREET

We will demonstrate the capabilities of our FEM code with a single example: the flow around a circular cylinder, starting from rest, at an ultimate Reynolds number of $\sim$100, at which vortex shedding is known to occur (Tritton, 1977). To our knowledge, this represents the first successful simulation of this flow phenomenon using the Galerkin FEM and the primitive variables; interestingly, it was (to our knowledge) only recently (Swanson and Spaulding, 1978) that the finite difference method has been used to compute vortex shedding using the primitive variables. Smith and Brebbia (1977) have previously computed vortex shedding using the FEM and the stream function-vorticity approach, while Glowinski et al. (1978) and Bristeau et al. (1978) have briefly presented some vortex shedding results using the primitive variables and a least-squares FEM. The most accurate and detailed results which we have found for this problem are those by Jordan and Fromm (1972), who used finite differences (and > 12000 nodes!) and the stream function-vorticity equations.

As in a laboratory experiment, our actual control of Re is not perfect; since we cannot impose $u = u_w$ = constant across the inlet at $t = 0$ with an otherwise motionless fluid (it violates mass conservation, since $C^T u_o \neq 0$ in that case), we must drive the flow with an imposed pressure gradient, or by moving the top and bottom walls at $u = u_w$, in order to initiate the flow (in fact, we employed both driving forces simultaneously). Only as $t \to \infty$ do we know the actual Re of our simulation. To obtain a reasonably appropriate pressure gradient, we cheated somewhat (from a laboratory simulation) by first running our steady-state code at Re = 100 with $u = u_w$ across the inlet (which is legal for steady flow calculations) and along the top and bottom walls. The resulting solution (although not a physically realizable flow since it is unstable) was used to drive our time-dependent code by using the computed inlet pressures as a normal traction BC (as discussed earlier). We simply hope that these BC's will ultimately lead to the desired Re (as it turned out, our 'final' Re was ~110).

Figure 3 shows the mesh of 196 "quadratic" elements (element 2 of an earlier section) and the BC's used for the simulation. The computational domain extends about four cylinder diameters above, below, and upstream of the cylinder and about 15 diameters downstream. The normal traction force at the inlet is seen to be slightly larger in the upper region than in the lower. The reason for this appears to be the imperfectly symmetrical mesh obtained from the FEM mesh generator code (asymmetry was present in the 3rd-4th decimal place, both in nodal coordinates and in the results from the steady code). It is very likely that this small asymmetry was instrumental in initiating the vortex shedding, since a triggering mechanism of some kind is always required. (In retrospect, we might have done better in triggering the flow — if quasi-steady vortex shedding were the main goal -- since the 'gestation time' turned out to be quite long.) Figure 3b shows the mesh details near the cylinder (with some artistic license; e.g. the two layers of elements nearest the cylinder actually have curved boundaries), showing the finest zoning in the downstream region where the vortices will be formed; the cylinder boundary was interpolated via the quadratic basis functions (i.e., the surface nodes were placed on the circle). The grid contains 850 nodes, which leads to a total of 1929 equations (1700 velocities, 229 pressures — in clear violation of the optimum constraint ratio referred to earlier -- nevertheless, the results appear to be quite accurate; probably an optimum 'quadratic' element (element 4 or 7) would generate equivalent accuracy on a coarser mesh). The diameter (D) of the cylinder, the wall speed (u_w) and the viscosity are 1.0, and the density is 100 (units are arbitrary). We used a 4 x 4 Gaussian quadrature rule to construct the element matrices and used ε = .001 for the error control and time step selection algorithm. The initial time step was 2.0, based on our previous experience which indicated that an initial step size of a few percent

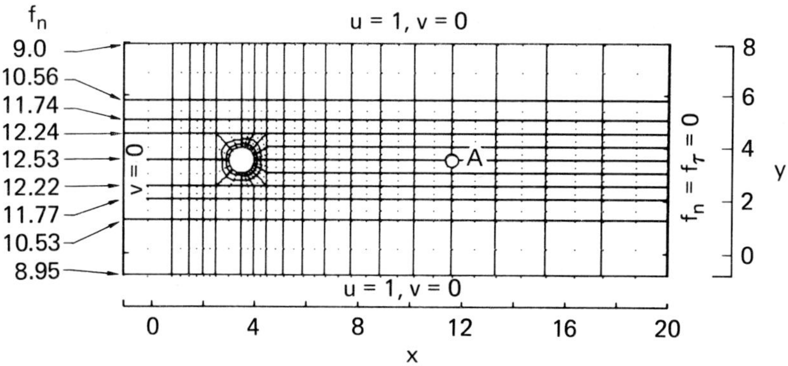

3a. Overall domain, mesh, and boundary conditions.

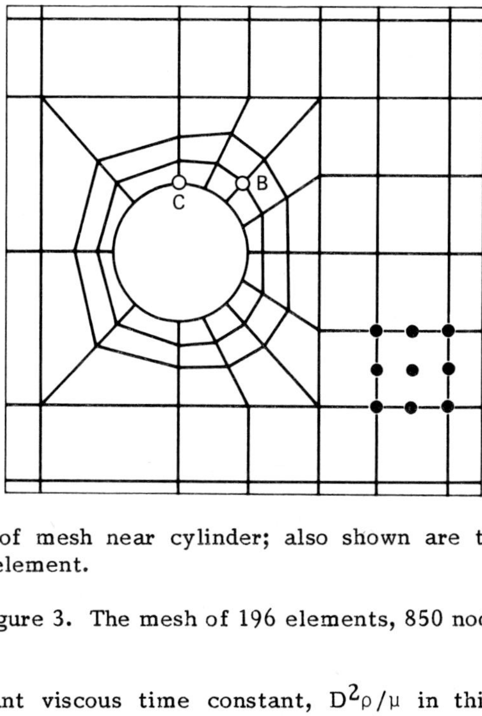

3b. Details of mesh near cylinder; also shown are the nodes for a typical element.

Figure 3. The mesh of 196 elements, 850 nodes.

of the relevant viscous time constant, $D^2\rho/\mu$ in this case, usually satisfies our accuracy criterion for flows starting from rest.

The first result we present is the time step history (Figure 4) since this information actually gives significant insight into the "physics" of the flow. The rapid increase in Δt (or in the time-scale of viscous diffusion) up to $t \cong 20$ is typical of linear viscous flow. At this time, however, some new "physics" appears, since Δt stops increasing. The reversal of Δt followed by the slower growth from $t \cong 40$ onward corresponds to the formation of the separated flow regions behind the cylinder. The time step then grows monotonically while the eddies continue to grow and elongate. From a time of ~ 150 to

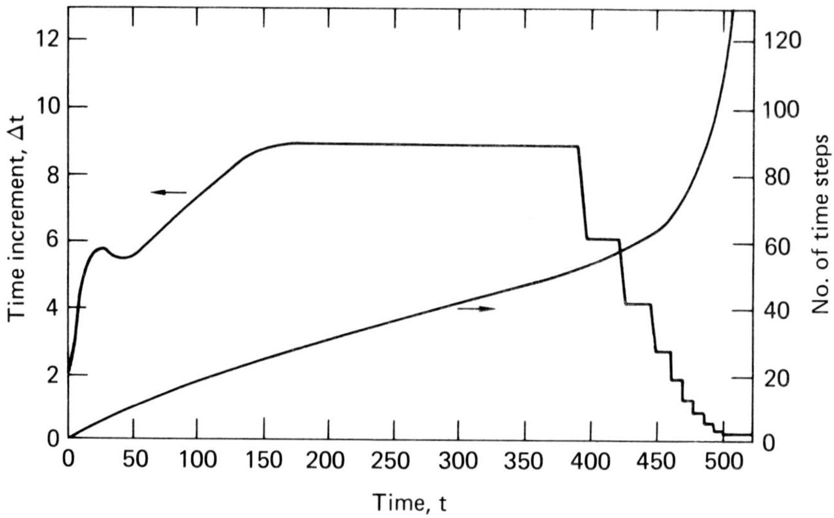

Figure 4. Time step history and number of time steps.

$\sim$350 or so, an approximately constant Δt is sufficient to follow this growth. The reductions in Δt, beginning at $t \cong 400$, signal the beginning of a new dynamic phenomenon and correspond, of course, to the oscillations which are caused by the inherent instability of the flow. The time step decreases (in a staircase manner, a result of using $\gamma = .7$ in the Δt-algorithm) from a maximum of $\sim$8.9 to $\sim$0.37 as the Karman vortices gain in strength and set a final time-scale for the flow. The number of time steps as a function of time is also included in the figure; the growth is linear in the last portion, $495 < t < 520$, ending at $\sim$170 total steps.

It may be of interest to compare these time steps with those based on certain stability requirements if simple, explicit time integration methods had been employed (the manner in which such schemes are employable is discussed in a subsequent section). The diffusive stability limit is on the order of $\Delta t \leq \rho \Delta x^2/4\mu$ for 2-D flows with $\Delta x \cong \Delta y$ (Roache, 1976); since the minimum Δx (and Δy) $\cong .05$, the stability limit would be $\Delta t \leq \sim .06$, which we exceed by a factor of $\sim$6 (during vortex shedding) and $\sim$150 (prior to shedding). Also, the cell Reynolds number, $Re^c \equiv \rho u \Delta x/\mu$, should not exceed $\sim$2 (ibid.) for stability; our maximum Re^c is about 170. Finally, the "Courant" stability limit is given by $\Delta t < \sim \Delta x/u$, which is as small as $\sim$.25 during our simulation. These comparisons help to point out some of the differences in time step sizes between those based on stability and those based on accuracy. During vortex shedding, our 'final' Δt was $\sim .34 \pm .02$; the final shedding period was $\sim$6.2, which results in about 18 steps per cycle, a reasonable number for TR to 'follow' a periodic solution. Finally had we used the implicit BE scheme with its artificial kinematic viscosity of $O(u^2 \Delta t/2)$, our time step would again need to be very small (e.g., if Δt were $\sim$.02, the numerical diffusion

$(u^2 \Delta t/2)$ would be approximately equal to the physical diffusion (μ/ρ); and in this case, the requirement for small time steps is not caused by stability considerations.

The next set of figures presents time histories of several selected nodal values. These are conveniently presented in three segments for each nodal value and correspond to the manner in which we performed the computations (we used two "restarts"). The first segment, $t = 0 \to 200$, covers the period in which the eddies form and grow on the rear of the cylinder. The next segment, $t = 200 \to 500$, covers the period in which the eddy growth gives way to the oscillatory instability and the final period, $t = 495 \to 520$, shows several cycles of vortex shedding.

Figures 5 and 6 show the u and v time history for node B in Figure 3b, which is located 45° from the centerline and ∿.15D from the cylinder surface. The formation of the downstream eddy ($t \cong$ 20-40) is reflected in the u and v plots in Figures 5a and 6a, as is the ensuing growth. Figures 5b and 6b show the initiation and growth of the oscillations quite clearly, and Figures 5c and 6c depict the details of the approach to quasi-steady vortex shedding (clearly we have not yet achieved a quasi-steady oscillation).

Figures 7 through 9 present the time history for u, v, and P at node A of Figure 3a, which lies on the centerline about 7.65D downstream. In Figure 8a we see a harmless manifestation of the numerical oscillations associated with the TR; it is harmless because the oscillations are so small ($< .001$) and therefore insignificant and because, as Figures 8b and 8c reveal, they are no longer visible (they are $< \varepsilon$) when the magnitude of v increases to a significant level (presumably all nodes exhibit similar TR "noise"). Also noteworthy here is the almost explosive growth of the instability beginning at t = 450-475. The Strouhal number (dimensionless frequency of vortex shedding, $S = fD/u_o$) is ∿0.156, which means that $\tau = 1/f \cong 6.4$ is the cycle time required to shed a pair of vortices (alternately from the "top" and "bottom" of the cylinder); at later times yet, τ decreased to ∿6.2, giving S = .161. We see that both u and P oscillate at twice the shedding frequency, as they respond to the passage of each vortex, whereas v responds at the shedding frequency since the vertical component of velocity is upward (say) during one-half the passage of two successive vortices.

The final time history, in Figure 10, shows the pressure at the top surface (90° from the centerline) of the cylinder (node C in Figure 3b), which oscillates at the shedding frequency. The diametrically opposed pressure also oscillates at this frequency and amplitude, but is 180° out of phase.

We next present some contour plots to further elucidate the flow field. Figure 11 depicts some streamlines and pressure contours at "small" time ($t \cong 20$), just before the flow separates. The pressure field shows only a small departure, except near the cylinder, from the essentially linear dependence on x caused by the "ΔP" boundary

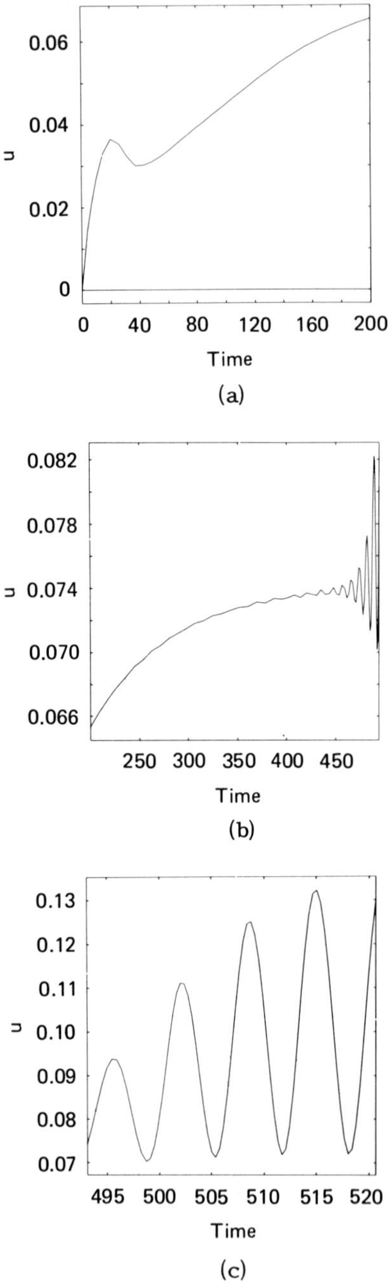

Figure 5. Horizontal velocity at node B of Figure 3b.

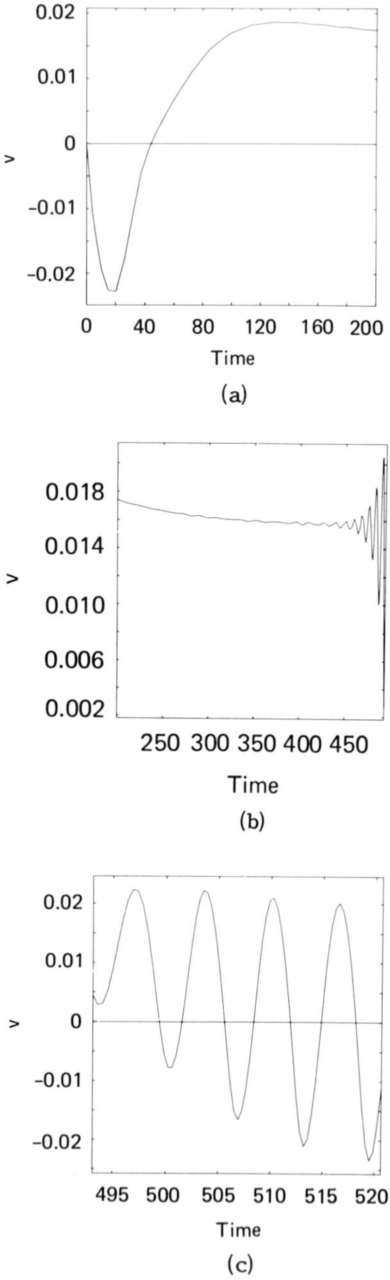

Figure 6. Vertical velocity of same node.

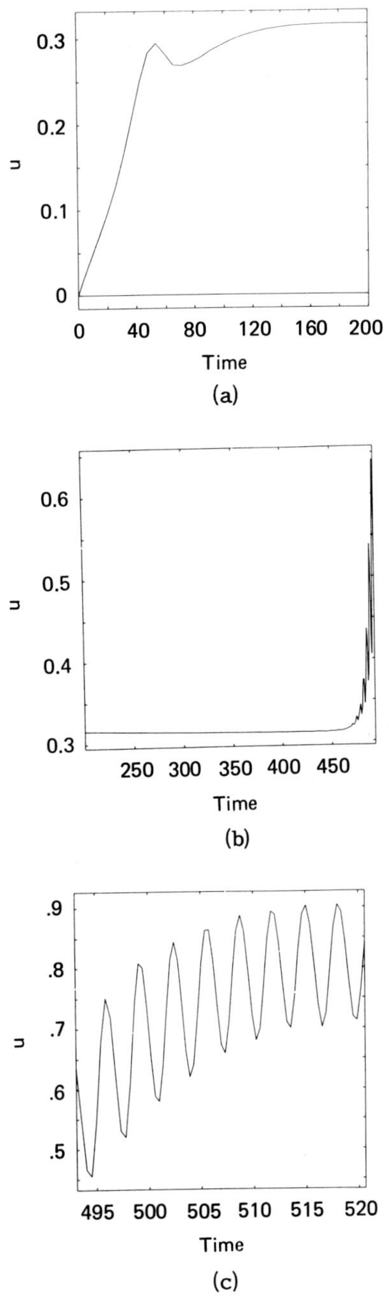

Figure 7. Horizontal velocity at node A of Figure 3a.

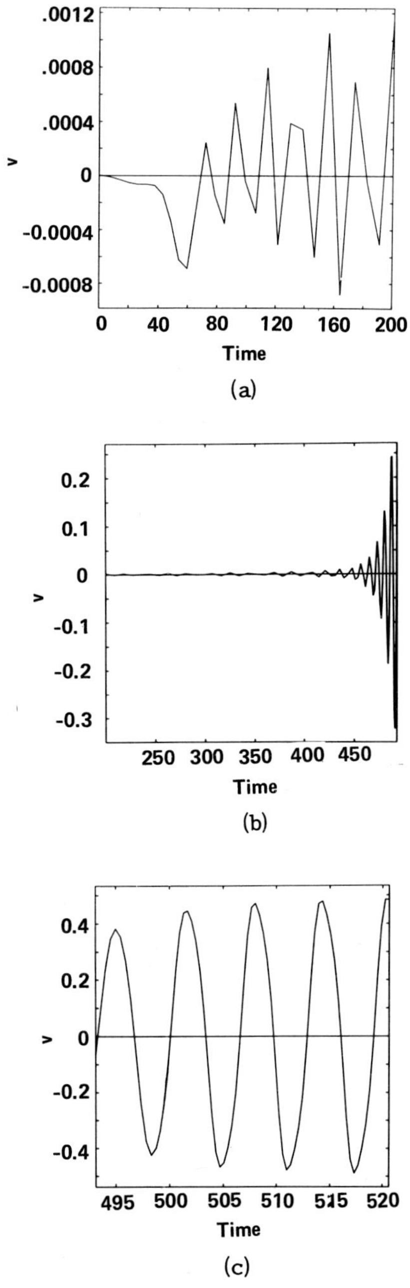

Figure 8. Vertical velocity of same node.

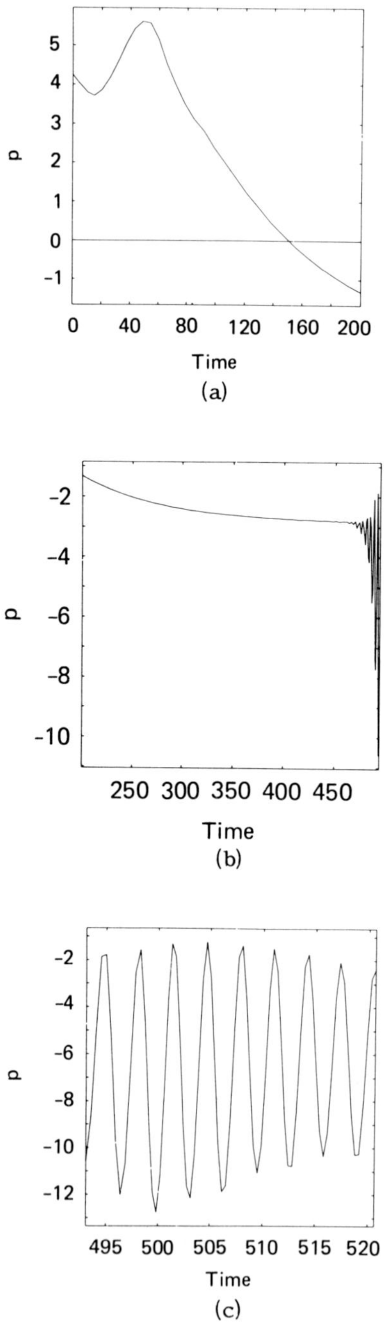

Figure 9. Pressure at node A.

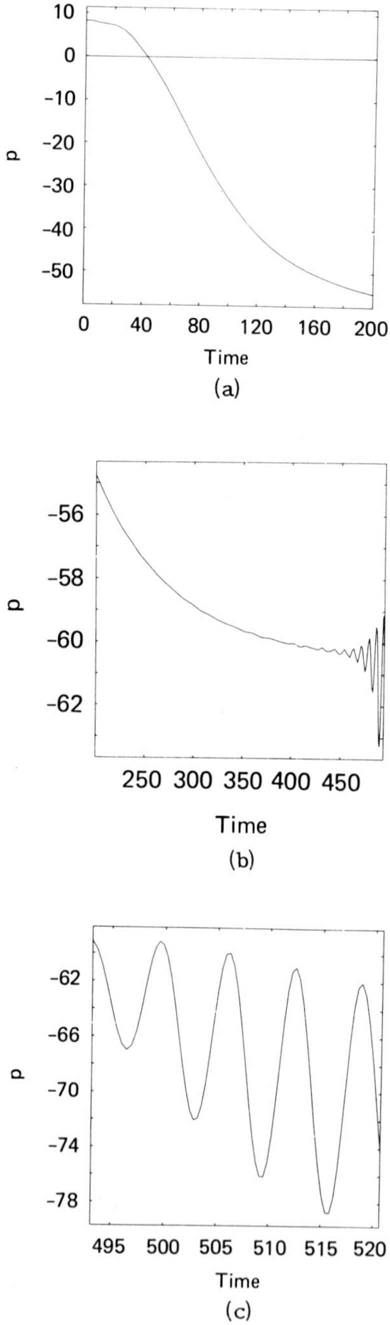

Figure 10. Pressure at node C of Figure 3b.

(a)

(b)

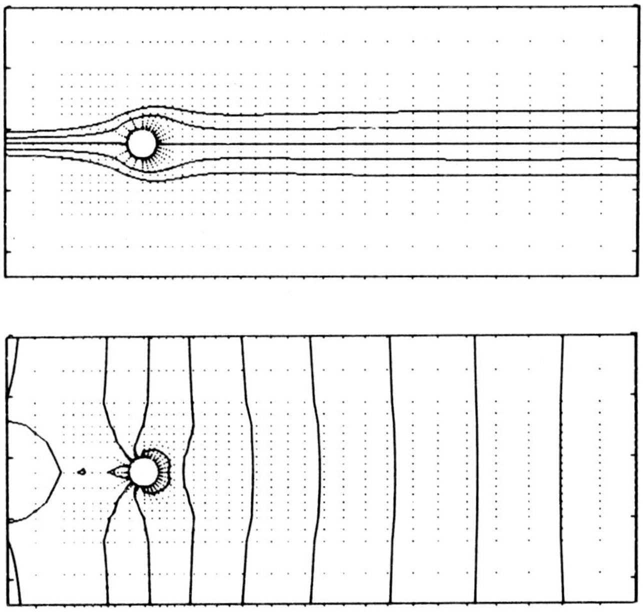

Figure 11. Contour plots at early time (t ≅ 20): (a) streamlines, (b) pressure.

(a)

(b)

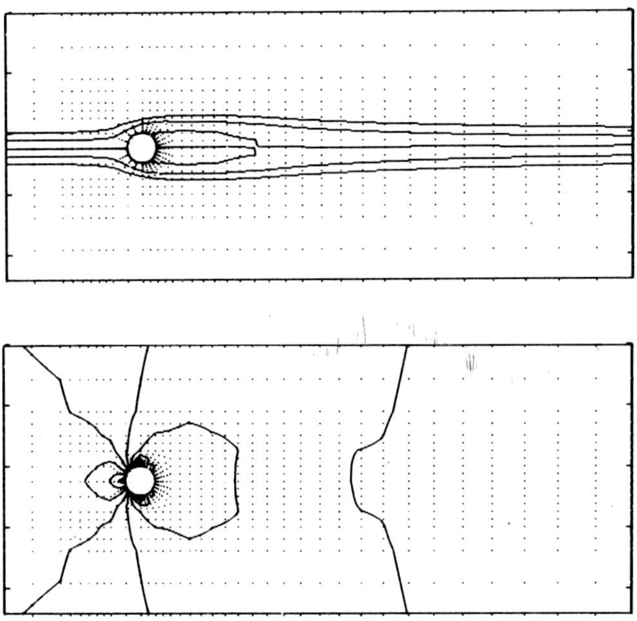

Figure 12. Contour plots near maximum eddy length (t ≅ 200): (a) streamlines, (b) pressure.

condition. These variables are shown at a later time in Figure 12, wherein the pressure distribution is much different and the two nearly symmetric, elongated eddies are clearly visible. Apparently these eddies must grow to a significant length before the instability can begin: once the oscillations start, these eddies shrink back in size until one of them is shed.

The next set of figures depicts the overall solution during one complete cycle (t = 532 to 538.2; shown at the beginning, 1/4, 1/2, and 3/4 of a cycle later). The overall vector field is shown in Figure 13, and the streamlines are presented in Figure 14; the streamlines in Figure 13d, for example, appear to be in excellent agreement with those presented by Swanson and Spaulding (1978) -- their Figure 7b. In Figure 15 are shown snapshots of streamline contours as seen by an observer who is moving to the right at the wall velocity. The closed streamlines clearly reveal the existence of the classic Karman vortex street as seen in this moving reference frame. It is noteworthy, in this and earlier figures, that the traction-free outflow BC's seem to work quite well in that this artificial boundary does not appear to degrade the quality of the solution as the vortices leave the grid. The associated isobars are shown in Figure 16 and reveal the near upstream symmetry and downstream asymmetry. The vorticity contours in Figure 17 are not well portrayed since they are concentrated near the front surface of the cylinder where vorticity is being generated; downstream advection, while apparent, only shows up for one or two contour values (the rest are diffused by viscosity).

In Figures 18-21, we focus on the detailed flow patterns in the interesting near-wake region, where the vortex street is formed; again, the results are shown at the same times during a shedding cycle. Figures 22-24 show the time histories for the same three nodes presented earlier; at this time the flow is closer to being quasi-steady.

Finally, Figure 25 presents the lift and drag coefficients over the same cycle. While the lift coefficient, which oscillates at the shedding frequency, exhibits an essentially steady oscillation, the drag coefficient, which is dominated by the pressure component ($\sim$2/3 of the total) and oscillates at twice the shedding frequency, clearly shows that we have not yet attained a "steady-state". These coefficients are computed via Gaussian quadrature (line integrals) of the velocity and pressure results along the surface of the cylinder (Γ_c) according to the following equations:

$$C_L = (F_y^P + F_y^V)/\tfrac{1}{2}\rho u_w^2 \quad , \tag{22}$$

and

$$C_D = (F_x^P + F_x^V)/\tfrac{1}{2}\rho u_w^2 \quad , \tag{23}$$

where F^P is the pressure force and F^V is the viscous force on the surface of the cylinder; in particular,

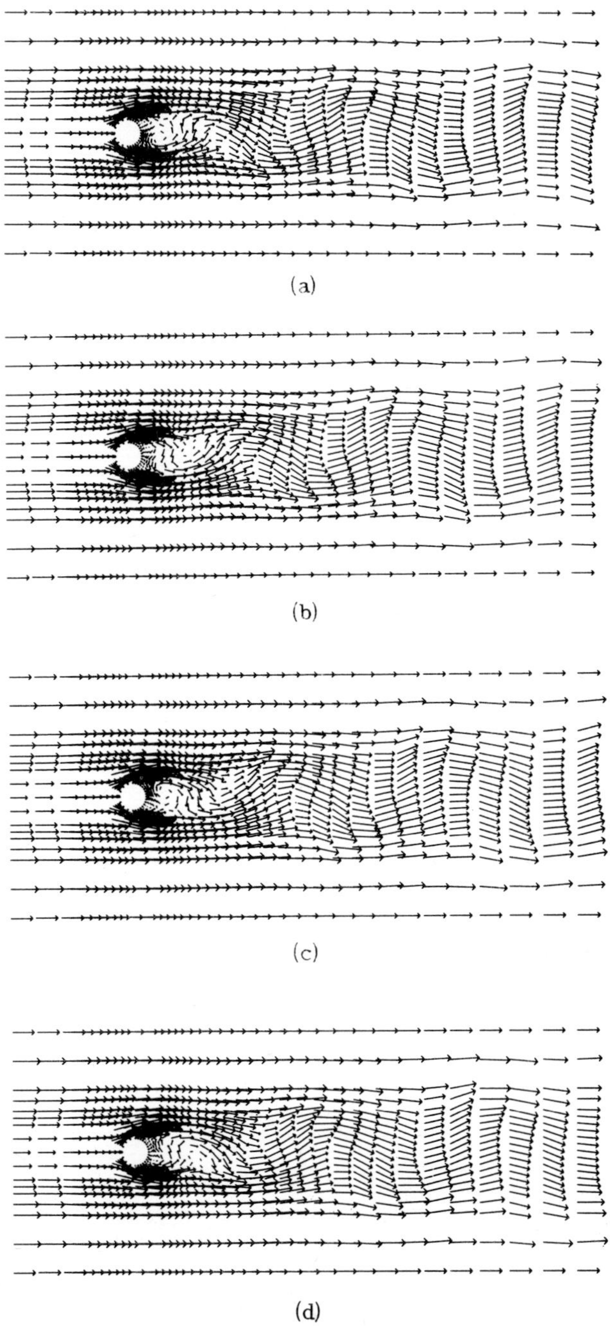

Figure 13. Velocity vectors during vortex shedding: (a) t = 532.00, (b) t = 533.55, (c) t = 535.10, (d) t = 536.65.

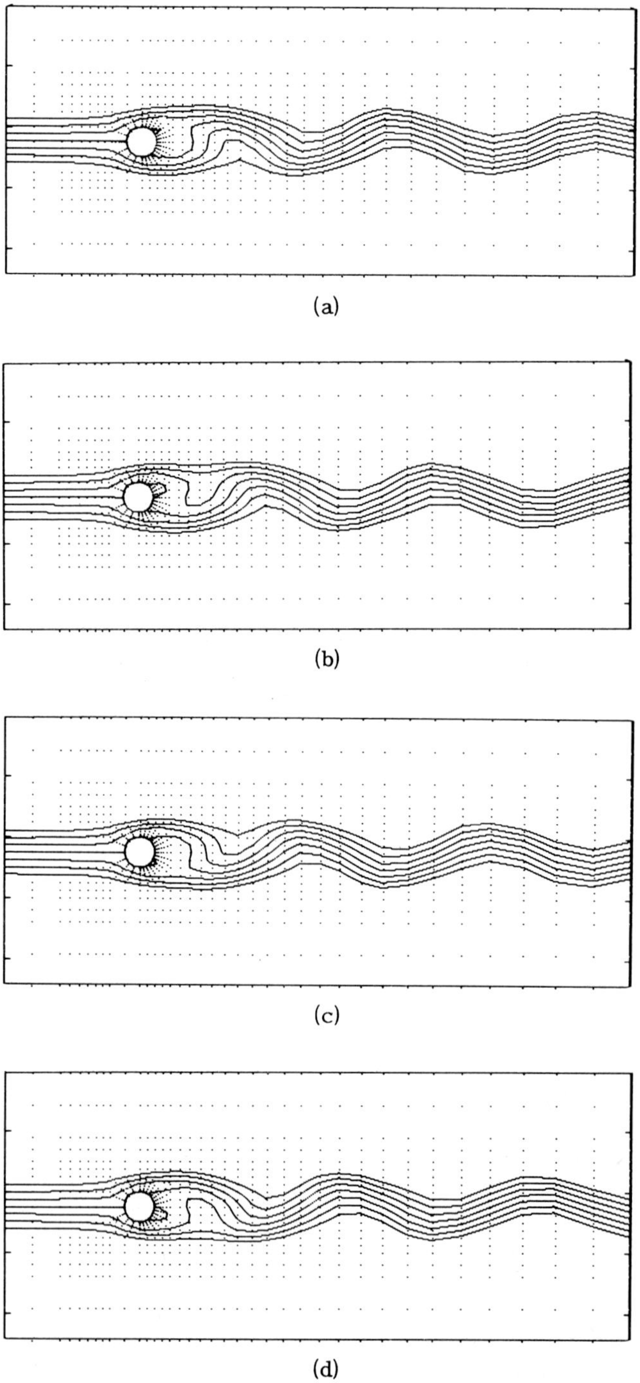

Figure 14. Streamlines during the same shedding cycle as in Fig. 13.

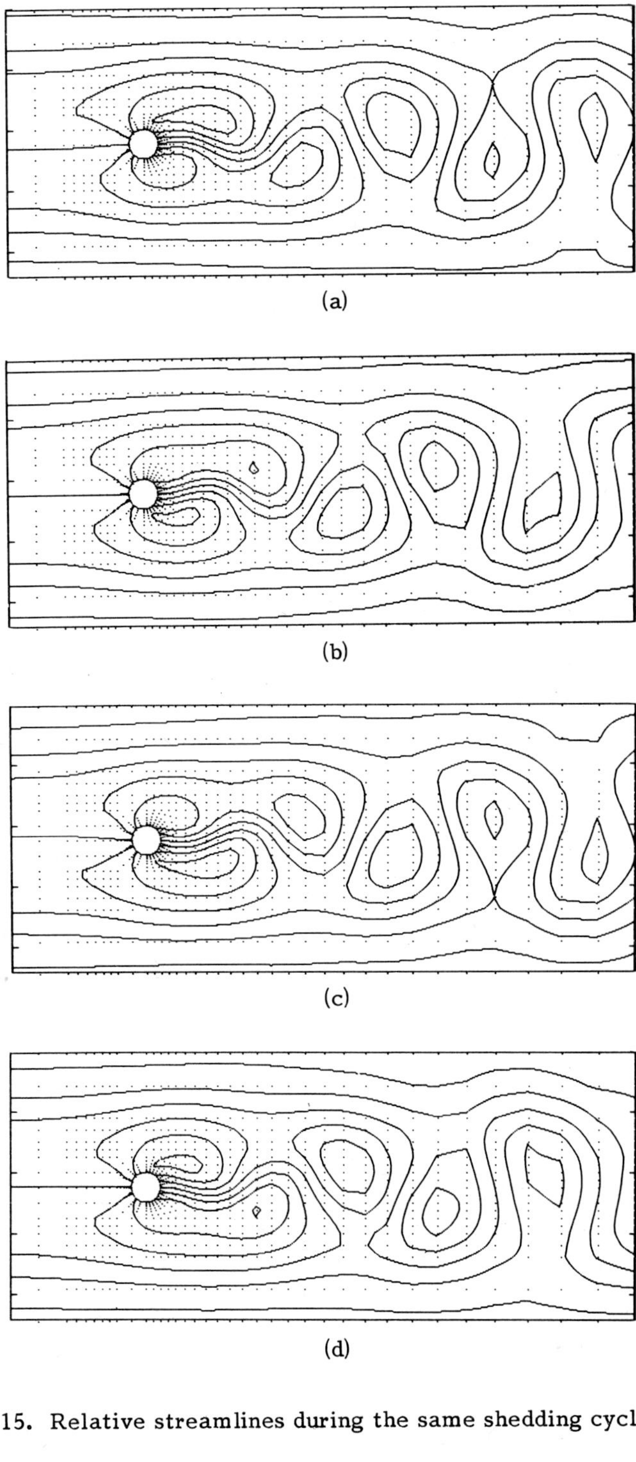

(a)

(b)

(c)

(d)

Figure 15. Relative streamlines during the same shedding cycle.

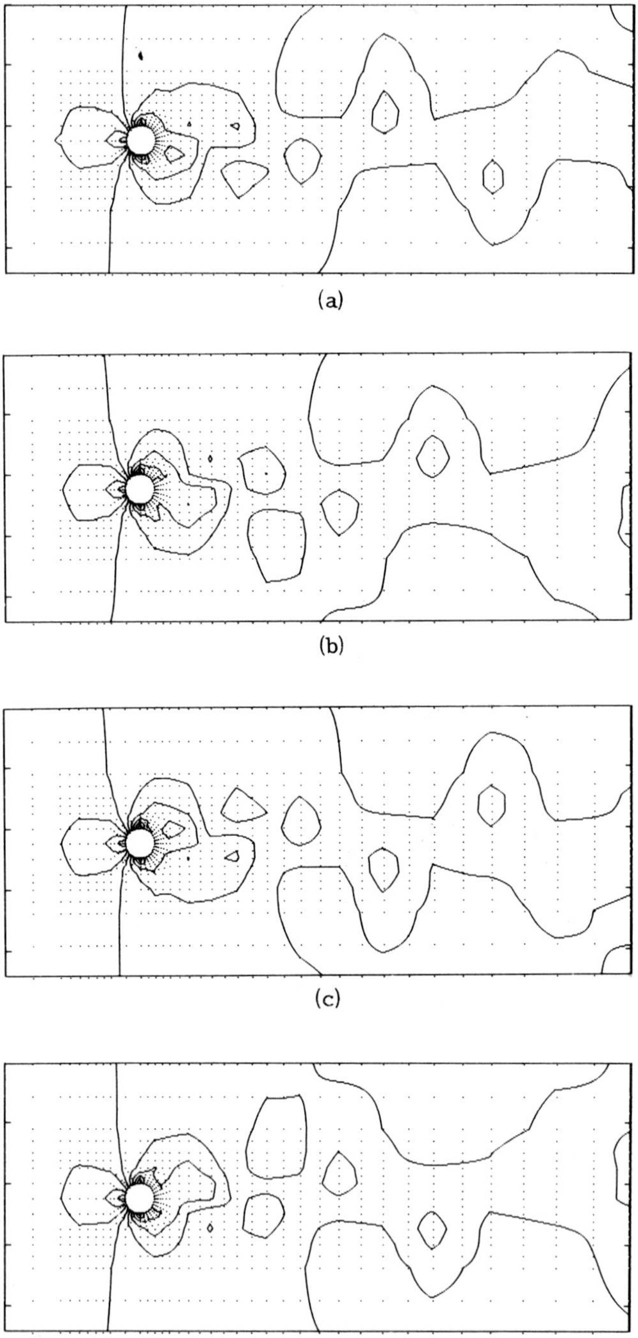

(a)

(b)

(c)

(d)

Figure 16. Pressure contours during the same shedding cycle.

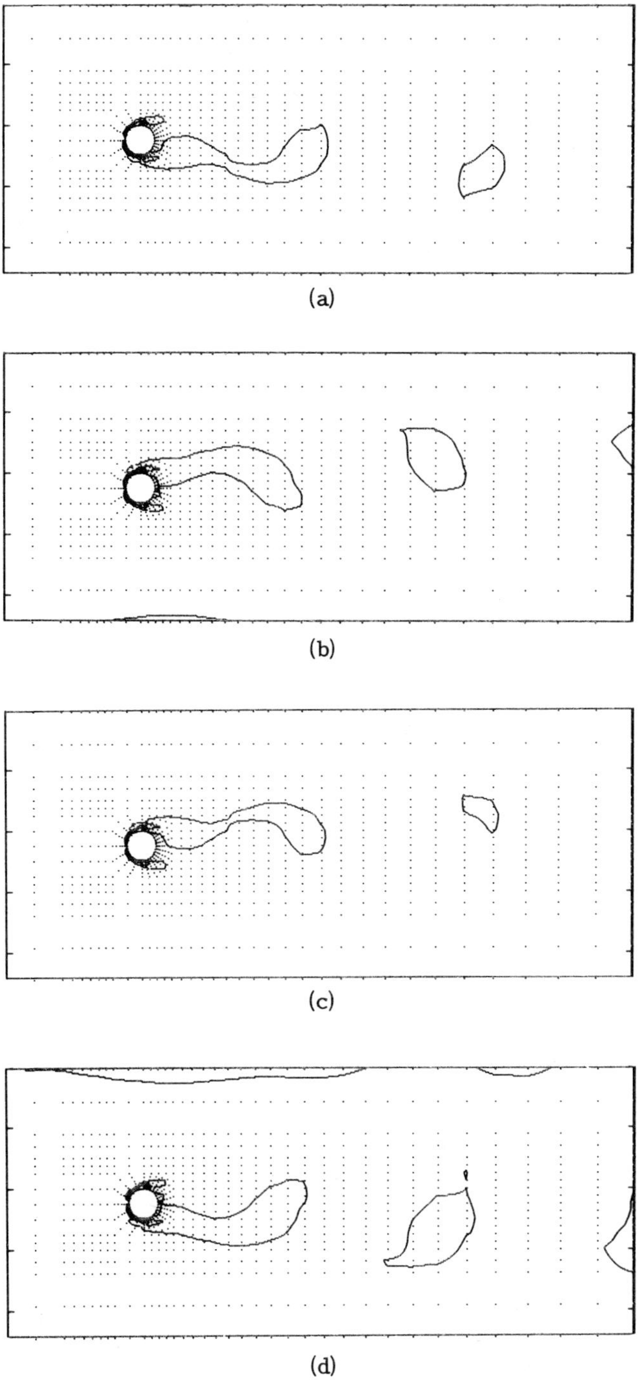

Figure 17. Vorticity contours during the same shedding cycle.

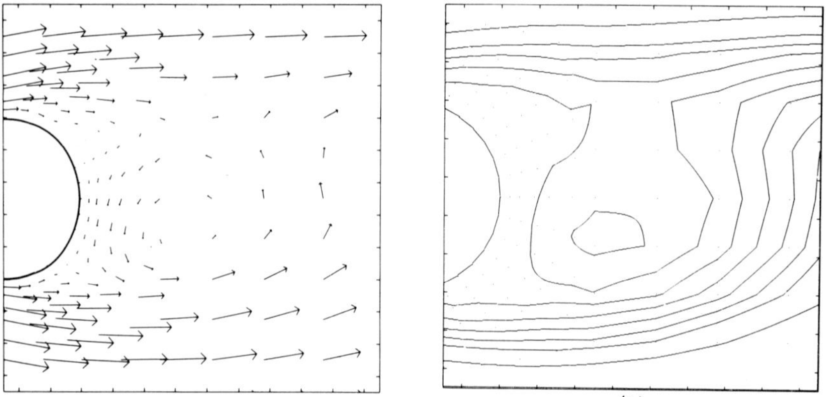

(a) (b)

Figure 18. Vectors (a) and streamlines (b) near the cylinder at t = 532.00.

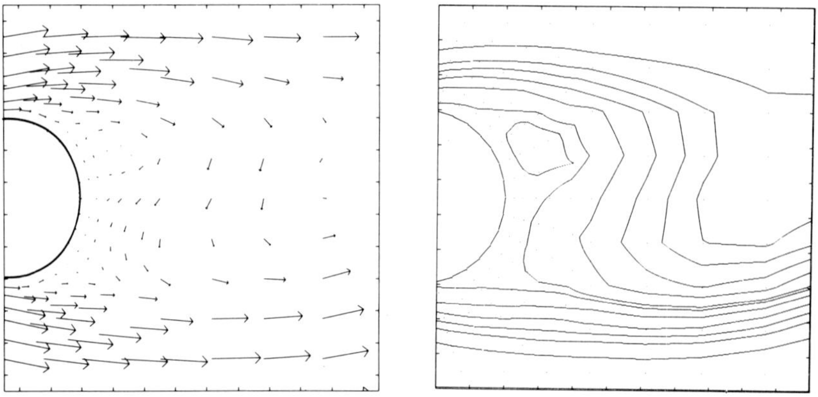

(a) (b)

Figure 19. Same as Figure 18 except at t = 533.55.

$$F_x^P = - \int_{\Gamma_c} n_x P \, d\ell \quad , \tag{24a}$$

$$F_y^P = - \int_{\Gamma_c} n_y P \, d\ell \quad , \tag{24b}$$

$$F_x^V = \mu \int_{\Gamma_c} \left[2n_x \frac{\partial u}{\partial x} + n_y \left(\frac{\partial u}{\partial y} + \frac{\partial v}{\partial x} \right) \right] d\ell \quad , \tag{24c}$$

and

$$F_y^V = \mu \int_{\Gamma_c} \left[n_x \left(\frac{\partial u}{\partial y} + \frac{\partial v}{\partial x} \right) + 2n_y \frac{\partial v}{\partial y} \right] d\ell \quad . \tag{24d}$$

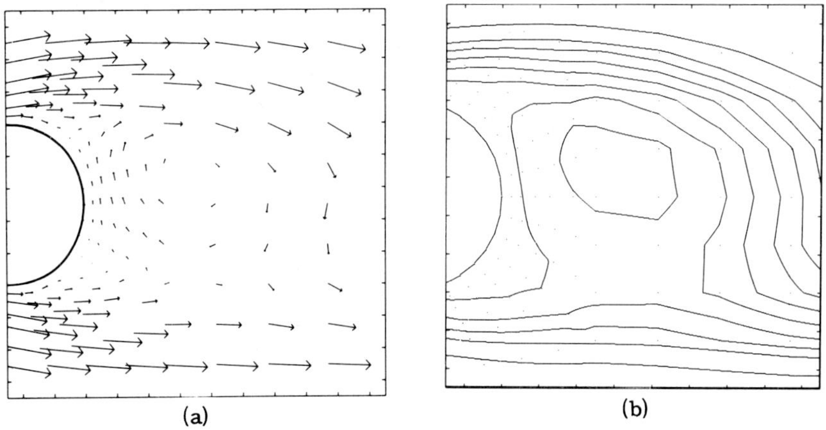

Figure 20. Same as Figure 18 except at t = 535.10.

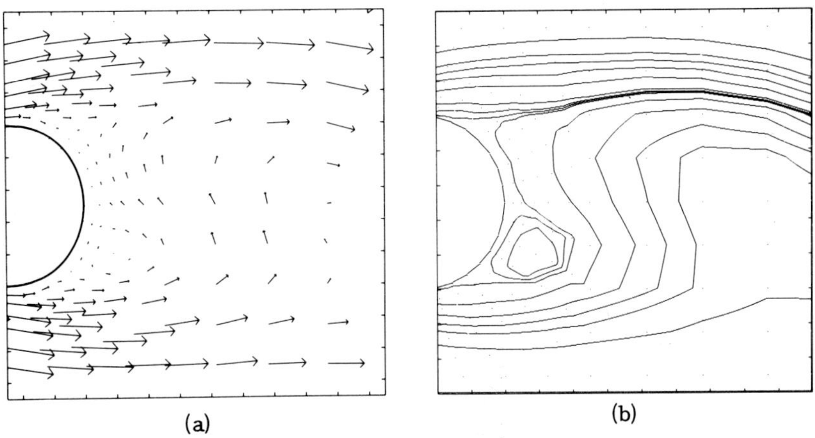

Figure 21. Same as Figure 18 except at t = 536.65.

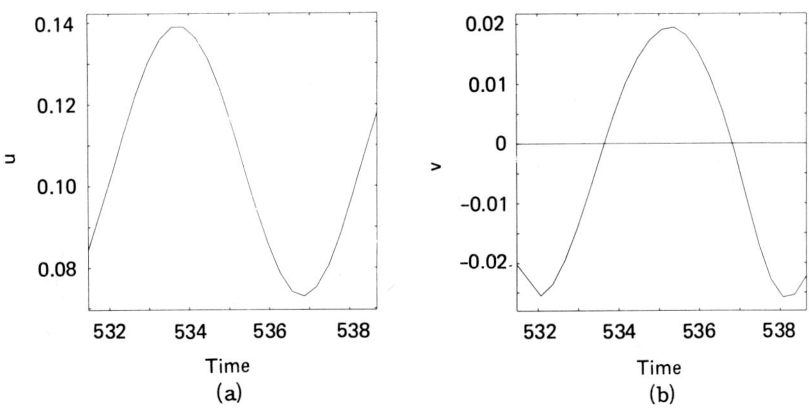

Figure 22. Velocity at node B during a shedding cycle: (a) horizontal, (b) vertical.

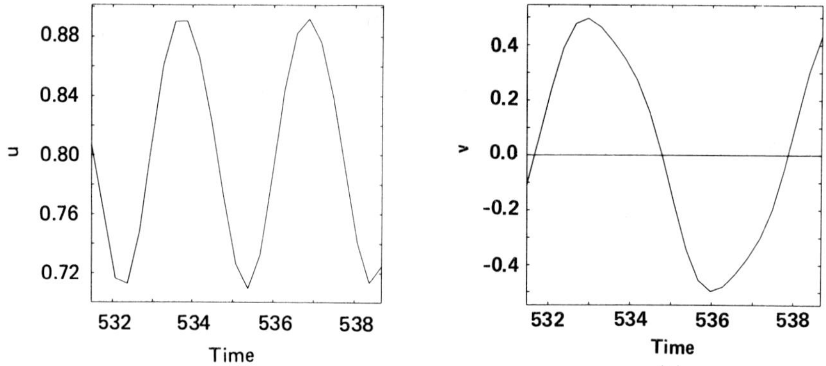

(a) (b)

Figure 23. Same as Figure 22 except at node A.

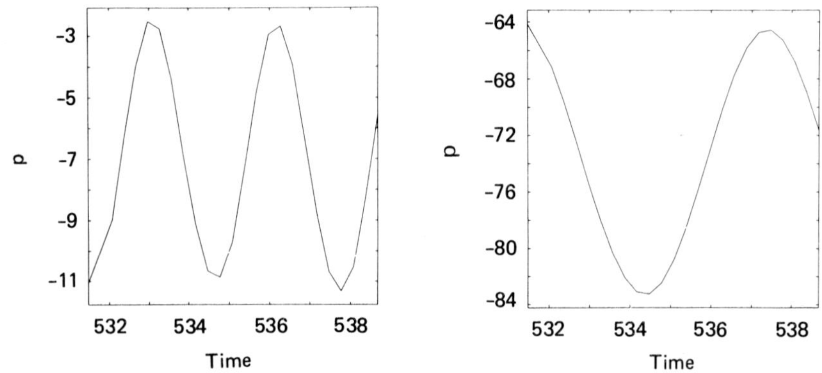

(a) (b)

Figure 24. Pressures during a shedding cycle: (a) node A, (b) node C.

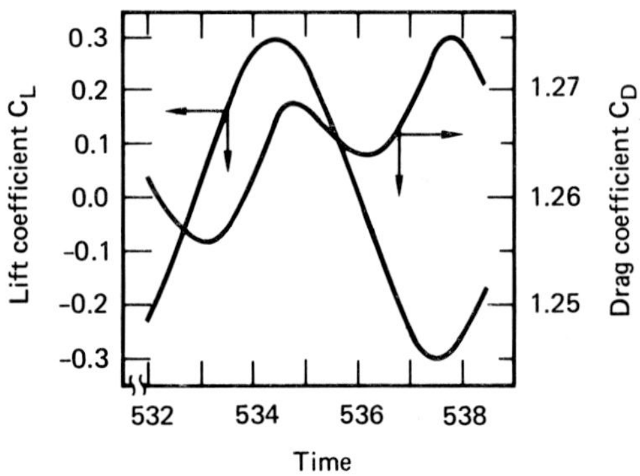

Figure 25. Lift and drag coefficients during a shedding cycle.

In Table 1 we compare some of our results with those of others; the agreement is seen to be quite good.

We repeated the last time segment of this simulation using a lumped mass matrix (row sum at element level) and were pleasantly surprised at the results; the frequency and phase were virtually identical and the amplitudes were within 10% of those using consistent mass. Although this good agreement may be partly due to a "very fine" grid, it does corroborate our earlier result wherein we found little error caused by lumping the mass on a pure advection (rotating cone) problem (Gresho et al., 1978) with this element. This interesting result seems to hold only for the 9-node element. Although we have not yet tested the linear element on this problem, we found that mass lumping with this element seriously degraded the phase accuracy on the rotating cone problem (ibid.). These results seem to indicate that mass lumping and explicit time integration (as implemented according to method described in the next section) may, under some circumstances (e.g. three-dimensional flow) be a viable option provided the full quadratic velocity approximation is employed (this seems to imply that the 27-node brick is required for equivalent accuracy in three-dimensional flows).

Table 1. Comparison of Computational and Experimental Results at Re $\cong$ 100.

Reference	Drag Coefficient Average	Drag Coefficient Peak-to-Peak	Lift Coefficient Peak-to-Peak	Strouhal No.
This study (Re=110*)	∿1.27 (and increasing)	.012	.60	.161
Jordan and Fromm (1972)	1.28	.012	.54	.16
Swanson and Spaulding (1978)	1.39	.02	.69	.174
Smith and Brebbia (1977; Table 1, scheme 2)	1.43	.02	.50	.166
Tritton (1959) (experimental)	∿1.26	—	—	.157 (.163 at Re=110)
Tritton (1977) (experimental)	∿1.33	—	—	—

*Obtained via $Re = \frac{D\rho}{\mu h}\int_0^h u\,dy$ at the channel inlet (t = 532).

Although the results obtained for this simulation are believed to be of excellent quality, the cost of the computations was fairly high. Using a very general 'research-version' of our code, each time step cost $\sim$20 seconds on a CDC-7600 computer; current estimates from our more efficient code (although still not truly optimized(are 10-12 seconds per step, which would give a cost of $\sim$3 minutes per shedding cycle. About 70% of the total 'charge time' was CPU and the remainder was "input/output" (I/O). We attempted to minimize the cost of I/O by using binary "READS" and "WRITES", and by employing very large buffers in large-core memory; three-fourths of available memory was, in fact, allocated to these buffers. Of the total CPU time, about 30% was spent in matrix factorizations, about 9% in I/O-related CPU charges, and about 6% in the construction of the nonlinear portions of the Jacobian matrix and the RHS vector.

THREE-DIMENSIONAL FLOW

In the near future we will be developing FEM methods directed toward our ultimate goal -- the time-dependent simulation of the 3-D wind field, temperature, and pollutant concentration in the atmospheric boundary layer. A logical step in this direction is the extension of our 2-D Navier-Stokes simulations to 3-D, toward which we limit subsequent discussion. It is obvious a priori that any adequate 3-D simulation will require at least an order of magnitude more node points than that for 2-D. This fact, coupled with the additional degrees of freedom (another velocity component, plus temperature for buoyancy-coupled flows and, utlimately, an additional equation for pollutant conservation), will require on the order of 10,000 nodes and anywhere from 30,000 to 60,000 equations. Fortunately we will at least be using the most powerful computer currently available -- the CRAY-1, which is a several-fold improvement, in both speed and memory, over our CDC-7600's. Even so, the magnitude of the problem is so large that we must carefully plan and select an appropriate (hopefully optimum) strategy, including equation formulation, time integration method, equation solution methods, and the 'best' elements. It is toward these issues that this section is devoted; we will discuss potentially viable techniques, ultimately focusing on those we believe to be most promising and will further develop.

Based on our experience in 2-D on a wide range of problems (which includes thermally-coupled flows), we will a priori select two elements which we believe have the best chance of success; these are the 'optimum constraint' elements referred to earlier in 2-D: (1) the 8-node isoparametric brick, using piecewise constant pressure and trilinear approximation for all other variables and (2) the 27-node isoparametric brick, using trilinear discontinuous pressure approximation (at the $2 \times 2 \times 2$ Gauss points) and triquadratic approximation for all other variables. The former, being somewhat simpler, will be the first element employed (also, we can confidently filter the spurious pressure mode for this element).

Coupled equations approaches

A direct extension of proven methods in 2-D will generally require the three momentum equations and the continuity equation to be solved simultaneously. Toward this end, we have considered (and rejected) three alternatives.

Direct methods The simplest technique is a direct extension of what we (and many others) have done in 2-D; viz, use implicit time integration techniques and Gaussian elimination methods (the profile, or skyline solver appears to be the most promising - see the chapter on solution methods by R. Taylor in Zienkiewicz (1977); see also Bercovier (1977), Hasbani and Engelman (1979), and Bercovier and Engelman (1979)) applied to the full, coupled system. While such a straightforward approach is 'guaranteed to work', we believe that it would be so expensive as to be limited to very small problems; for this reason, we exclude these simplest techniques from further consideration.

Indirect (iterative) methods Iterative methods for solving linear systems are known to become competitive for certain large problems because of the excessive storage and execution costs associated with direct methods. We have, and are still, considering iterative methods in 3-D -- we will discuss them in more detail in a subsequent section; here we dismiss them early on because of our "Basic-limitation-of-iterative-methods-hypothesis". Based on discussions (some of them quite detailed) with a number of mathematicians who qualify as experts in iterative solution methods (e.g. Richard Varga, Lou Hageman), and on the results from some of our own limited numerical experiments on the steady and time-dependent Stokes equations, we offer the following hypothesis: there is currently no iterative solution method which can be guaranteed to work on the discretized, primitive-variable incompressible NS equations, as depicted, for example, by Equations (3) or (11). The main reason for this "fear" is the diagonal block of zeros associated with the discretized continuity equation, and the associated indefiniteness of the matrix. Another serious limitation stems from the unsymmetric part of the matrix generated by the advection terms, which are especially important at large Re. The remaining difficulty, well-known to those who have used iterative methods even on symmetric, positive definite matrices associated with elasticity theory, is the irregular matrix structure which can result when the FEM is used to model geometrically complex regions. Thus, we also reject iterative methods as applied to the coupled equations.

Penalty methods The penalty method applied to Equation (1) yields a set of ODE's of the form (analogous to Equation (3))

$$M\dot{u} + (K + \lambda B)u + N(u)u = f \quad , \qquad (25)$$

where B is the penalty matrix (it corresponds to the operator $\nabla(\nabla \cdot \underline{u})$) and λ is the large 'penalty' parameter (e.g. 10^6). This method is listed under "coupled equations" since the penalty matrix necessarily couples all three velocity components and does not appear amenable to any uncoupled approach. While we have stated earlier that this method may be optimum (most cost-effective) in 2-D, we believe that it too will be limited to small problems in 3-D for the following reasons: (1) implicit time integration methods are virtually mandatory because λ is very large (the λ-terms mimic the pressure, which is inherently implicit), (2) the resulting matrix problem is unattractive for both direct and iterative methods. Direct methods are judged to be simply too expensive (even without a pressure variable), and for iterative methods, our fear is that the difficulties attendant to the pressure (i.e., incompressibility) are still 'present' (in a disguised form) and therefore, the convergence of iterative methods may well be a strong function of the magnitude of $\lambda\Delta t$ (even though B is symmetric). If, in fact, it turns out that $\lambda\Delta t = O(1)$ is required, then clearly the time step required for convergence would be prohibitively small. Although these ideas are somewhat speculative, we will let someone else prove us wrong and (for now at least) reject penalty methods for 3-D flow simulation.

Our 'bottom line' on techniques using the coupled equations is as simple as this: none of them appear to be viable for our purposes.

Uncoupled equations via the Poisson equation for pressure

It is well-known that if the pressure can be made to satisfy the appropriate Poisson equation and BC's, then the velocity field will automatically satisfy the appropriate continuity equation. The virtue of this strategy is that it opens the door to solving the NS equations in a less intensively-coupled manner.

'Continuum approaches' The appropriate continuum Poisson equation is (5), and it might appear that this is an appropriate starting point for the FEM solution. That this is not so is quite well-known in finite difference discretizations (Roache, 1976) and the reason there is the same one which will apply to FEM; viz, the utilization of $\nabla \cdot \underline{u} = 0$ is inherent in the derivation of Equation (5), but its discretized counterpart, even though leading to a "classic-type" matrix problem (positive-definite, symmetric, etc.), is not sufficiently compatible with the associated (discretized) momentum equations (and, the corresponding discretized form of $\nabla \cdot \underline{u} = 0$ is, to our knowledge, not even definable). One 'solution' to this 'incompatibility problem', as summarized by Roache (ibid) for the finite difference approach and implemented by Olson and Tuann (1978) using the FEM, is to carry the time-derivative of $\nabla \cdot \underline{u}$ when deriving the Poisson equation; this leads to

$$\nabla^2 P = -\rho\left[\frac{\partial(\nabla \cdot u)}{\partial t} + \nabla \cdot (\underline{u} \cdot \nabla \underline{u})\right] \quad . \tag{26}$$

When this equation is discretized in space and time, there results a term like $(D_{n+1}-D_n)/\Delta t$, where D is the 'appropriate' (or an estimate thereof) discretized version of $\nabla \cdot \underline{u}$. The trick is to retain the D_n term in the final discretized system, but to drop the D_{n+1} term (to try to enforce $D = 0$ at t_{n+1}, even though it might not have been zero at t_n). This device has proven most useful and should probably be imitated in any scheme which uses a discretized version of Equation (26); i.e., Equation (5) should probably not be used.

In spite of this trick, we are still fearful of employing Equation (26), for the following reasons (which we share with Chorin (1968)):

1. It is a higher-order, derived equation. This means first, that a continuous (C^0) pressure approximation is required, in contrast to that from the weak form of NS equations, which can tolerate discontinuous (C^{-1}) pressure fields (and this would seem to immediately preclude the use of those elements which we have identified as 'optimum'). Also, the higher order leads directly to the second reason.

2. As noted previously, the BC's for such a Poisson equation must be derived from the original momentum equations (1a), and this introduces the necessity of computing second spatial derivatives of the velocity at the boundary; e.g. for the simplest case of a rigid wall, the BC is

$$\frac{\partial P}{\partial n} = \mu \frac{\partial u_n}{\partial n^2} \quad . \tag{27}$$

Such a BC is very difficult to satisfy with C^0 velocity approximations, especially on complex boundaries. Short of employing a smoother velocity approximation (e.g., C^1 basis functions, which are also not required in the weak form of the NS equations), such as that used by Olson and Tuann (1976) for the stream function, there appears to be no simple way to satisfy Equation (27), especially using <u>linear</u> approximation for velocity (perhaps the special wall element designed by Hutton and Smith (1979) would work here; they claim some success by employing cubic velocity variation for the normal velocity in the first row of elements along a solid wall - they use quadratic velocity elsewhere via the serendipity element and do not form a Poisson equation). This is not to imply that it is hopeless, however. Finite difference practitioners approximate such difficult BC's in a variety of ways, such as polynomial extrapolation. These notions could probably be carried over to FEM by some clever device (they must be, if Equation (27) is to be satisfied; especially if simple elements are employed).

3. Finally, since the continuity equation is introduced in a rather subtle and indirect manner, one is never sure exactly what discretized form of $\nabla \cdot \underline{u} = 0$ is actually contained in the simulation.

Hence, we discard another approach, in favor of a more direct and (we believe) consistent approach, described next.

<u>Consistent discretized approach</u> We return now to the original discretized NS system of Equation (3) except that we replace

Equation (3b) by its time derivative, $C^T\dot{u} = 0$, which is legitimate for reasons previously discussed. Since M is non-singular, the acceleration may be eliminated to give the appropriate discretized Poisson equation for the pressure (which includes the appropriate BC's, permits C^{-1} pressure representations, and enforces a specific discretized continuity equation — $C^T u = 0$):

$$(C^T M^{-1} C)P = C^T M^{-1} [f - Ku - N(u)u] \quad . \tag{28}$$

There is, of course, a price to be paid for these alleged advantages; viz, it appears hopeless to be able to form either $C^T M^{-1}$ or the 'Laplacian' matrix ($C^T M^{-1} C \equiv L$ is, in fact, an appropriate representation of ∇^2) when the (consistent) mass matrix is employed. Hence, the unfortunate practical consequence of this approach is the need to "lump the mass" (i.e., convert M to a diagonal form, by techniques discussed, for example in Zienkiewicz (1977) or Lee et al. (1979)); this permits the simple construction of M^{-1}, while retaining the principal advantages enumerated above. While we are optimistic that mass lumping is not too deleterious using quadratic velocity approximation (as discussed earlier for our vortex shedding simulation), we are still concerned that it may cause a significant loss of phase accuracy using linear velocity approximation. Even with the lumped mass approximation, the construction of the L matrix will be nontrivial, since M^{-1} is a global vector and it appears that global construction techniques may be required (only once per problem, however; or even once 'per grid' if the BC's are not changed).

It should be emphasized that if P satisfies Equation (28) and u satisfies Equation (3a), that Equation (3b) will indeed be 'automatically' satisfied, and it can be removed from the system, thus leading to the possibility of an uncoupled approach. To further uncouple the equations, we also invoke the following new 'ground rule': rather than starting with the stress-divergence form of the NS equations, we will replace Equation (1a) by its equivalent (in the continuum at least) form,

$$\rho\left(\frac{\partial \underline{u}}{\partial t} + \underline{u} \cdot \nabla \underline{u}\right) = - \nabla P + \mu \nabla^2 \underline{u} \quad , \tag{29}$$

which will have two effects in the discretized form:

1. The viscous terms are uncoupled in the velocity components (the stress-divergence form leads to an additional viscous term, $\mu\nabla(\nabla \cdot \underline{u})$, which, while zero in the continuum, couples all velocity components in the FEM discretization; i.e., it is not zero in the discretized form).

2. The natural BC's associated with the weak form are changed from those in Equation (4) to

$$f'_n = - P + \mu \frac{\partial u_n}{\partial n} \quad , \tag{30a}$$

and

$$f'_\tau = \mu \frac{\partial u_\tau}{\partial n} \quad , \tag{30b}$$

where the prime is a reminder that these expressions are no longer interpretable as traction forces (this loss of physical significance seems to be much less relevant in fluid mechanics than in solid mechanics, however -- at least at significant Re; in fact, there are even situations where Equation (30b) is a preferred outflow BC to Equation (4b) -- cf. Hutton and Smith (1979)). With this new "ground rule", the only remaining term which couples the three momentum equations is the advective operator, $\underline{u} \cdot \nabla$; we will later present methods by which even these terms may be uncoupled.

Since the discretized form of Equation (29) is still represented by Equation (3a) (the details of the K matrix are changed; it is now block diagonal), the final discretized system we consider for 3-D flow is given by Equations (3a) and (28), where M is now a diagonal matrix. It remains only to define a solution method which is able to utilize the ostensibly uncoupled formulation, and this is disussed next.

Proposed solution methods The technique we plan to use on Equations (3a) and (28) is equivalent to that devised and demonstrated by Chorin (1968, 1969) using finite difference methods, and is basically as follows, given a solution at time t_n:

1. Omit the pressure gradient terms (CP) from Equation (3a), discretize the resulting equation in time (in any way you like, which includes explicit, implicit, or semi-implicit schemes), and advance the solution by one time step. The resulting 'velocity', called $\tilde{u}_{n+1}$ is a provisional quantity, since it does not include the effect of the pressure gradient on the acceleration.

2. Solve Equation (28) for P, using $\tilde{u}_{n+1}$ to evaluate the terms on the RHS; call the result P_{n+1} (the new pressure).

3. Obtain the final velocity, $\tilde{u}_{n+1}$, by re-introducing the pressure gradient terms into Equation (3a); an 'implicit version' of the time integration scheme used to obtain $\tilde{u}_{n+1}$ must be used here, as it must involve the new pressure, P_{n+1}.

The claim, which has more or less been proven by Chorin (ibid), is that the final results from this sequential (uncoupled) strategy are the same (within a multiple of Δt^2; i.e., of the local time truncation error of a first order time integration scheme) as those which would obtain using the same (implicit) time integration scheme on the fully coupled system, Equations (3a) and (3b) or their equivalent, Equations (3a) and (28). It is important to note that the time integration scheme must be one which treats the pressure and continuity equation implicitly.

It may be useful to clarify these ideas; we will attempt this in two ways, both employing the simplest (first-order accurate) Euler time integration schemes for simplicity of presentation (we will also consider schemes based on TR and AB, in our implementation).

In the simplest approach, the provisional velocity field is obtained purely explicitly, via

1. $\tilde{u}_{n+1} = u_n + \Delta t M^{-1}[f - Ku_n - N(u_n)u_n]$ (31)

2. This interim result is inserted into the RHS of Equation (28) to obtain P_{n+1}.

3. The final velocity is obtained from

$$u_{n+1} = \tilde{u}_{n+1} - \Delta t M^{-1} C P_{n+1} \tag{32}$$

If implicit integration is exclusively employed, the first step is changed to:

1. Solve for $\tilde{u}_{n+1}$ from

$$\tilde{u}_{n+1} = u_n + \Delta t M^{-1}[f - K\tilde{u}_{n+1} - N(\tilde{u}_{n+1})\tilde{u}_{n+1}] \;, \tag{33}$$

which is a large nonlinear system.

Steps (2) and (3) are the same as those above using FE. Chorin has conjectured that the stability of the overall scheme will be the same as that used to obtain $\tilde{u}_{n+1}$ (i.e. Step (1)); this would make the former scheme conditionally stable and the latter unconditionally stable.

While the above steps describe the essence of the procedure, the variant employed by Chorin (1968) is computationally more efficient, with results which appear to be equivalent to within a power of Δt (i.e., to within $O(\Delta t^2)$), and this is the approach we will use; viz, after solving for $\tilde{u}_{n+1}$, from Equation (31) or (33), the equivalent Poisson equation for P_{n+1} is formed via the equations

$$u_{n+1} = \tilde{u}_{n+1} - \Delta t M^{-1} C P_{n+1} \quad , \tag{34a}$$

and

$$C^T u_{n+1} = 0 \quad . \tag{34b}$$

This leads to

$$LP_{n+1} = C^T \tilde{u}_{n+1}/\Delta t \quad , \tag{35}$$

which is solved for P_{n+1}, after which u_{n+1} is obtained from Equation (34a). The advantages of this approach are: (1) the RHS vector for the Poisson equation is simpler to construct and (2) the continuity equation at t_{n+1} appears to be more rigidly enforced; this may be especially useful if iterative methods are employed.

Having obtained what appears to be a legitimate and useful method for uncoupling the equations, we now examine some of the subsequent options in more detail.

Solution of the Poisson equation Since the L matrix is symmetric and constant (it depends only on the FEM mesh, the BC's on velocity, and, of course, the basis functions), we believe that it is most appropriate to employ a disk-based Gaussian elimination technique (e.g. banded or profile solver). The matrix would be formed (probably globally, for simpler programming), factored, and stored in factored form on disk in a 'pre-processor' step; subsequent pressure solutions would be obtained rather efficiently via one forward-reduction and back-substitution per time step. Iterative methods could be employed, and would probably work well (the matrix is of the 'right type'), but we believe they would generally be less cost-effective (and more difficult to program) in this case.

The remaining discussion is devoted to the options available for computing the interim velocity, $\tilde{u}_{n+1}$, which will probably be the most expensive part of the algorithm; again, we limit the current discussion to variants of the first-order Euler methods.

Explicit method Since the CRAY-1 is an efficient machine in vector mode (e.g. matrix-vector scalar products), it is conceivable that this simplest of approaches may be viable. For example, it has recently been reported (Hallquist, 1979) that an explicit 3-D dynamics code (solid mechanics) has achieved an efficiency of "2 minutes per million mesh-cycles"; this translates into, for example, 50 time-steps per minute of CPU time on a mesh with 10^4 nodes and three variables per node. If we could achieve something close to this speed for the discretized NS equations, it would appear that explicit methods might be competitive in spite of severe time step restrictions (diffusive and advective stability limits will usually restrict Δt to very small values).

Implicit methods Fully-coupled implicit methods necessitate the simultaneous solution of '3N' coupled nonlinear equations (even when discretized from Equation (29) because of the advection operator). We believe such an approach is impractical (it violates our ground rule) and drop it from further consideration; i.e., we will not solve Equation (33), except, perhaps, approximately via one-step Newton, as discussed below.

Semi-implicit methods In order to achieve the desired uncoupling of the three momentum equations, yet gain some stability advantages over explicit methods, the advection terms must be treated in some sort of semi-implicit manner; we will consider several such schemes, as outlined below.

1. Implicit treatment of viscous terms. If we write the equation for the provisional velocity as

$$\tilde{u}_{n+1} = u_n + \Delta t M^{-1} [f - K\tilde{u}_{n+1} - N(u_n)u_n] \quad , \tag{36}$$

the time step restriction associated with the viscous terms is removed, but we must solve three linear systems (each of size N, owing to the discretized form of Equation (29) — with P omitted of course) whose symmetric matrix is $(I + \Delta t M^{-1} K)$. This approach could be attractive if the critical time step associated with the diffusion terms is exceedingly small compared to that for advection (Courant condition limit) when purely explicit methods are considered.

2. Implicit diffusion, semi-implicit advection. Here we write

$$\tilde{u}_{n+1} = u_n + \Delta t M^{-1} [f - K\tilde{u}_{n+1} - N(u^*)\tilde{u}_{n+1}] \quad , \tag{37}$$

where u^* is presumed to be known (it could either be u_n or a predictor value of $\tilde{u}_{n+1}$). In this approach, the matrix is unsymmetric (it is given by $I + \Delta t M^{-1}(K + N(u^*))$ and therefore the solution of the three linear systems will be more expensive. The increased stability of the advection terms (the 'amount' of which is currently unknown) could conceivably make this approach useful in certain situations.

3. One-step Newton. Finally, it may be possible to extend some of our successful experience with one-step Newton to the uncoupled systems in 3-D (albeit not rigorously). For example, we could form the one-step Newton system corresponding to the coupled equation for $\tilde{u}_{n+1}$ (33), and then split the 3N system into three systems of size N by simply dropping all 'coupling terms' from the Jacobian. This scheme, or a similar one, might even be the most stable and efficient of the semi-implicit methods.

Solution of linear systems In any of the semi-implicit schemes discussed above, one of the most important cost considerations (besides those caused by Δt stability limits) is related to the solution of large (of order $N = 10^4$ or so) systems of linear algebraic equations. Additional issues relate to variable vs. 'fixed' (or occasionally variable) Δt algorithms and symmetric vs unsymmetric systems. For example, if the simplest semi-implicit scheme is chosen, and a more-or-less fixed Δt strategy is adopted, the solution of the symmetric linear systems associated with Equation (36) might be quite efficiently processed by a direct solution method, since the matrix can be stored in factored form. On the other hand, if Δt were to be varied often and/or if a semi-implicit scheme were adopted (e.g. Equation (37)) which required an update of the unsymmetric coefficient matrix at each time step, some of the inherent advantages of direct methods may be lost relative to iterative schemes (note that the problems referred to earlier, related to the continuity equations and iterative solvers, do not arise in this uncoupled approach).

As mentioned earlier, we have been considering the possibility of solving linear systems via iterative methods, a brief summary of which is presented in this section. Direct solvers, being less mysterious and generally more robust, are also sufficiently well-developed that we will not discuss them here except to say that we will consider only disk-based profile or skyline solvers of the type described by R. Taylor in Zienkiewicz (1977) or Hasbani and Engelmann (1979). The iterative methods we believe are worth considering will be briefly discussed, as will their remaining most serious (potential) limitation.

Iterative methods considered All iterative methods considered (and those not considered) should have the property that little or no storage of zero coefficients is required (which implies a minimum of multiplications by zero), and their convergence rates should be "high enough" to indeed make them faster than any direct solver. It is also desirable, but probably not assuredly attainable, that they be able to solve systems containing unsymmetric matrices.

We have considered three types of iterative solvers as being more-or-less appropriate to at least some of the matrices in question. All three are guaranteed to converge for symmetric, positive-definite matrices. These are: block successive overrelaxation (SOR), Chebyshev polynomial semi-iterative (CSI), and conjugate gradient (CG). We cannot hope to describe these techniques in this paper, but will only attempt to indicate a few of their properties and features. For more information on SOR and CSI methods, the reader is referred to the books by Varga (1962), Young (1971), Wachspress (1966), and Hageman and Young (1980)). For CG references, it is best to consult the literature, starting with the classic paper by Hestenes and Stiefel (1952); other useful references are Jennings and Malik (1978), Fried (1970), and Metzler and Fried (1977), the latter two being applied to FEM matrices.

The SOR (or block SOR) method has the advantage of having a fairly broad base of theoretical knowledge and it is known to work well on certain types of matrices. The disadvantages of SOR include: (1) the convergence rate is often quite sensitive to the accurate estimate of the relaxation parameter, for which good estimates are only available for symmetric, positive-definite matrices which are structured in the so-called 2-cyclic form, (2) the global matrix must be available in a known, structured form.

The CSI method has the advantage of being somewhat more generally applicable than SOR, yet also suffers from the need to estimate convergence parameters. Also, CSI converges at about one-half the rate of the optimum SOR method.

The CG method has some inherent advantages, especially for FEM applications: (1) often the global matrix need never be assembled, since only matrix-vector multiplications are involved, (2) there are no parameters which need to be optimized in order to accelerate convergence, and (3) it seems to converge as rapidly as other schemes. The principal disadvantage of CG is related to the less broad base of experience with it; e.g. in certain cases, it is not yet clear when the iterations should be stopped.

Potential limitation of iterative methods The convergence of all iterative methods becomes somewhat uncertain when addressing non-symmetric matrices, which occur (and are more dominant at large Re) when the advection operator is placed on the LHS and treated semi-implicitly. Since the flows are time-dependent, however, there is a potentially saving feature in that the identity matrix, which is positive-definite and symmetric, is (effectively) multiplied by $1/\Delta t$, so that iterative methods are almost assured of converging if Δt is small enough. Of course, if these Δt requirements are anywhere near as stringent as the basic stability limits associated with explicit methods, they are obviously not cost-effective.

On the other side of the ledger, however, is a successful experience we have had using "point SOR" on a purely hyperbolic problem (convergence of SOR is not guaranteed). When we used this method on the "rotating cone problem", in which we solved

$$\frac{\partial T}{\partial t} + u\frac{\partial T}{\partial x} + v\frac{\partial T}{\partial y} = 0 \qquad (38)$$

in a regular domain, where u(x,y) and v(x,y) describe a purely rotational flow, using 8-node serendipity elements, we were able to solve about 800 time-dependent equations in the same (or less) time as with a banded direct solver. The reason for this was ascribed to the desirable properties of the mass matrix (positive-definite, symmetric) and a sufficiently small time step, which was required for temporal accuracy (the Courant number was $\sim$.15). The method did not converge when Δt was too large, in which case the "skew-symmetric" advection matrix became too "important".

With these admonitions in mind, we must surely be careful about selecting problems and domains, treatment of the advective terms, and selection of time-steps to be used with iterative methods. Proper regard to these items may lead to successful flow simulations via iterative methods. Depending on the methods selected, we may or may not be able to (or want to) invoke some form of automatic time step control; at the time of this writing, all we can say is that if such a technique is deemed worthwhile, it will be necessary more ad hoc than the 'rigorous' algorithm developed for 2-D flows. In closing this section on 3-D flows, it may even be appropriate to offer a final speculation; viz, that the 'optimum' method finally developed for 3-D may even be cost-effective when applied to very large 2-D problems.

SUMMARY AND CONCLUSIONS

We have carefully described and demonstrated a useful, accurate, honest, and robust technique for solving the time-dependent NS equations in 2-D. We believe that the solutions obtainable from our 2-D code (and the Boussinesq version thereof — see Gresho et al. (1979) will rival those of any other with respect to accuracy and honesty in the simulation of actual physical flows.

We have also outlined an analogous method (FE/BE) which we feel would be quite useful for obtaining steady-state solutions to the NS or Boussinesq equations.

We have also discussed some of the numerous and complex issues to be faced in 3-D FEM simulations and have outlined our plans in this area. We will carefully compare the relative merits of explicit vs. semi-implicit approaches.

Finally, we have suggested some optimum elements and solution strategies for both 2-D and 3-D flows. Those elements with the optimum constraint ratio are judged to be superior in both 2-D and 3-D; in the former, the penalty approximation is probably a more cost-effective technique than 'mixed-interpolation', while in 3-D the converse appears to be true (in conjunction with the suggested uncoupled equation approaches).

ACKNOWLEDGMENTS

The time integration scheme was developed with the assistance of Dr. A. C. Hindmarsh, whose many contributions were crucial to the ultimate success of the method. Dr. N. Madsen assisted us in both the theory and implementation of iterative solution methods. T. W. Stullich provided much of the programming support, and C. Upson provided the recently incorporated 'zoom' capability. Finally, we are grateful to Dr. G. Goudreau and M. Gerhard for providing us with their mesh generator code and guidance in using it. All of the above are at LLL.

This work was performed under the auspices of the U. S. Department of Energy by the Lawrence Livermore Laboratory under contract No. W-7405-Eng-48.

REFERENCES

Aris, R. (1962) Vectors, Tensors, and the Basic Equations of Fluid Mechanics. Prentice-Hall, N. J., p. 129.

Bercovier, M. (1977) A Family of Finite Elements with Penalization for Numerical Solution of Stokes and Navier-Stokes Equations, in Information Processing 77, B. Gilchrist (ed.), North Holland Publishing Co.

Bercovier, M. and M. Engelman (1979) A Finite Element for Incompressible Viscous Fluid Flows. J. Comp. Phys., 30, 181-201.

Bristeau, M., R. Glowinski, J. Periaux, P. Perrier, O. Pironneau, and G. Poirier (1978) Application of Optimal Control and Finite Element Methods to the Calculation of Transonic Flows and Viscous Incompressible Flows, Research Report No. 294, Institut de Recherche d'Informatique et d'Automatique, Le-Chesuay, France.

Byrne, G., A. Hindmarsh, K. Jackson, and H. Brown (1977) A

Comparison of Two ODE Codes: GEAR and EPISODE. Comp. and Chem. Eng., 1, 133-147.

Chorin, A. J. (1968) Numerical Solution of the Navier-Stokes Equations. Math. Comp., 745-762.

Chorin, A. (1969) On the Convergence of Discrete Approximations to the Navier-Stokes Equations. Math. Comp., 23, 341-353.

Fried, I. (1970) A Gradient Computational Procedure for the Solution of Large Problems Arising from the Finite Element Discretization Method. Int. J. Num. Meth. Eng., 2, 447-494.

Gartling, D. R. (1974) Texas Fluid Analysis Program — User's Manual. TICOM Report 75-2, Texas Institute for Computational Mechanics, Austin, TX.

Gear, C. W. (1971) Numerical Initial Value Problems in Ordinary Differential Equations. Prentice-Hall, N. J.

Glowinski, R., B. Mantel, J. Periaux, and O. Pironneau (1978) H^{-1} Least Squares Method for the Navier-Stokes Equations, in Numerical Methods in Laminar and Turbulent Flow, Proceedings of the First International Conference held at University College, Swansea, U.K., July 17-21, 1978; Pentech Press, London.

Gresho, P. M. and R. L. Lee (1979) Don't Suppress the Wiggles — They're Telling You Something, in Proc. Special Symposium on Finite Element Methods for Convection-Dominated Flows, Winter Annual ASME Meeting, December 2-7, 1979, N. Y.

Gresho, P. M., R. L. Lee and R. L. Sani (1978) Advection-Dominated Flows, with Emphasis on the Consequences of Mass Lumping. Finite Elements in Fluids, v. 3, J. Wiley & Sons, N. Y.

Gresho, P. M., R. L. Lee, S. T. Chan, and R. L. Sani (1979) Solution of the Time-Dependent Incompressible Navier-Stokes and Boussinesq Equations Using the Galerkin Finite Element Method; in Proceedings of the IUTAM Symposium on Approximation Methods for Navier-Stokes Problems, held in Paderborn, W. Germany, September 9-15, 1979, Springer-Verlag, Berlin.

Gresho, P. M., R. L. Lee, S. T. Chan, and J. M. Leone (1980) A New Finite Element for Incompressible or Boussinesq Fluids, in Proc. Third International Conference of Finite Elements in Flow Problems, June 10-13, 1980, Banff, Canada.

Hageman, L. and D. M. Young (1980) Applied Iterative Methods, Academic Press, N. Y. (to appear).

Hallquist, J. O. (1979) Private Communication, Lawrence Livermore Laboratory, Livermore, CA.

Hasbani, Y. and M. Engelman (1979) Out-of-Core Solution of Linear Equations with Non-Symmetric Coefficient Matrix. Comp. and Fluids, 7, 13-31.

Hestenes, M. and E. Stiefel (1952) Method of Conjugate Gradients for Solving Linear Systems. Nat. Bur. Standards J. of Res., 49, 409-436.

Hindmarsh, A. (1972) Construction of Mathematical Software, Part III: The Control of Error in the GEAR Package for Ordinary Differential Equations. Lawrence Livermore Laboratory Report UCID-30050, Part 3, Livermore, CA.

Hood, P. (1976) Frontal Solution Program for Unsymmetric Matrices. Int. J. Num. Meth. Eng., 10, 379-399.

Hughes, T. J. R., W. T. Liu, and A. Brooks (1979) Finite Element Analysis of Incompressible Viscous Flows by the Penalty Function Formulation. J. Comp. Phys., 30, 1-60.

Hutton, A. and R. Smith (1979) The Prediction of Laminar Flow over a Downstream-Facing Step by the Finite Element Method, Report RD/B/N3660, Central Electricity Generating Board, London, England.

Huyakorn, P. S., C. Taylor, R. L. Lee and P. M. Gresho (1978) A Comparison of Various Mixed-Interpolation Finite Elements in the Velocity-Pressure Formulation of the Navier-Stokes Equations. Comp. and Fluids, 6, 25-35

Jennings, A. and G. Malik (1978) The Solution of Sparse Linear Equations by the Conjugate Gradient Method. Int. J. Num. Meth. Eng., 12, 141-158.

Jordan, S. K. and J. E. Fromm (1972) Oscillatory Drag, Lift, and Torque on a Circular Cylinder in Uniform Flow, The Physics of Fluids, 15, 371-376.

Lee, R. L., P. M. Gresho, and R. L. Sani (1979) Smoothing Techniques for Certain Primitive Variable Solutions of the Navier-Stokes Equations. Int. J. Num. Meth. Eng. (in press).

Lee. R. L., P. M. Gresho, S. T. Chan, and R. L. Sani (1980) A Comparison of Several Conservative Forms for Finite Element Formulations of the Incompressible Navier-Stokes or Boussinesq Equations, in Proc. Third International Conference on Finite Elements in Flow Problems, June 10-13, 1980, Banff, Canada.

Leone, J. M. (1980) Finite Element Simulations of Stratified Flow over Simple Geometrical Obstructions and Arbitrarily Complex Terrain. Ph.D. Thesis, Iowa State University, Ames, Iowa.

Leone, J. M. and P. M. Gresho (1979) Finite Element Simulations of Steady, Two-Dimensional, Viscous Incompressible Flow over a Step. (Submitted to J. Comp. Phys.)

Marshall, R., J. Heinrich, and O. C. Zienkiewicz (1978) Natural Convection in a Square Enclosure by a Finite Element Penalty Function Method, Using Primitive Fluid Variables. Num. Heat Transfer, 1, 315-330.

Metzler, J. and I. Fried (1977) The Conjugate Gradient Method with Finite Elements, in Proc. 2nd IMACS Int. Symp. Computer Mech. for Partial Differential Equations, Lehigh University.

Olson, M. D. (1977) Comparison of Various Finite Element Solution Methods for the Navier-Stokes Equations, in Finite Elements in Water Resources. Pentech Press, London.

Olson, M. D. and S. Y. Tuann (1978) Primitive Variables Versus Stream Function Finite Element Solutions of the Navier-Stokes Equations. Finite Elements in Fluids, v. 3, J. Wiley & Sons, N. Y.

Roache, P. J. (1976) Computational Fluid Dynamics, 2nd Ed., Hermosa Press, Albuquerque, N. M.

Sani, R. L., P. M. Gresho, and R. L. Lee (1978) The Imposition of Incompressibility Constraints via Variational Adjustment of Velocity Fields, in Proc. First International Conference on Numerical Methods in Laminar and Turbulent Flow, Pentech Press, London.

Sani, R. L., P. M. Gresho, and R. L. Lee (1980) On the Spurious Pressures Generated by Certain FEM Solutions of the Incompressible Navier-Stokes Equations, in Proc. Third International Conference on Finite Elements in Flow Problems, June 10-13, 1980, Banff, Canada.

Shampine, L. and M. Gordon (1975) Computer Solution of Ordinary Differential Equations: The Initial Value Problem, W. Freeman & Co., San Francisco.

Smith, S. L. and C. A. Brebbia (1977) Improved Stability Techniques for the Solution of Navier-Stokes Equations, Appl. Math. Modeling, v. 1, 227-234.

Swanson, J. C. and M. L. Spaulding (1978) Three-Dimensional Numerical Model of Vortex Shedding From a Circular Cylinder, in Nonsteady Fluid Dynamics, ASME, N. Y.

Tritton, D. (1959) The Flow Past a Circular Cylinder at Low Reynolds Number, J. Fl. Mech., 6, 547-567.

Tritton, D. (1977) Physical Fluid Dynamics, Van Nostrand Reinhold Co., N. Y.

Varga, R. S. (1962) Matrix Iterative Analysis, Prentice-Hall, N. J.

Wachspress, E. (1966) Iterative Solutions of Elliptic Systems, Prentice-Hall, N. J.

Young, D. M. (1971) Iterative Solution of Large Linear Systems, Academic Press, N. Y.

Zienkiewicz, O. C. (1977) The Finite Element Method, 3rd ed., McGraw-Hill, London.

APPENDIX 1

Weak form and natural boundary conditions

Converting to tensor notation and employing the Einstein summation convention permits Equation (1) to be restated as

$$\rho\,(\dot{u}_\alpha + u_\beta u_{\alpha,\beta}) = \tau_{\alpha\beta,\beta} \quad , \tag{A1}$$

$$u_{\gamma,\gamma} = 0 \quad , \tag{A2}$$

and

$$\tau_{\alpha\beta} = -P\,\delta_{\alpha\beta} + \mu(u_{\alpha,\beta} + u_{\beta,\alpha}) \quad , \tag{A3}$$

where the dot means partial differentiation with respect to time and a comma followed by an index indicates partial differentiation with respect to the spatial coordinate indicated by the index. All Greek indices range from 1 to DN, the dimension of the physical space.

To obtain the weak, or Galerkin form of Equation (A1), it is first multiplied by any velocity basis function, ϕ_i; $i = 1,2,\ldots,N$, where N is the dimension of the finite dimensional space for the velocity approximation, and integrated over the spatial domain Ω:

$$\int_\Omega \rho(\dot{u}_\alpha + u_\beta u_{\alpha,\beta})\,\phi_i = \int_\Omega \tau_{\alpha\beta,\beta}\,\phi_i \quad ; \tag{A4}$$

the RHS of which can be rewritten as

$$\int_\Omega (\tau_{\alpha\beta}\,\phi_i)_{,\beta} - \int_\Omega \tau_{\alpha\beta}\,\phi_{i,\beta} \quad .$$

Applying the divergence theorem to the first of the above integrals introduces boundary integral terms (and natural BC's) and leads to the final weak form of the momentum equation; viz.

$$\int_\Omega \rho\,(\dot{u}_\alpha + u_\beta u_{\alpha,\beta})\phi_i = \int_{\partial\Omega} \phi_i\, n_\beta \tau_{\alpha\beta} - \int_\Omega \tau_{\alpha\beta}\,\phi_{i,\beta} \quad , \tag{A5}$$

where $\partial\Omega$ is the boundary of Ω and n_α is the α-component of the outward unit normal vector on $\partial\Omega$. Similarly, the vector $n_\beta\tau_{\alpha\beta}$ is the α-component of the surface traction force, $f_\alpha \equiv n_\beta\tau_{\alpha\beta}$. Note that the weak formulation has relaxed the differentiability requirements on $\tau_{\alpha\beta}$ (derivatives have been shifted to ϕ_i); this is the reason that C^{-1} approximations can be employed for velocity derivatives and pressure (ala Equation (A3)), since $\tau_{\alpha\beta}$ must now be merely square integrable. To complete the Galerkin equations, Equation (A2) (which implicitly defines the pressure) is multiplied by any pressure basis function, ψ_i; $i = 1,2,\ldots,M$, where M is the dimension of the finite dimensional space used for pressure approximation, and integrated over Ω; viz.

$$\int_\Omega u_{\gamma,\gamma}\,\psi_i = 0 \quad . \tag{A6}$$

For convenience and ease of interpretation, we now write the Galerkin equations in terms of Cartesian coordinates in 2-D (i.e., u_1 = u, u_2 = v, etc.):

$$\int_\Omega \rho\left(\frac{\partial u}{\partial t}+u\frac{\partial u}{\partial x}+v\frac{\partial u}{\partial y}\right)\phi_i = \int_{\partial\Omega}\phi_i f_x - \int_\Omega\left(\tau_{xx}\frac{\partial\phi_i}{\partial x}+\tau_{xy}\frac{\partial\phi_i}{\partial y}\right) \quad , \tag{A7}$$

$$\int_\Omega \rho\left(\frac{\partial v}{\partial t}+u\frac{\partial v}{\partial x}+v\frac{\partial v}{\partial y}\right)\phi_i = \int_{\partial\Omega}\phi_i f_y - \int_\Omega\left(\tau_{yx}\frac{\partial\phi_i}{\partial x}+\tau_{yy}\frac{\partial\phi_i}{\partial y}\right) \quad , \tag{A8}$$

and

$$-\int_\Omega\left(\frac{\partial u}{\partial x}+\frac{\partial v}{\partial y}\right)\psi_i = 0 \quad , \tag{A9}$$

where the minus sign is introduced into Equation (A9) to preserve symmetry,

$$\left.\begin{aligned}
\tau_{xx} &= -P + 2\mu\frac{\partial u}{\partial x} \quad , \\
\tau_{xy} = \tau_{yx} &= \mu\left(\frac{\partial u}{\partial y}+\frac{\partial v}{\partial x}\right), \\
\tau_{yy} &= -P + 2\mu\frac{\partial v}{\partial y} \quad ,
\end{aligned}\right\} \tag{A10}$$

and the x- and y-components of the surface traction force are

$$\left.\begin{aligned}
f_x &\equiv n_x\tau_{xx} + n_y\tau_{xy} = n_x\left(-P + 2\mu\frac{\partial u}{\partial x}\right)+n_y\mu\left(\frac{\partial u}{\partial y}+\frac{\partial v}{\partial x}\right), \\
f_y &\equiv n_x\tau_{yx} + n_y\tau_{yy} = n_x\mu\left(\frac{\partial u}{\partial y}+\frac{\partial v}{\partial x}\right)+n_y\left(-P + 2\mu\frac{\partial v}{\partial y}\right).
\end{aligned}\right\} \tag{A11}$$

If the portions of $\partial\Omega$ on which tractions are to be specified are aligned with the x- and y-directions, then f_x and f_y can be immediately identified with normal and tangential traction components, and this is the approach we have taken. A more general approach would require a local transformation of coordinates into normal and tangential components. In the remainder of this discussion we therefore restrict ourselves to the Cartesian case with natural BC's (tractions) only imposed on portions of $\partial\Omega$ which are parallel to either x or y.

Matrix definitions

The next step is to expand the solution for u, v, and P according to Equation (2); i.e., into the basis set. When this is done, Equations (A7) through (A9) become, using Equation (A10) and the summation convention,

$$\begin{aligned}
&\left[\int_\Omega\rho\phi_i\phi_j\right]\dot u_j + \left[\int_\Omega\mu\left(2\frac{\partial\phi_i}{\partial x}\frac{\partial\phi_j}{\partial x}+\frac{\partial\phi_i}{\partial y}\frac{\partial\phi_j}{\partial y}\right)\right]u_j + \left[\int_\Omega\mu\frac{\partial\phi_i}{\partial y}\frac{\partial\phi_j}{\partial x}\right]v_j \\
&+\left[u_k\int_\Omega\rho\phi_k\phi_i\frac{\partial\phi_j}{\partial x}+v_k\int_\Omega\rho\phi_k\phi_i\frac{\partial\phi_j}{\partial y}\right]u_j - \left[\int_\Omega\frac{\partial\phi_i}{\partial x}\psi_j\right]P_j = \int_{\partial\Omega}\phi_i f_x \;;
\end{aligned} \tag{A12}$$

for i = 1,2,...N,

$$[\int_\Omega \rho\phi_i\phi_j]\dot{v}_j + \left[\int_\Omega \mu \frac{\partial\phi_i}{\partial x}\frac{\partial\phi_j}{\partial y}\right]u_j + \left[\int_\Omega \mu\left(\frac{\partial\phi_i}{\partial x}\frac{\partial\phi_j}{\partial x} + 2\frac{\partial\phi_i}{\partial y}\frac{\partial\phi_j}{\partial y}\right)\right]v_j$$

$$+\left[u_k\int_\Omega \rho\phi_k\phi_i\frac{\partial\phi_j}{\partial x} + v_k\int_\Omega\rho\phi_k\phi_i\frac{\partial\phi_j}{\partial y}\right]v_j - \left[\int_\Omega \frac{\partial\phi_i}{\partial y}\psi_j\right]P_j = \int_\Omega \phi_i f_y; \qquad \text{(A13)}$$

for i = 1,2,...,N, and

$$-\left[\int_\Omega \psi_i\frac{\partial\phi_j}{\partial x}\right]u_j - \left[\int_{\partial\Omega}\psi_i\frac{\partial\phi_j}{\partial y}\right]v_j = 0; \quad i=],2,\ldots,M \quad . \qquad \text{(A14)}$$

These are the expanded form of Equation (3), from which the matrix identifications follow immediately:

$$M_{ij} = \left[\begin{array}{c|c} \int_\Omega \rho\phi_i\phi_j & 0 \\ \hline 0 & \int_\Omega \rho\phi_i\phi_j \end{array}\right], \qquad \text{(A15)}$$

$$K_{ij} = \left[\begin{array}{c|c} \int_\Omega \mu\left(2\frac{\partial\phi_i}{\partial x}\frac{\partial\phi_j}{\partial x} + \frac{\partial\phi_i}{\partial y}\frac{\partial\phi_j}{\partial y}\right) & \int_\Omega \mu\frac{\partial\phi_i}{\partial y}\frac{\partial\phi_j}{\partial x} \\ \hline \int_\Omega \mu\frac{\partial\phi_i}{\partial x}\frac{\partial\phi_j}{\partial y} & \int_\Omega \mu\left(\frac{\partial\phi_i}{\partial x}\frac{\partial\phi_j}{\partial x} + 2\frac{\partial\phi_i}{\partial y}\frac{\partial\phi_j}{\partial y}\right) \end{array}\right], \qquad \text{(A16)}$$

$$N_{ij}(u) = \left[\begin{array}{c|c} u_k\int_\Omega \rho\phi_k\phi_i\frac{\partial\phi_j}{\partial x} + v_k\int_\Omega\rho\phi_k\phi_i\frac{\partial\phi_j}{\partial y} & 0 \\ \hline 0 & u_k\int_\Omega \rho\phi_k\phi_i\frac{\partial\phi_j}{\partial x} + v_k\int_\Omega\rho\phi_k\phi_i\frac{\partial\phi_j}{\partial y} \end{array}\right], \qquad \text{(A17)}$$

$$C_{ij} = -\left[\begin{array}{c} \int_\Omega \frac{\partial\phi_i}{\partial x}\psi_j \\ \hline \int_\Omega \frac{\partial\phi_i}{y}\psi_j \end{array}\right], \qquad \text{(A18)}$$

and finally,

$$f_i = \left[\begin{array}{c} \int_{\partial\Omega} \phi_i f_x \\ \hline \int_{\partial\Omega} \phi_i f_y \end{array}\right] \qquad \text{(A19)}$$

While the basis functions (and hence, the matrices) are usually defined globally, the matrix assembly is performed element-wise in the conventional way; i.e., these matrix definitions can also be interpreted as element matrices if the basis functions are re-interpreted as being locally defined (on one element).

Jacobian matrix

The Jacobian matrix of the vector A(X)X is given by

$$J_{ij} = \frac{\partial}{\partial X_j}\,[A_{ik}(X)X_k] = A_{ij}(X) + \left[\frac{\partial A_{ik}(x)}{\partial X_j}\right]X_k \quad .$$

Applying this formula to the LHS of Equation (11) gives (dropping the n+1 subscript on u and P for convenience), with $X = (u \quad P)^T$, where u contains both u_j and v_j,

$$J_{ij} = \left[\begin{array}{c|c} \frac{2}{\Delta t_n} M_{ij} + K_{ij} + N_{ij}(u) + J'_{ij} & C_{ij} \\ \hline C^T_{ij} & 0 \end{array}\right] \quad , \tag{A20}$$

where

$$J'_{ij} \equiv \left[\begin{array}{c|c} \frac{\partial N_{ik}(u)}{\partial u_j} u_k & \frac{\partial N_{ik}(u)}{\partial v_j} u_k \\ \hline \frac{\partial N_{ik}(u)}{\partial u_j} v_k & \frac{\partial N_{ik}(u)}{\partial v_j} v_k \end{array}\right] \quad . \tag{A21}$$

Inserting (A17) into (A21) gives, finally,

$$J'_{ij} = \left[\begin{array}{c|c} u_k \int_\Omega \rho\phi_i\phi_j \frac{\partial \phi_k}{\partial x} & u_k \int_\Omega \rho\phi_i\phi_j \frac{\partial \phi_k}{\partial y} \\ \hline v_k \int_\Omega \rho\phi_i\phi_j \frac{\partial \phi_k}{\partial x} & v_k \int_\Omega \rho\phi_i\phi_j \frac{\partial \phi_k}{\partial y} \end{array}\right] \quad . \tag{A22}$$

Note that only the J_{ij} matrix includes all of the variables — its size is (2N + M) x (2N + M); the size of the other matrices in Equations (A15) through (A22) is 2N x 2N except C_{ij}, which is 2N x M.

APPENDIX 2

Euler integration formulas and one-step Newton

The FE scheme (the predictor), applied to the system of equations $\dot{y} = f(y)$ gives

$$y^p_{n+1} = y_n + \Delta t_n f(y_n) = y_n + \Delta t_n \dot{y}_n \tag{B1}$$

and the BE (corrector) scheme is

$$y_{n+1} = y_n + \Delta t_n f(y_{n+1}) \quad , \tag{B2}$$

whose inversion can yield $\dot{y}_{n+1}$ for the more efficient implementation of Equation (B1); i.e.,

$$\dot{y}_{n+1} = (y_{n+1} - y_n)/\Delta t \quad . \tag{B3}$$

The local truncation errors of the two schemes are required next:

$$y^p_{n+1} - y(t_{n+1}) = -\frac{\Delta t_n^2}{2}\ddot{y}_n + O(\Delta t^3) \tag{B4}$$

and

$$y_{n+1} - y(t_{n+1}) \equiv d_{n+1} = \frac{\Delta t_n^2}{2}\ddot{y}_n + O(\Delta t^3) \quad , \tag{B5}$$

which leads directly to

$$d_{n+1} = \frac{1}{2}(y_{n+1} - y^p_{n+1}) + O(\Delta t^3) \tag{B6}$$

as an estimate of the error in a single BE step. This is used to compute the next step size, similar to the AB/TR method, via

$$\Delta t_{n+1} = \Delta t_n(\varepsilon/|d_{n+1}|)^{1/2} \quad . \tag{B7}$$

Finally, applying the one-step Newton method to Equation (B2) gives

$$J_p(y_{n+1} - y^p_{n+1}) = y_n + \Delta t\, f(y^p_{n+1}) - y^p_{n+1} \quad , \tag{B8}$$

where

$$J_p \equiv \left(I - \Delta t \left.\frac{\partial f}{\partial y}\right|_{y^p_{n+1}}\right) \quad .$$

The application of this algorithm to the spatially-discretized NS equation (Equation (3)) can be made in the same manner as presented in the text for the AB/TR scheme.

CHAPTER 3

THE SOLUTION OF THE LAMINAR BOUNDARY LAYER EQUATIONS BY A VARIABLE ORDER SELFADAPTIVE DIFFERENCE METHOD

W. Schönauer, G. Glotz, K. Raith

Rechenzentrum der Universität Karlsruhe,
D-7500 Karlsruhe 1, Germany

Summary: A selfadaptive difference method for the solution of parabolic equations is presented. The estimates of the discretization errors are used to adapt the step size and order of accuracy to a prescribed accuracy. The iteration of the nonlinear equations is also adapted to the discretization error. The application to the incompressible laminar boundary layer equations, to the hypersonic boundary layer equations with nonequilibrium chemical reactions and to the 3-D incompressible boundary layer equations is discussed.

1. INTRODUCTION

The paper describes the results of a research program which has been performed by the Forschungsgruppe Numerische Strömungsmechanik at the computing center of the University of Karlsruhe and which was sponsored by the Stiftung Volkswagenwerk and in part by the Deutsche Forschungsgemeinschaft and Gesellschaft für Weltraumforschung.

1.1 Historical development

The starting point of the research program was the bad experience gained with a computer program for the solution of the hypersonic boundary layer (bdl) equations in chemical nonequilibrium (Schönauer et al. 1974). The extremely stiff (by chemical reactions) and nonlinear (by complicated thermo-chemical functions) system of bdl equations was solved by a fully implicit difference method with a difference star of the order $O(\Delta x)$ and $O(\Delta y^2)$ in the direction along and normal to the surface. At each station x_i

the nonlinear difference equations were solved by one iteration step of a Newton-Raphson procedure. The solution of one test case required several hours of CPU-time on a Univac 1108 and at the end nothing was known about the accuracy of the solution.To test the accuracy, further jobs with different step sizes Δx and Δy and different number of iterations had to be run. The difficulty of the problem is illustrated by the fact that Blottner (1970) in a similar problem could not apply a Crank-Nicolson method because of the stiffnes of the equations. It was clear that the large computation time could only be reduced by higher order methods. It could be deduced from the experience of Blottner that these methods could use only backward difference formulae in the x-direction.

So we decided to develop a method of higher order, using as test case the incompressible bdl equations (Grüning, Schönauer 1975). At first we gradually augmented the order of the method separately in the x- and y-direction and computed the solution with these (fixed) difference stars. But improving the error e.g. in the x-direction was useless when in the y-direction a large error was present and vice versa. And all improvements of the discretization errors were needless if the iteration of the nonlinear difference equations was not continued until the iteration error was lower than the discretization error. But if it was too small, computation time was wasted. From these experiences we learned that only a selfadaptive method could be an efficient method.

1.2 Necessary properties of a selfadaptive difference method

The prime requirement for a selfadaptive difference method is some knowledge of the discretization error in the x- and y-direction and of the error of the iteration procedure. If one of these errors is large, e.g. 0.1%, and the other ones are small, e.g. 0.001%, then the large error will determine the accuracy of the solution and the small ones only waste computing time. So these errors must be balanced. We soon learned that higher order might not be better than lower order, depending on the desired accuracy. So one has to have some knowledge of the discretization error of the neigbouring orders, too. It is very essential never to use higher order without such a knowledge. The number of iterations largely determines the overall amount of computing time. So it is necessary to have an efficient and reliable iteration method. Therefore we chose the Newton-Raphson procedure. But this means simultaneous solution of all equations of a system. For an efficient iteration a very good guess is needed. So we extrapolated an initial guess with an interpolation polynomial of the same order as the difference formula in the x-direction and in consequence needed only one iteration in most of the x-stations. This gives a predictor-corrector type of procedure.

2. ERROR ESTIMATES FOR FAMILIES OF DIFFERENCE FORMULAE

2.1 Families of difference formulae

The bdl equations are of parabolic type, i.e. IBVP (initial-boundary value problem), with the IVP in the x-direction with derivative of type u_x and with the BVP in the y-direction with derivatives of type u_y an u_{yy}. As we need difference formulae of "arbitrary" order we derive these formulae from a Newton interpolation polynomial which allows easily the addition of further points. In Figure 2.1 there is depicted the family of nonequidistant backward difference formulae for u_x with arbitrary order p, with $\overline{\Delta x}$ some "mean value" of the Δx_j.

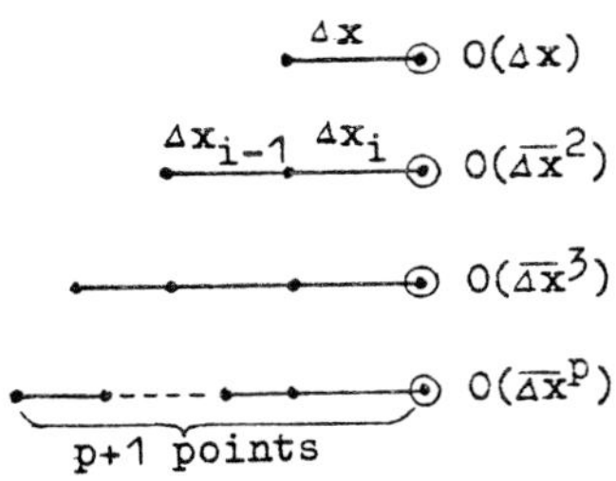

Figure 2.1 Difference formulae for u_x

For u_{yy} the family of equidistant difference formulae of Figure 2.2 with arbitrary order q (even) are used, but near the boundary unsymmetric formulae with one additional point (to maintain the order) are required until the symmetric formulae may be applied. For u_y the same number of points is used.

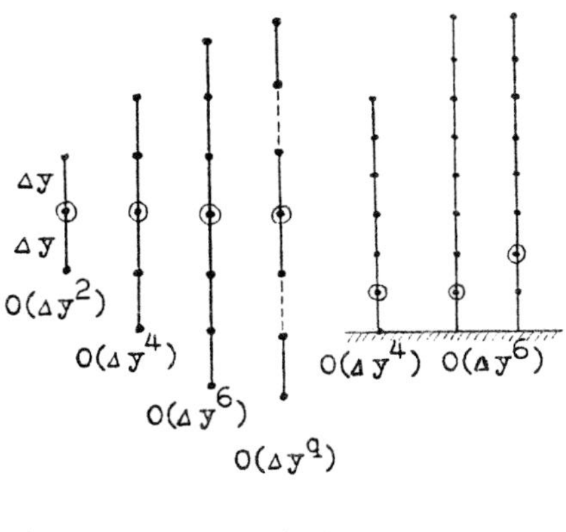

Figure 2.2 Difference formulae for u_{yy}

The derivation of the formulae is demonstrated for a function u(x), where x may be replaced by y. The n+1 values $u_i = u(x_i)$, $i=0(1)n$ are known. The Newton interpolation polynomial for these values is

$$P(x) = \underset{s_o}{[x_o]}\underset{P_o}{\cdot 1}+\underset{s_1}{[x_ox_1]}\underset{P_1}{(x-x_o)}+\underset{s_2}{[x_ox_1x_2]}\underset{P_2}{(x-x_o)(x-x_1)}+\ldots \qquad (2.1)$$

$$\ldots +\underset{s_n}{[x_ox_1\ldots x_n]}\underset{P_n}{(x-x_o)\ldots(x-x_{n-1})} .$$

The s_i are the edge values of the Newton scheme of divided differences

$$\begin{array}{lllll}
x_o & u_o = \overset{s_o}{[x_o]} & s_1 & & \\
 & & ---- [x_ox_1] = \dfrac{[x_o]-[x_1]}{x_o-x_1} & & \\
x_1 & u_1 & & & \\
\cdot & \cdot & & & s_n \\
\cdot & \cdot & ---------------- & & [x_ox_1\ldots x_n] \\
\cdot & \cdot & & & \\
x_n & u_n & & &
\end{array} \qquad (2.2)$$

For a difference formula the polynomial P should be ordered according to the u_i. The scheme (2.2), being linear in the u_i, is generated by the superposition of n+1 schemes. In (2.2) the vector $\underline{u} = (u_o,u_1,\ldots u_n)$ is replaced by the unit vector $\underline{e}_k = (0,0,.,0,1,0,\ldots 0)$ with 1 in the k-th position and zero elsewhere. The corresponding elementary Polynomial P is depicted in Figure 2.3, and the i-th divided difference is s_{ik}. Multiplication of the elementary polynomials for $\underline{e}_k$ by u_k and adding them results in the polynomial P of (2.1) but now represented in the following ordering

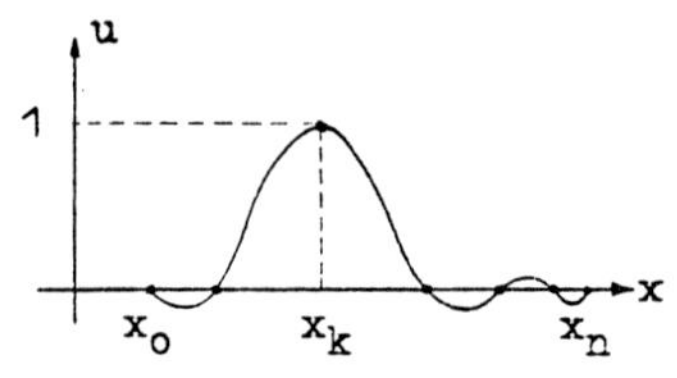

Figure 2.3 Interpolation polynomial for $\underline{e}_k$

$$
\begin{aligned}
P(x) = {} & u_o(\underline{s_{oo}P_o+s_{10}P_1}+\underline{s_{20}P_2}+ \dots + s_{no}P_n) + \\
& + u_1(\underline{s_{o1}P_o+s_{11}P_1}+\underline{s_{21}P_2}+ \dots + s_{n1}P_n) + \\
& + u_2(\underline{s_{o2}P_o+s_{12}P_1+s_{22}P_2}+ \dots + s_{n2}P_n) + \\
& \vdots \\
& + u_n(s_{on}P_o+s_{1n}P_1+s_{21}P_2+ \dots + s_{nn}P_n) \;.
\end{aligned}
\tag{2.3}
$$

This has the character of an interpolation formula, the parentheses being the "coefficients" of the function values u_i. Evaluation of the $P_i(x)$ gives the value of these coefficients. The underlined termes yield a linear interpolation formula, adding the dashed terms yields a quadratic interpolation formula. The interpolated value is $u_d(x) = P(x)$. The index d charcterizes the result of a difference formula.

Difference formulae are derived from (2.3) by differentiating (2.3) with respect to x:

$$u_{xd}(x) = P'(x) \;, \quad u_{xxd} = P''(x) \;, \tag{2.4}$$

which means differentiation of the P_i in (2.3). These P_i', P_i'' have a very special structure if $x = x_i$. So the formulae for the families of Figure 2.1 and 2.2 are easily derived. The coefficients are computed by special subroutines. There the properties of the scheme 2.2, that the ordering of the x_i is arbitrary, and that the adding or omitting of a point adds or drops an edge of the scheme (2.2) is used; the schemes for the unit vectors $\underline{e}_k$ are stored in the computer.

2.2 Error estimates for difference formulae

The basic ideas are shown up for the derivative u_x, but they hold equally for u_y, u_{yy} etc. Replacing a derivative u_x by a difference formula u_{xd} causes a discretization error d_x. We replace u_x by a higher and a lower order difference formula u_{xdh} and u_{xdl}, causing discretization errors d_{xh} and d_{xl}:

$$u_x = u_{xdh} + d_{xh} = u_{xdl} + d_{xl} \;. \tag{2.5}$$

Resolving for d_{xl} yields

$$d_{xl} \approx u_{xdh} - u_{xdl} \quad \{+ d_{xh}\} \;. \tag{2.6}$$

Neglecting d_{xh} we get an estimate (not a bound) for the lower order discretization error as the difference of the higher and lower order formulae, and the quality of the estimate is given by the discretization error of the higher order formula. For the formulae u_{xdh} and u_{xdl} two consecutive members of Figure 2.1 or 2.2 are used. Then one computes the error (2.6) directly from the interpolation polynomial (2.3), without explicitely computing the u_{xdh} and u_{xdl}. This can be seen in (2.3) where the dashed terms are just the difference of the higher and lower order approximation. So one computes together with the coefficients of the difference formulae directly the coefficients of the error formulae. This is included in the special subroutines for the difference formulae.

So from (2.3) we finally get for a derivative $u_{(\nu)}$ and its error estimate $d_{(\nu)}$ formulae of the usual form

$$u_{(\nu)d} = \sum_i a_i u_i \,, \qquad d_{(\nu)} = \sum_i b_i u_i \,. \tag{2.7}$$

2.3 Numerical test of the error estimate

The theory of the error estimate, equation (2.6), is rather trivial. The practical use may only be demonstrated by numerical experiments. We demonstrate the quality of the error estimates for the function $u = 1/x$ or $1/y$, see Figure 2.4, for the families of difference formulae of Figure 2.1 and 2.2. The backward and noncentral formulae are taken from $x = 2$ to the left and the central formulae are centered at $y = 2$. The exact and estimated errors of the difference formulae for different orders are computed for the step sizes Δx or $\Delta y = 0.1,\ 3 \cdot 10^{-2},\ 10^{-2}$.

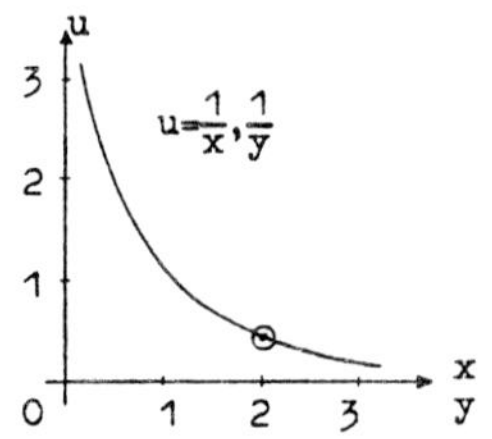

Figure 2.4 Test function 1/x

The exact error is the value of the derivative minus the result of the difference formula.

Figure 2.5 shows the <u>error</u> (not the value of the difference formula) for the backward formulae for u_x of Figure 2.1. For the step size $\Delta x = 0.1$ there is an optimum order of $p = 9$. If at this Δx we take a higher order, we get "irrelevant"points into the difference formula. At an error of 10^{-14} we are in

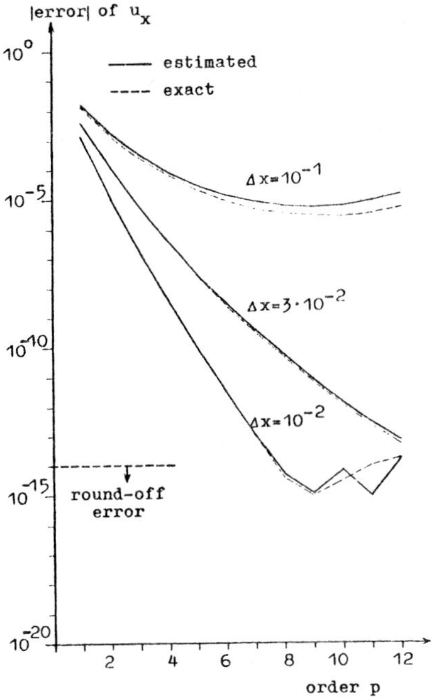

Figure 2.5 Error of u_x for the difference formulae of Figure 2.1 for $u = 1/x$ at $x = 2$.

the round-off error domain of the computer. Figure 2.6 and 2.7 show the errors for the central and unsymmetric boundary formulae (derivative at first interiour grid point) of Figure 2.2 for the derivative u_{yy} at y=2. It is clearly visible that the unsymmetric formulae have larger errors. For Δy=0.1 they have an optimum order at q=8. For the symmetric formulae below errors of 10^{-14} and the unsymmetric formulae below 10^{-12} we are in the domain of the round-off error. By these figures it is demonstrated that the error estimates computed according to (2.6) are extremely good estimates if the step size is small enough and the order not

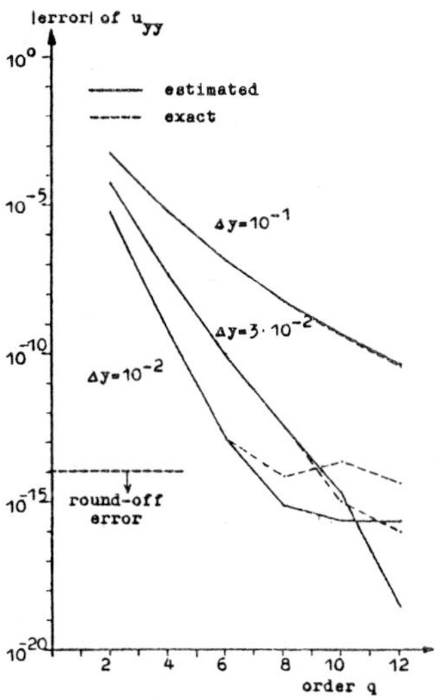

Figure 2.6 Error of u_{yy} for the central difference formulae of Figure 2.2 for u=1/y at y=2.

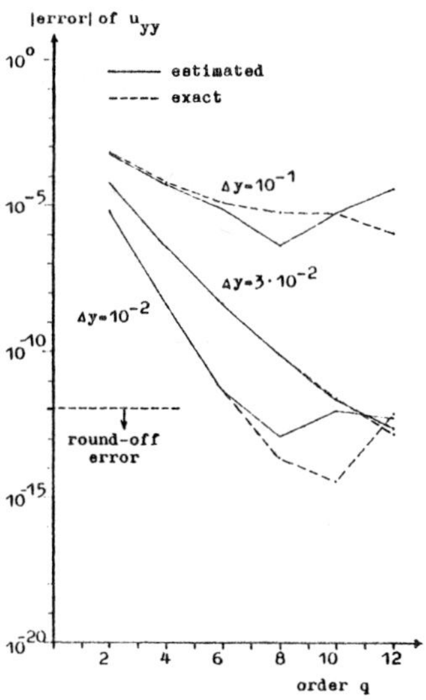

Figure 2.7 Error of u_{yy} for unsymmetric formulae of Figure 2.2 for u=1/y at y=2.

too high. Therefore high orders should not be used without an order control mechanism. The situation becomes still more difficult if we deal with the "rough" function values of a difference method and not with the smooth values of an analytic test function.

3. THE SOLUTION OF A MODEL EQUATION

3.1 Preparation of the equations

Blottner (1975) gives an excellent review of the state of the art in the numerical solution of the bdl equations. He states that for 3-D flows, complex reacting gas flows and turbulent flows with complex governing equations, prohibitive computing times can occur. He stresses the need for higher order methods and automatic techniques for finding optimum grid spacings. With an estimate of the discretization errors we now have a tool to develop a method which is selfadaptive with respect to order and step size.

The basic ideas of the selfadaptive solution technique will be demonstrated at the simple example of the incompressible 2-D bdl equations. The Prandtl bdl equations are used in nondimensional form, momentum and continuity equation:

$$uu_x + vu_y + p_x - u_{yy} = 0 \;, \quad u_x + v_y = 0 \;, \qquad (3.1)$$

with x and y as coordinates along and normal to the surface, y has been stetched by $\sqrt{Re}$. With $u_e(x)$ the inviscid velocity at the outer edge of the bdl we have $p_x = - u_e \, u_e'$. The boundary conditions are

$$u(x,o) = v(x,o) = 0 \;, \; u(x,y_e) = u_e \;,$$

$$u_y(x,y_e) = u_{yy}(x,y_e) = \ldots = 0 \;, \qquad (3.2)$$

where theoretically $y_e = \infty$, but is taken as a final value y_e for numerical calculations.

The equations (3.1) are not well suited for the numerical solution because they have a singularity at the stagnation point x=o,and the value y_e, i.e. the domain of solution in the y-direction, changes considerably with x. Therefore a similarity transformation in the form of a modified Falkner-Skan-transformation is introduced with
$x, y, u, v \rightarrow \xi, \eta, f, g$:

$$x = \xi \;, \; y = \delta(\xi)\phi(\eta) \;, \; u = u_e(\xi)\cdot f(\xi,\eta) \;,$$

$$v = \delta^{-1}(\xi)\cdot g(\xi,\eta) \;, \qquad (3.3)$$

with the functions depending on the inviscid velocity u_e

$$\delta^2(\xi) = 2\cdot\left(\int_0^{\xi} u_e(\xi^*)d\xi^*\right)/u_e^2(\xi) \ , \ A(\xi) = u_e(\xi)\delta^2(\xi),$$
$$\Lambda(\xi) = u_e'(\xi)\delta^2(\xi) \ . \qquad (3.4)$$

By the function

$$\phi(\eta) = C[c\eta+(1-c)\eta^N] \qquad (3.5)$$

with the constant $C = y_e/\delta$ (i.e. y_e proportional to δ) and the constants c,N which are determined to give an optimal distribution of mesh points, the domain $0 \leq y \leq y_e$ is transformed to $0 \leq \eta \leq 1$. The choice of C now must care for the asymptotic behaviour at $\eta = 1$. The value of C is the value of the outer edge coordinate in the classical Falkner-Skan-transformation, which is obtained from (3.5) for η=1.

With the transformation (3.3) - (3.5) the momentum equation (ME), contunuity equation (CE) and boundary conditions (BC) (3.1), (3.2) are

$$ME \equiv F_1 ff_\xi + F_2 ff_\eta + F_3 f^2 + gf_\eta + F_4 f_\eta + F_5 f_{\eta\eta} - F_3 = 0, \qquad (3.6a)$$

$$CE \equiv F_1 f_\xi + F_2 f_\eta + F_3 f + g_\eta = 0 \ , \qquad (3.6b)$$

$$BC \equiv \begin{cases} f(\xi,o) = 0 \ , \\ g(\xi,o) = 0 \ , \\ f(\xi,1) - 1 = 0 \ , \end{cases} \qquad (3.6c)$$

with the coefficient functions

$$F_1 = A\phi_\eta \ , \ F_2 = (\Lambda-1)\phi \ , \ F_3 = \Lambda\phi_\eta \ ,$$
$$F_4 = \phi_{\eta\eta}/\phi_\eta^2 \ , \ F_5 = -\ 1/\phi_\eta \ . \qquad (3.6d)$$

For

$$u_e = \xi^{\frac{\Lambda}{2-\Lambda}} \ , \quad \Lambda = \text{const.} \qquad (3.7)$$

the system (3.6) has similar solutions $u^*(\eta)$ which are not depending on ξ. This is the reason why the system (3.6) is far more suited for the numerical solution than the system (3.2). When for a general u_e the coefficient $\Lambda(\xi)$ ist known, an experienced engineer may roughly predict the behaviour of the bdl only by looking at this coefficient.

Taking ME, CE and BC as the corresponding operators of (3.6) we may abbreviate the whole system by the operator equation

$$P\underline{u} \equiv \begin{pmatrix} ME \\ CE \\ BC \end{pmatrix} = 0 \quad , \quad \underline{u} = \begin{pmatrix} f \\ g \end{pmatrix} \tag{3.8}$$

where $\underline{u}$ is the solution vector.

3.2 Basic solution technique

The classical way to solve the system (3.6) or (3.8) is to replace the derivatives by difference formulae, leading to a nonlinear algebraic system of difference equations which then is linearized and eventually iterated for the solution.

We chose the Newton-Raphson procedure for the iteration of the nonlinear equations because of its quadratic convergence (if one is near to the solution). We at first linearize the differential equations and then apply the difference method to the equations for the Newton-corrections. With the necessary care this leads to the same result as if we at first apply the difference method and then the Newton-Raphson procedure.

We replace $\underline{u}$ by the (ν+1)st iterate $\underline{u}^{(\nu+1)}$, which itself is the sum of the ν-th iterate $\underline{u}^{(\nu)}$ and of the correction $\Delta\underline{u}^{(\nu+1)}$:

$$\underline{u} \Leftarrow \underline{u}^{(\nu+1)} = \begin{pmatrix} f^{(\nu+1)} \\ g^{(\nu+1)} \end{pmatrix} = \underline{u}^{(\nu)} + \Delta\underline{u}^{(\nu+1)} = \begin{pmatrix} f^{(\nu)} + \Delta f^{(\nu+1)} \\ g^{(\nu)} + \Delta g^{(\nu+1)} \end{pmatrix} . \tag{3.9}$$

We omit the upper indices ν and ν+1 in the following, recalling that $\underline{u}$ and $\Delta\underline{u}$ has the upper index ν and ν+1. For the first term of (3.6a) e.g. we get

$$F_1 f f_\xi \Leftarrow F_1(f+\Delta f)(f_\xi+\Delta f_\xi) = F_1 f f_\xi + F_1 \Delta f f_\xi + F_1 f \Delta f_\xi + F_1 \Delta f \Delta f_\xi . \tag{3.10}$$

Then we linearize, dropping the term $\Delta f \Delta f_\xi$. The same we do with all terms of (3.6) and order the equations, so that at the left hand sides there are the corrections and at the right hand sides the original operators:

$$QME \equiv - F_1 \Delta f f_\xi - F_1 f \Delta f_\xi - F_2 \Delta f f_\eta - F_2 f \Delta f_\eta - 2F_3 f \Delta f - \Delta g f_\eta - g \Delta f_\eta - - F_4 \Delta f_\eta - F_5 \Delta f_{\eta\eta} = ME \;, \tag{3.11a}$$

$$QCE \equiv - F_1 \Delta f_\xi - F_2 \Delta f_\eta - F_3 \Delta f - \Delta g_\eta = CE \;, \tag{3.11b}$$

$$QBC \equiv \begin{cases} \Delta f(\xi,o) = f(\xi,o) \;, \\ \Delta g(\xi,o) = g(\xi,o) \;, \\ \Delta f(\xi,1) = f(\xi,1) - 1 \;. \end{cases} \tag{3.11c}$$

In this case the right hand sides of (3.11c) are zero, but if the boundary conditions instead of (3.6c) would be themselves nonlinear like (3.6a,b), then in (3.11c) there would be equations like (3.11a). The system (3.11) is a system of parabolic partial differential equations for the correction functions Δf, Δg. Corresponding to (3.8) we abbreviate this system by the operator equation

$$Q\underline{\Delta u} \equiv \begin{pmatrix} QME \\ QCE \\ QBC \end{pmatrix} = \begin{pmatrix} ME \\ CE \\ BC \end{pmatrix} = P\underline{u} \;. \tag{3.12}$$

Looking at (3.8) we could conclude that the right hand side of (3.12) is zero,but in (3.8) $\underline{u}$ is the solution of the system whereas in (3.12) $\underline{u}$ is really an iterate $\underline{u}^{(\nu)}$, for which $P\underline{u}^{(\nu)} \neq 0$. The value $P\underline{u}^{(\nu)}$ is the residual or defect of $P\underline{u}$.

Now we want to solve (3.11) or (3.12) for the correction $\underline{\Delta u}$ by the difference method. The step size should be determined only from accuracy requirements. Therefore we choose a fully implicit method: ξ-derivarives are replaced by a member of the family of backward difference formulae of Figure 2.1 which gives the best attainable stability. These formulae allow a nonequidistant step size $\Delta\xi$. The η-derivatives are replaced by difference formulae of the family of Figure 2.2, which are central equidistant formulae in the interiour of the domain and unsymmetric formulae near the boundaries (until central formulae can be used). The difference grid is shown in Figure 3.1. The differential equation is approximated at ξ_i, η_k; one of the possible difference stars is indicated in Figure 3.1, it has the order $O(\overline{\Delta\xi^3}, \Delta\eta^4)$. But according to (2.5), (2.7) we now replace a derivative by its difference formula and its estimate for the discretization error, e.g. the first term in the right hand side of (3.11a), which is the first term in (3.6a), yields

$$F_1 ff_\xi \Leftarrow F_{1,i,k}\, f_{i,k}\, (f_{\xi,d,i,k} + d_{f_\xi,i,k}) \;, \qquad (3.13a)$$

where $f_{\xi,d,i,k}$ is the difference formula for f_ξ at ξ_i, η_k and $d_{f_\xi,i,k}$ its error estimate, according to (2.7). In this way all derivatives in the right hand side of (3.11) or (3.12) are replaced. The first term of the left hand side of (3.11a) yields

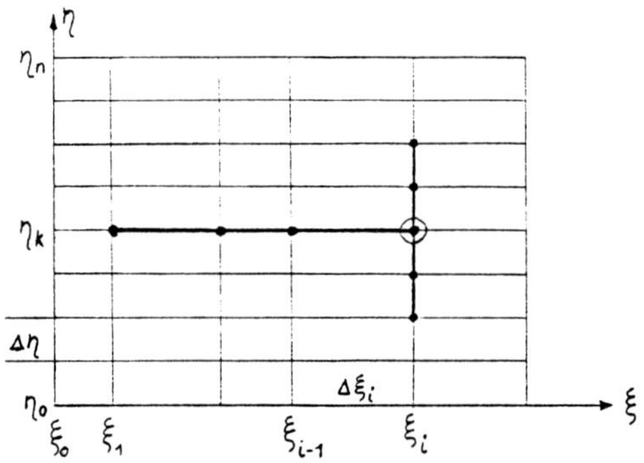

Figure 3.1 Difference grid with one possible difference star

$$F_1 \Delta ff_\xi \Leftarrow F_{1,i,k}\, \Delta f_{i,k}\, (f_{\xi,d,i,k} + d_{f_\xi,i,k}) \simeq$$

$$\simeq F_{1,i,k}\, \Delta f_{i,k}\, f_{\xi,d,i,k} \;, \qquad (3.13b)$$

where the product of the correction and the discretization arror has been dropped as small of higher order. The second term in (3.11a) yields

$$F_2 f \Delta f_\xi \Leftarrow F_{2,i,k}\, f_{i,k}\, (\Delta f_{\xi,d,i,k} + d_{\Delta f_\xi,i,k}) \simeq$$

$$\simeq F_{2,i,k}\, f_{i,k}\, \Delta f_{\xi,d,i,k} \qquad (3.13c)$$

where the discretization error of the correction also has been dropped as small of higher order. One should remember that the corrections quadratically approach zero with increasing number of iteration steps. So we can neglect the discretization errors when discretizing the left hand side of (3.11) or (3.12).

There is a special remark concerning Δf_ξ (i.e. the ξ-derivative of a correction). We compute the solution on lines ξ_i = const, see Figure 3.1. If we compute the solution iteratively at ξ_i, the solution profiles for ξ_j, $j<i$, are given and remain fixed. The

difference formula for Δf_ξ has according to (2.7) and Figure 3.1 the form

$$\Delta f_{\xi,d,i,k} = a_i \, \Delta f_i \, \left(+a_{i-1} \, \Delta f_{i-1} + \dots\right) , \qquad (3.13d)$$

but as there are no corrections at the foregoing profiles, $\Delta f_j = 0$ for $j<1$ and there remains only the first term in (3.13d).
So we finally get from (3.12)

$$Q_d \underline{\Delta u} = (P\underline{u})_d + D_\xi + D_\eta , \qquad (3.14)$$

where Q_d is the matrix of the linear system of equations for $\underline{\Delta u}$ on the grid line ξ_i, $(P\underline{u})_d$ is the discretized defect vector on the grid line, D_ξ and D_η are vectors resulting from the discretization errors in ξ- and η-direction. The matrix Q_D is a (q+1)-diagonal block matrix, if we use q-th order formulae (Figure 2.2) in the η-direction, the "elements" are 2x2-matrices because we have a system of two equations in each grid point. But at the top and bottom of Q_d there are additional elements, see the shaded triangles in Figure 3.2, which result from the unsymmetric formulae at the boundaries.

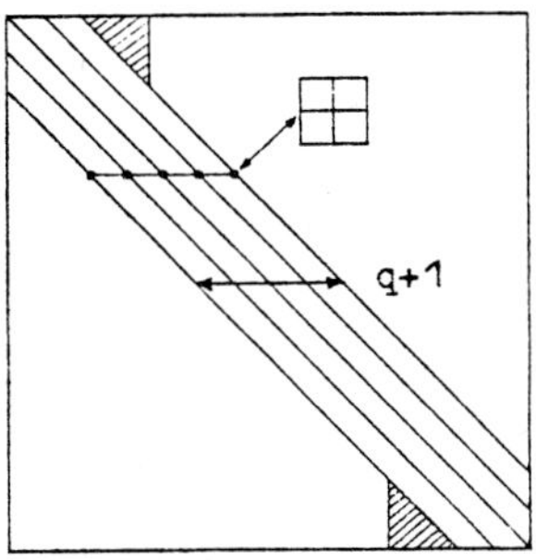

Figure 3.2 Structure of the matrix Q_d

The numerical solution of (3.14) by the Gauss method is done by a special subroutine which takes care of the structure of Q_d. The formal solution is

$$\underline{\Delta u} = \underline{\Delta u}_{Pu} + \underline{\Delta u}_{D_\xi} + \underline{\Delta u}_{D_\eta} = Q_d^{-1}[(P\underline{u})_d + D_\xi + D_\eta] . \qquad (3.15)$$

Here the correction $\underline{\Delta u}$ is split up into the three separate errors which correspond to the three terms at the right hand side: the error $\underline{\Delta u}_{Pu}$ due to the nonlinearity which is reduced by the number of iteration steps, and the errors $\underline{\Delta u}_{D_\xi}$ and $\underline{\Delta u}_{D_\eta}$ due to the discretization errors which depend on the step sizes and orders in the ξ- and η-direction.
For the iteration of the Newton-correction

$$\underline{\Delta u}_{Pu} = Q_d^{-1}(P\underline{u})_d \qquad (3.16)$$

is computed (neglecting the discretization errors) and according to (3.9) a new iterate $\underline{u}^{(\nu+1)}$ is computed.

But there is a difficulty in solving (3.15) or (3.16), because in the equation (3.11b) there is only Δg_η and no Δg. If one orders the unknowns in the form Δf_o, Δg_o, Δf_1, Δg_1,... and the equations QBC of Δf, QBC of Δg, QME_1, QCE_1,...., i.e. the momentum equation is taken as typical Δf-equation and the continuity equation as typical Δg-equation, the system cannot be solved (without pivoting and thus destructing the structure of the matrix Q_d). Therefore the continuity equation QCE, (3.11b), is fulfilled at k-1/2 (instead of k), which requires special difference formulae for u_y, shown in Figure 3.3; near the boundaries onesided formulae must be used.

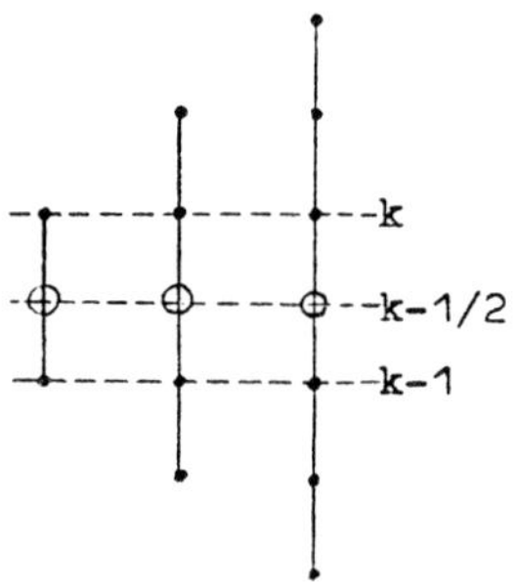

Figure 3.3
Difference formulae for the continuity equation

Also the function values are interpolated to k-1/2 with the same order of accuracy. For compressible equations this special trick is not necessary. But this method only results from the ancient "uncoupling" of the equations. If we would reverse the ordering of the unknowns, i.e. (g,f) instead of (f,g) in (3.8) (a sort of pivoting), then the system would be regular with the usual difference formulae of Figure 2.2. The development of the method had been made with the formulae of Figure 3.3 for the continuity equation, the examples of chapter 6 are computed with the formulae of Figure 2.2 for the continuity equation, using the "reversed" ordering. There the continuity equation at $\eta=1$ acts a "boundary condition".

3.3 Determination of the η-grid and η-order

The questions to be answered are: what is the adequate grid spacing $\Delta\eta$ for a given accuracy and what is the optimum order q, i.e. what member of the family of difference formulae of Figure 2.2 should be used? First publications of these problems are in Grüning and Schönauer (1975), Däubler et al. (1976) and Schönauer et al. (1976).

At the stagnation point the coefficient A of (3.4) is zero, therefore $F_1=0$ in (3.6d) and it can be shown that the system (3.6) or (3.11) reduces to a system of ordinary differential equations in η. The solution of this system is the initial condition for (3.6) at $\xi=o$. The computation of this solution can easily be incorporated in the general program, if one takes care that the coefficients (3.4) get their right (limiting) values for ξ=o. Then in the error equation (3.15) D_ξ=0 because we have no ξ-derivatives. If we assume that by an appropriate number if iterations $(P\underline{u})_d \approx 0$, then only the η-discretization error

$$\underline{\Delta u}_{D_\eta} = Q_d^{-1} D_\eta \tag{3.17}$$

remains. We now compute for an initial step size $\Delta\eta$ and an order q=2 the solution and the error $\|\underline{\Delta u}_{D_\eta}\|$, with $\|\cdot\|$ as maximum absolute value over k, then from the error law $\Delta u_{D_\eta} = O(\Delta\eta^2)$ we compute the step size $\Delta\eta$, which we need for a desired accuracy,

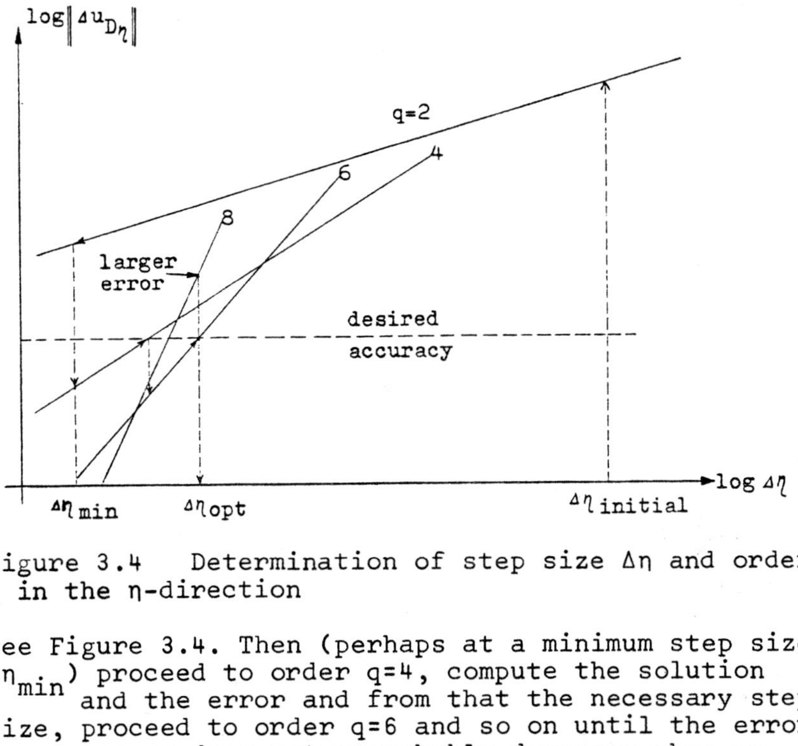

Figure 3.4 Determination of step size $\Delta\eta$ and order q in the η-direction

see Figure 3.4. Then (perhaps at a minimum step size $\Delta\eta_{min}$) proceed to order q=4, compute the solution and the error and from that the necessary step size, proceed to order q=6 and so on until the error increases or does not remarkably decrease when we proceed to higher order. The last combination of $\Delta\eta$ and q then yields the optimum values. From Figure 3.4 it can easily be detected that low accuracy leads to low order, high accuracy to high order. The error law of the high order prevails only when the step size is small enough because of the worse error coefficients of the higher order formulae and of the necessary unsymmetric formulae at the boundaries. Figure 3.4 is only schematic, the errors for the guessed $\Delta\eta$ must be checked by calculation.

Now we make use of the property of the similarity-transformed bdl equations, that the character of the bdl profiles does not change considerably with ξ. We therefore determine the η-grid at the stagnation point and leave this grid unchanged for the downstream calculation.

3.4 Stop of the iteration and determination of the ξ-grid and ξ-order

In the error equation (3.15) there are three partial errors in the bracket. If one of them is large, it will determine the accuracy and the other ones only waste computing time. The η-discretization error D_η now is prescribed by the η-grid and the solution, it is the key error. During the iteration only $\underline{\Delta u}_{Pu}$ from (3.16) is computed. In order to adapt the iteration error, we want

$$\|(P\underline{u})_d\| < 0.1\,\|D_\eta\|_{i-1} \tag{3.18}$$

to hold, where $\|\cdot\|$ is again the maximum absolute value over k. We use $D_{\eta,i-1}$ because $D_{\eta,i}$ is computed after the stopping of the iteration. Only during the iteration of the initial profile $D_{\eta,i}$ must be calculated at each iteration step.
The ξ-discretization error D_ξ is adapted to D_η by the choice of the step size $\Delta\xi$. We want to hold $\|D_\xi\| \simeq 0.3\,\|D_\eta\|$, so we determine the new step size from

$$\Delta\xi_{i+1} = \Delta\xi_i \sqrt{\frac{1}{3}\,\frac{\|D_\eta\|_i}{\|D_\xi\|_i}} \quad . \tag{3.19}$$

Here we use the squareroot instead of the error law because the error law holds for a "medium" step size $\Delta\xi$ which is quite near to the maximum step size covered by the nonequidistant difference formula. As $\Delta\xi_{i+1}$ is computed from information at ξ_i, the guess (3.19) may lead to an error with $\|D_\xi\| > \|D_\eta\|$. If this holds then the whole solution is dropped and with the information of ξ_{i+1} a new $\Delta\xi_{i+1}$ is determined and we recompute the solution.
Now there remains only to determine the order p of the ξ-derivative, i.e. what member of the family of Figure 2.1 should be used. For this purpose we compute the discretization errors D_ξ for the orders p-1, p and p+1, see Figure 3.5. If the situation (a) holds p remains unchanged, if (b) holds, p is augmented and if (c) holds p is reduced.

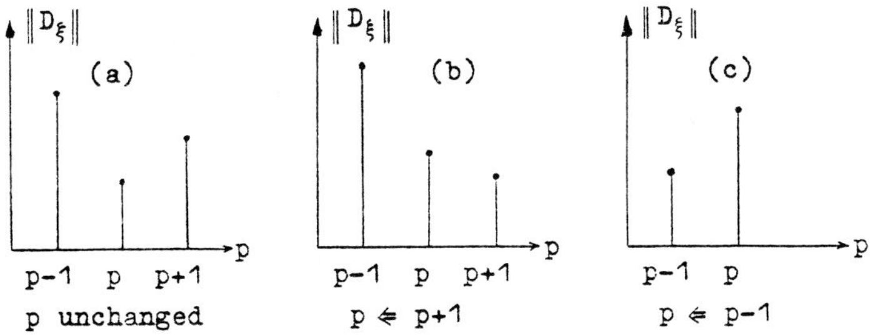

Figure 3.5 Determination of order p in the ξ-direction

3.5 Additional considerations

If one computes the new solution at a station ξ_i, see Figure 3.1, it is essential to have a good guess of the new solution. Therefore we extrapolate with a polynomial of the order p new values from the foregoing profiles. Thus in many cases only one iteration step is necessary at each ξ_i. The extrapolation acts like a predictor, so we have a predictor-corrector type method. The procedure is selfstarting: after three steps with order p=1 the order control of Figure 3.5 can begin. At the first step ξ_1 the ξ-discretization error is determined by comparison to a Hermite formula (instead to a three point formula).
The stability of the method has been tested numerically by the exact computation of the spectral radius of the error transfer matrix, Raith (1976), which indicates that the method is stable for orders $p \leq 6$. There is a strong correspondence to the backward difference methods for ordinary differential equations which are stable for $p \leq 6$ and are A-stable for $p \leq 2$.
The implicit mesh stretching function ϕ, (3.5), determines how the mesh points, which are equally in the η-scale, are distributed in the y-scale. For different accuracies the parameters c and N have been determined to give a minimum $\|\Delta u_{D_\eta}\|$, (3.17). Thus the knowledge of the discretization error is used for the implicit mesh spacing. Good values are c=0.5,N=2 for incompressible bdl, but for hypersonic bdl c=0.1, N=4 are to be preferred. The value $C=y_e/\delta$ which is the value of the outside coordinate in the "classical" Falkner-Skan similarity scale is automatically determined during the iteration of the initial profile. The aim is that tan α in Figure 3.6, with α as the angle in the y/δ-scale, is nearly equal to the desired accuracy of f.

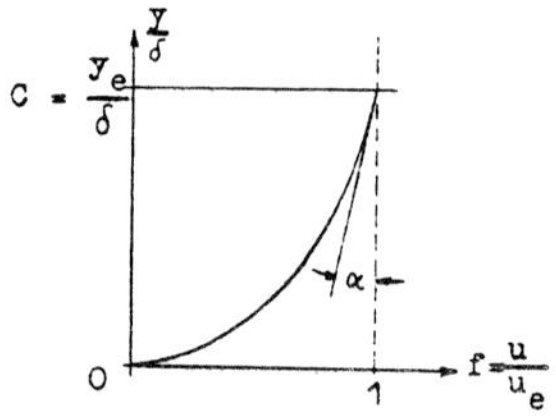

Figure 3.6 Determination of C in equation (3.5)

If tan α is too large or too small then $y_e/\delta = C$ is augmented or reduced. Here again it is very essential that by the similarity transformation the character of the profiles does not change considerably with ξ.

The error $\underline{\Delta u}$ is a local error and nobody can in general predict how the summation of all the local errors at all ξ_i add up to a global error $\underline{\Delta u}_g$. In the differential equation (3.11) and (3.12) for the error function $\underline{\Delta u}$ we did not distinguish between a local and global error. But when we transformed it to a difference equation, we used (3.13d) which states that there is only a (local) error at the actual station ξ_i. If we use instead of (3.13d) the full difference formula for $f_{\xi,d,i,k}$, i.e. also the terms in the parentheses of (3.13d) and thus include the errors of the foregoing profiles, we get instead of (3.15)

$$\underline{\Delta u}_g = Q_d^{-1}[(P\underline{u})_d + D_\xi + D_\eta + s] , \qquad (3.20)$$

where s results from the additional terms in the parentheses of (3.13d). By (3.20) we have the difference approximation for the global error, which is computed together with the solution in the general purpose program of paragraph 6.
In this paragraph no results are presented as the system of incompressible bdl equations here only served as an example to demonstrate the solution techniques. In paragraph 6 these equations are used as an example for the application of standard software and there results will be presented.

4. HYPERSONIC BOUNDARY LAYER EQUATIONS

4.1 Numerical solution

In this paragraph we have only to present the differential equations and boundary conditions and then to apply the solution mechanism of the preceding paragraph 3.
In hypersonic flow, e.g. in a reentry problem, the air behaves as a chemically reacting mixture whose composition is variable and is described by the mass fraction $\omega_i = \rho_i/\rho$ for each species i, with ρ_i the

density of the species i and ρ the mean density. The unknowns and necessary equations are for hypersonic bdl flow

Unknown		Equation
u	velocity along surface	global continuity equ.
v	velocity normal to surface	momentum equation
T	temperature	energy equation
$\omega_i=\rho_i/\rho$	mass fraction $i=1(1)N$	species continuity equations, $i=1(1)N$, but <u>one</u> replaced by $\sum_{i=1}^{N} \omega_i = 1$.

The pressure p is given by the inviscid flow at the outside of the bdl at $y=y_e$, and $\rho=\rho(p,T)$ is given by the equation of state. The corresponding original bdl equations are not written down here. In these equations all quantities are made dimensionless with reference values, e.g. L_∞, u_∞, T_∞, ρ_∞,... which may be the free stream values. The corresponding equations may be found in Däubler et al. (1976a) and Glotz et al. (1975). Then we apply a similarity transformation, which now must take care of the compressibility. We use a modified Falkner-Skan- and simplified Howarth-Dorodnycin-transformation. If x,y,u,v are nondimensional quantities in the boundary layer scale (y-coordinate stretched by $\sqrt{Re_\infty}$) then by

$$x=\hat{x} \ , \ y=\frac{\delta}{\rho_e}\hat{y} \ , \ u=u_a f \ , \ v=\frac{1}{\delta\rho_e} g \qquad (4.1)$$

we go from (x,y,u,v) to the similarity system $(\hat{x},\hat{y},f,g)$ and by

$$\hat{x}=\xi \ , \ \hat{y}=\hat{y}_e(\xi)[c\eta+(1-c)\eta^{N^*}] \ , \qquad (4.2)$$

from $(\hat{x},\hat{y})$ to the numeric coordinates (ξ,η), where the index e means outer edge of the boundary layer and where the relations (3.4) hold for δ and the coefficients A and Λ appearing in the equations. This is the compressible analogy of (3.3) to (3.5). The coefficients c and N^* in (4.2) are determined to give an optimum distribution of mesh points in the $\hat{y}$-scale. The coordinate η being between 0 and 1, the outside coordinate $\hat{y}_e$ must guarantee the adequate asymptotic behaviour of Figure 3.6. The resulting equations are presented here in the nomenclature of Däubler et al. (1976a), omitting ^ (i.e. y means $\hat{y}$ in the following equations:

Global continuity equation

$$GCE \equiv \left(F_1 \frac{\rho_\xi}{\rho} + F_2 \frac{\rho_\eta}{\rho} + F_4 + \varepsilon F_1 \frac{r_\xi}{r}\right) f + F_1 f_\xi + F_2 f_\eta + \\ + \frac{\rho_\eta}{\rho} g + g_\eta = 0 \qquad (4.3)$$

with ε=o for plane flow and ε=1 for axisymmetric flow with r as the contour radius.

Species continuity equations

$$SCE \equiv \rho A f \omega_{i\xi} + (F_2 f + g) \frac{\omega_{i\eta}}{y_\eta} \rho - \delta^2 M_i \sum_{r=1}^{R} \nu_{ir} w_r - \\ - Z_1 y_\eta \frac{\partial}{\partial \eta} \left[\frac{D_i^T T_\eta}{y_\eta T} - \sum_{\substack{k=1 \\ k \neq i}}^{N} \frac{\rho M_i D_{ik}}{y_\eta M^2} (\omega_k M)_\eta \right] = 0, i=1(1)N, \qquad (4.4a)$$

where <u>one</u> is replaced for a <u>nonvanishing</u> species i by

$$\sum_{i=1}^{N} \omega_i - 1 = 0 \ . \qquad (4.4b)$$

Momentum equation

$$ME \equiv f(F_1 f_\xi + F_2 f_\eta + F_4 f) + g f_\eta + \frac{\rho_a^2}{\rho y_\eta} \left(\frac{\mu \, y_{\eta\eta}}{y_\eta} - \mu_\eta \right) f_\eta - \\ - \frac{\rho_a^2 \mu}{\rho y_\eta} f_{\eta\eta} - F_4 \frac{\rho_a}{\rho} = 0 \ . \qquad (4.5)$$

Energy equation for chemical nonequilibrium

$$EE \equiv \rho c_p A f T_\xi + (F_2 f + g) \rho c_p \frac{T_\eta}{y_\eta} - Y_6 \frac{\partial}{\partial \eta} \left\{ \left[\frac{\lambda}{Pr_\infty} + \right. \right. \\ \left. + Y_4 \sum_{i=1}^{N} \sum_{k=1}^{N} \frac{M\omega_i \omega_k}{\rho M_i M_k \mathcal{D}_{ik}} \left(D_i^{red} - D_k^{red} \right)^2 \right] \frac{T_\eta}{y_\eta} + \frac{2 Y_4 T}{y_\eta} \times \\ \times \left. \sum_{i=1}^{N} \frac{D_i^{red}}{M \, M_i} (M\omega_i)_\eta \right\} - Y_6 \frac{T_\eta}{Sc_\infty} \sum_{i=1}^{N} c_{p i} \left\{ \frac{D_i^T \, T_\eta}{T \, y_\eta} - \sum_{\substack{k=1 \\ k \neq i}}^{N} \times \right. \\ \times \left. \frac{\rho M_i D_{ik} (M\omega_k)_\eta}{M^2 y_\eta} \right\} + \delta^2 \sum_{r=1}^{R} w_r H_r + \\ + Y_5 \left(\Lambda f - \rho_a \, \mu \frac{f_\eta^2}{y_\eta^2} \right) = 0 \ . \qquad (4.6)$$

These equations have been written down here only to illustrate the complexity of the system of equations. Functions with capital letters F_1, F_2,... are given functions of ξ. But all thermochemical coefficients like thermodiffusion-coefficient D_i^T or $D_i^{red} = D_i^T/\omega_i$, diffusion coefficient D_{ik} (binary $\mathcal{D}_{ik}$), mean molecular

weight M of the mixture, specific heat c_p, dynamic viscosity μ, heat conduction coefficient λ and density ρ are mostly very complicated nonlinear functions of the unknowns T and ω_i. Especially the chemical reaction velocity w_r of the r-th chemical reaction is a function which may change its value by several orders of magnitude if one mass fraction ω_i is changed by a fraction of one percent. On the other hand the ω_i may have values from 10^{-40} to the order of 1. This makes the system extremely stiff.
The boundary conditions for a noncatalytic wall are (no chemical reaction at the wall, i.e. species diffusion is zero):

$$\text{BCW} \equiv \begin{cases} f(\xi,o) = 0 \\ g(\xi,o) = 0 \\ T(\xi,o) - T_w = 0 \\ \left[D_i^T \dfrac{T_\eta}{T} + \sum\limits_{\substack{k=1\\k\neq i}}^{N} \dfrac{\rho M_i D_{ik}}{M^2} (M\omega_k)_\eta\right]_w = 0, \quad i=1(1)N \,, \\ \text{where } \underline{\text{one}} \text{ is replaced for a nonvanishing species by} \\ \qquad \sum\limits_{i=1}^{N} \omega_i = 0 \end{cases} \tag{4.7}$$

and the boundary conditions at the outer (inviscid) edge of the bdl are

$$\text{BCE} \equiv \begin{cases} f(\xi,1) - 1 = 0 \\ \text{GCE} = 0 \\ T(\xi,1) - T_e = 0 \\ \omega_i(\xi,1) - \omega_{i,e} = 0 \,, \quad i=1(1)N \,, \end{cases} \tag{4.8}$$

where T_e, $\omega_{i,e}$ result from the inviscid flow.
Now we have a system of parabolic equations like equation (3.8), which is now

$$\underline{Pu} \equiv \begin{pmatrix} \text{BCW} \\ \text{GCE} \\ \text{ME} \\ \text{EE} \\ \text{SCE}(i=1) \\ \vdots \\ \text{SCE}(i=N) \\ \text{BCE} \end{pmatrix} = 0 \,, \qquad \underline{u} = \begin{pmatrix} f \\ g \\ T \\ \omega_1 \\ \vdots \\ \omega_N \end{pmatrix} . \tag{4.9}$$

The whole solution procedure described in paragraph 3

is now applied to the system (4.9): a Newton-ansatz like (3.9) leads to a system of linear parabolic equations for corrections $\underline{\Delta u}$ like equation (3.12), which further yields a difference equation like (3.14). The corresponding matrix Q_d now has "elements" which are (3+N)×(3+N)-matrices. But for the application of the difference method some care must be taken for terms like the y-derivative of the heat conduction coefficient, which for simplicity here is taken only to be a function of temperature,

$$\lambda_y(T) \rightarrow \lambda_{y,d} + (\lambda_{y,T}\,\Delta T)_d \tag{4.10}$$

in order to yield the correct Newton-Raphson-equations for the nonlinear difference equations. It is a very tedious work to write down the expansions (3.12) for this nonlinear system (4.9).

Here are some additional remarks concerning the solution of such a complicated system. For the accuracy control according to Figure 3.4, 3.5 and equations (3.18), (3.19) only the norm of the temperature error has been used, e.g. to compute T with 1% accuracy relative to the maximum temperature. It would be meaningless to compute all species, also those of very small value (e.g. 10^{-10}), to the same relative accuracy.

The examples in the reports given in the references have been computed with f, g reversed in (4.9), leading to difficulties in some cases, so in (4.8) GCE=0 had been replaced by $g_{\eta,w}=o$ (resulting from 4.3). But with the remarks at the end of paragraph 3.2, the ordering of (4.9) should present no difficulties.

By the discretization errors near the wall negative mass fractions ω_i in the order of the discretization error (instead of zero) could occur. By the thermochemical functions these negative mass fractions led to physically impossible values of the reaction velocities and destroyed the solution. Therefore the species continuity equation SCE(i)=0 was replaced by $\omega_i=o$ whenever $\omega_i<o$ occured.

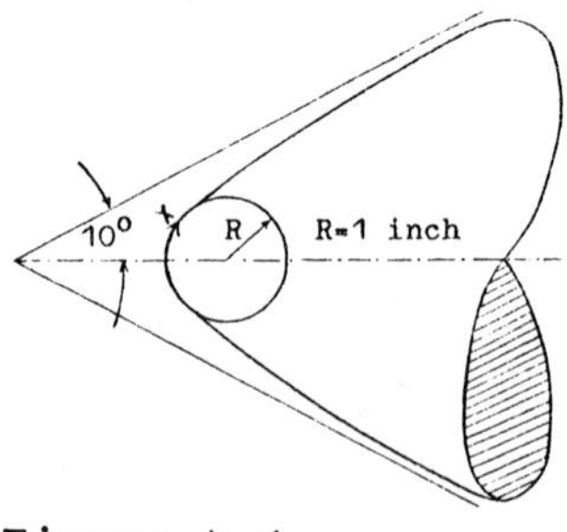

Figure 4.1
AGARD-hyperboloid

A general purpose program has been established with a variable number of components and chemical reactions. For different chemical models only the corresponding subroutines for the thermochemical functions have to be changed. The inviscid data are computed from the system of ordinary differential equations in ξ which results for $y=y_e$

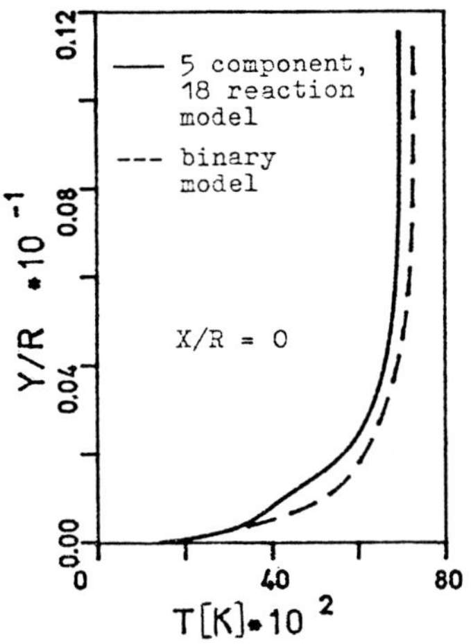

Figure 4.2 Temperature profiles at stagnation point for different chemical air-models

from the general system; the pressure was taken from the modified Newton theory, see Däubler et al. (1976a) and Glotz et al. (1975).

The examples have been calculated for the AGARD-hyperboloid of Figure 4.1. For the AGARD-case A with flight conditions 30.48 km altitude, 6.096 km/sec velocity and a wall temperature of 1400K, giving $Ma_\infty \approx 20.1$, $Re_\infty \approx 1.6 \cdot 10^5$ (depending on the chemical model for the air) the temperature profiles computed with 0.1% accuracy are depicted in Figure 4.2 for a 5 component/18 reaction model and a binary model for the air; the stagnation temperatures are 6973K and 7281K. In Figure 4.3 the Stanton

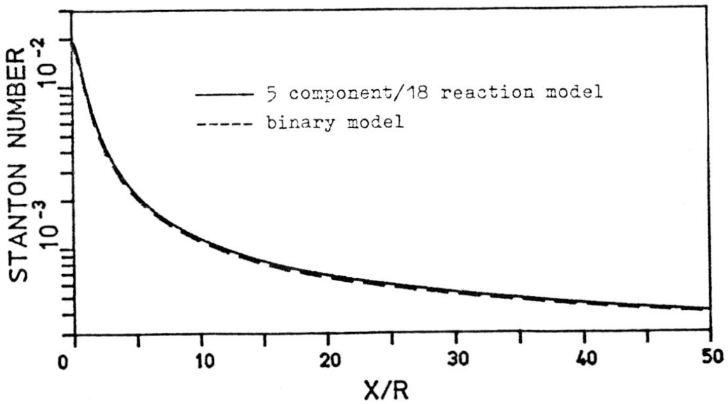

Figure 4.3 Stanton number for different chemical air-models

number for these two air-models is depicted (1% accuracy) over $\xi = x/R$ up to a value of 50 nose radii. It is interesting to note that the order p in the ξ-direction mostly was 2 and never exceeded 3 because of the stiffnes of the equations. The step size $\Delta\xi$ varied from 0.1 up to 10. Most of the computing time is used for the calculation of the thermodynamics.

4.2 Overall accuracy considerations

If we prescribe a numerical accuracy e.g. of 0.1% for Figure 4.2, we do not know how good the result really is. There are errors caused by the mathematical model of the physical reality (bdl equations), but in real gas flow main sources of errors are the thermochemical coefficients which are only approximately known at the relevant high temperatures. These questions have been discussed in Schönauer et al. (1979), Däubler (1977), Glotz and Schönauer (1977) and Glotz (1979). Here a brief summary of the results will be given.

The description of the thermodynamic properties (specific heat, enthalpy), of heat and mass transfer and of the chemical reactions is based on experimentally determined coefficients. These coefficients, above all for high temperatures, are not exact values. How do errors of these coefficients influence the result of a hypersonic bdl calculation? For the engineer the surface heat transfer is the most important quantity. Therefore a nonlinear numerical analysis has been carried out for coefficients varying in the range of their uncertainty and the change in the heat transfer has been calculated. The distinction between numerical errors and changes by the variation of the coefficients has been made possible only by the selfadaptive solution method. Most tests were made for the stagnation point profiles which reflect the typical behaviour, but some calculations were made for the whole hyperboloid and they confirmed the stagnation point results.

If one considers the possible situations of a reentry trajectory, for the temperature the following statements hold:

i) The thermodynamic properties generate errors of about 1%.

ii) The chemical reaction coefficients generate errors of about 3% if a 5 component/18 reaction model for air is used, although their own uncertainties are very much larger.

iii) The transport coefficients generate errors of about 10%. They are the most uncertain coefficients and limit severely the attainable overall accuracy.

The next question was: What chemical model of the air should be used and what boundary conditions? To answer these questions a series of different air models with different numbers of components and reactions has been used with different boundary conditions. With regard to iii) we should take a 5 component model (N_2, O_2, NO, N, O) with 3 reactions (Zeldovich-mechanism) and a fully catalytic wall (for security reason) if temperature and heat transfer are to be computed. If only the heat transfer should be computed a binary model (atoms and molecules) will be sufficient. In Figure 4.2 and 4.3 the difference between a 5 component and binary model can be seen. The difference between a 5 component/18 reaction model and a 5 component/3 reaction model would not be visible in the drawings. The most uncertain parameter in the chemical reaction mechanism is the boundary condition at the wall, where the real heterogeneous reactions should be known.

4.3 Transient equations

For a binary air model the transient hypersonic bdl equations have been solved, Däubler et al. (1977a) and (1977b). Here only the solution procedure will be described. To the hypersonic bdl equations (4.4)-(4.6) are added the instationary terms which were transformed in the same way as the equations themselves. But for a transient flow the free stream conditions may change considerably with time t, which results in a large variation of the bdl thickness and so destroys the advantage of the similarity transformation. Therefore all quantities are nondimensionalized with the _actual_ free stream values at the time t which has the consequence that the characteristic dimesionless numbers are time dependent, e.g. Reynolds number Re=Re(t). This leads independent of time to a nearly constant boundary layer tickness in the usual similarity coordinates. But the time dependence of the free stream quantities must be considered in the transformation of the instationary terms. We then derive with the Newton-Raphson method a system of differential equations of the type of equation (3.12).

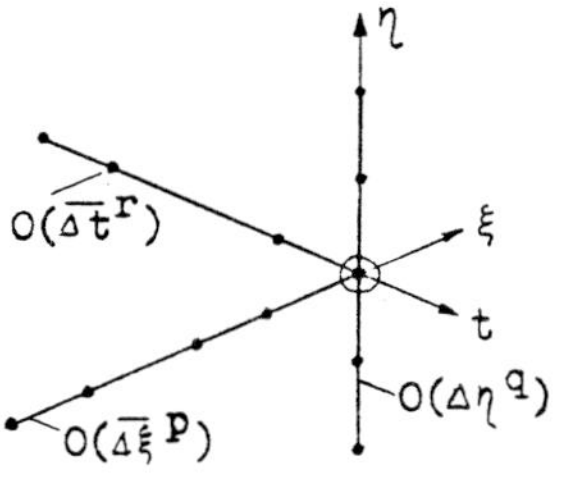

Figure 4.4 Possible difference star for transient equations

For the discretization of the time derivatives backward difference formulae of arbitrary order r of the type of Figure 2.1 are used which then results in a fully implicit difference star shown in Figure 4.4 with arbitrary orders r,p,q in t-, x-, y-direction.

From the time-discretization there results a discretization error vector D_t and the error equation corresponding to (3.15) is now with the nomenclature defined in (4.9)

$$\underline{\Delta u} = \underline{\Delta u}_{Pu} + \underline{\Delta u}_{D_\xi} + \underline{\Delta u}_{D_\eta} + \underline{\Delta u}_{D_t} = Q_d^{-1}[(Pu)_d + D_\xi + D_\eta + D_t]. \tag{4.11}$$

The error terms in the brackets must be balanced in order to yield an efficient solution method. Figure 4.5 shows the domain of solution. The difference grid is determined in the following way: For ξ=o (stagnation point) the initial profiles are computed for prescribed values $\Delta\eta$ and order q, with selfadapted step size Δt and order r. For these profiles the errors $\|\Delta u_{D_\eta}\|$ are determined, and then the time t_c which corresponds to the profile with largest η-discretization error is chosen as critical time, see Figure 4.5. For the conditions of time t_c a "normal" stationary bdl calculation is made by dropping the transient terms in the equations. The ξ- and η-grid of this calculation is stored, see Figure 4.5, and then the full transient equations are solved with this grid fixed. For each station ξ_i the order p in ξ-direction is determined according to Figure 3.5. The time steps Δt_j are determined by balancing $\|D_\eta\|$ and $\|D_t\|$ similarly to (3.19). The order r in t-direction is determined for t_j by comparison of the error of the order r to the errors of the orders r+1 and r-1 like Figure 3.5

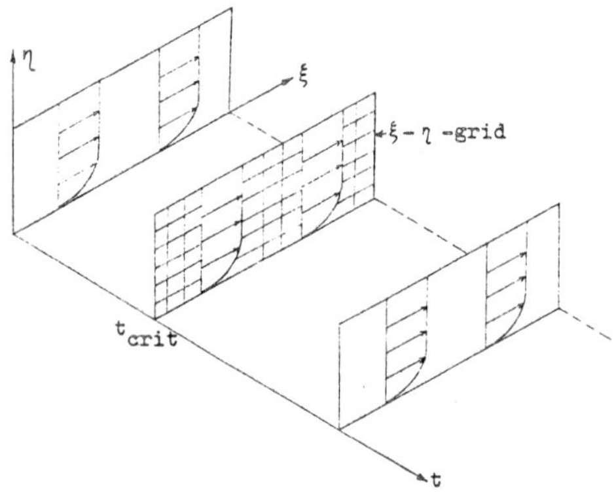

Figure 4.5 Domain of solution

For the numerical calculations the bdl for a reentry trajectory has been computed, but practically no instationary effects occured. Then the time scale has been reduced for the trajectory by a factor of 10^{-6} which corresponds to the timescale in a shock tube and clearly there were pronounced instationary effects. Details of the calculations and results are presented in Däubler et al. (1977b).

5. THREE DIMENSIONAL INCOMPRESSIBLE BOUNDARY LAYER

5.1 3-D boundary layer equations and similarity transformation

The program for the 3-D incompressible bdl equations is still under development. Preliminary reports were given by Schönauer et al. (1978) and (1977). The available space does not allow for a detailed presentation of the encountered problems, so rather a survey is given.

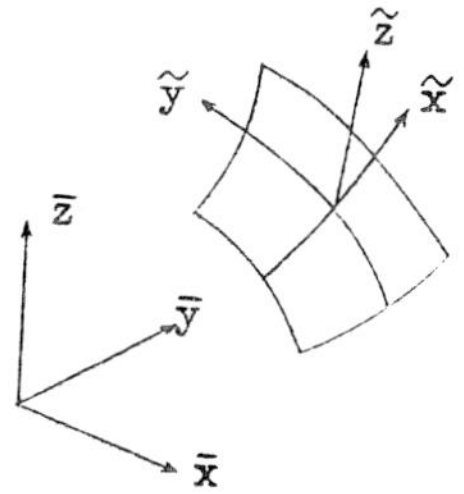

Figure 5.1
Coordinate systems

In a basic cartesian coordinate system $\bar{x},\bar{y},\bar{z}$ a general 3-D surface is given. On the surface there is a locally orthogonal coordinate system with surface coordinates $\tilde{x},\tilde{y}$ and the normal coordinate $\tilde{z}$, see Figure 5.1. The dimensionles 3-D bdl equations for this configuration are given in Schönauer et al. (1978) as equation (3). For an efficient numerical solution a similarity transformation is necessary which accounts for the history of the bdl, and for a selfadaptive method a proceeding of the solution only in positive coordinate direction is required. Both requirements are best met by streamline-potentialline coordinates, see Figure 5.2.

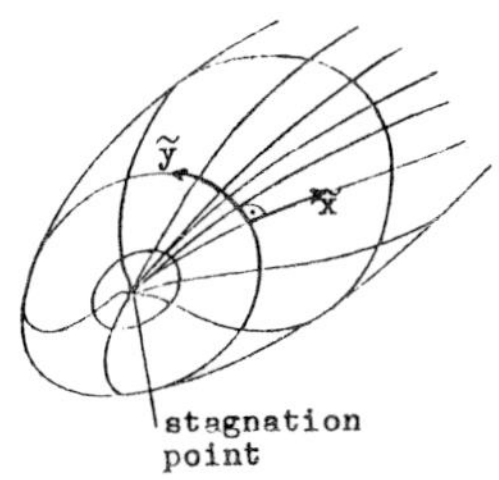

Figure 5.2
Streamline-potential line coordinates

The $\tilde{x}$-coordinate on streamlines is $\tilde{x}=\phi-\phi_o$, with ϕ as the inviscid velocity potential and ϕ_o its value in the front stagnation point. The $\tilde{y}$-coordinate on the closed potential lines is normalized to $o\leq\tilde{y}\leq 2\pi$. On the line $\tilde{y}=2\pi$ there is a jump to $\tilde{y}=o$.

We now generalize the modified Falkner-Skan transformation (3.3) to the 3-D flow with the inviscid velocity $u_e(\tilde{x},\tilde{y})$, but we have to stretch here the normal coordinate $\tilde{z}$. So we go from the bdl coordinates $(\tilde{x},\tilde{y},\tilde{z})$ to the similarity coordinates (x,y,z):

$$\tilde{x}=x\ ,\ \tilde{y}=y\ ,\ \tilde{z}=\delta(x,y)\Phi(x,y,z)\ ,\quad 0\leq z\leq 1\ ,$$
$$\Phi = (\tilde{z}_a/\delta)[cz+(1-c)z^N]\ , \tag{5.1}$$

where Φ cares for an optimum distribution of mesh

points by the choice of c and N and for the asymptotic behaviour by the choice of $\tilde{z}/\delta$, which may depend on x,y. The dependent variables are transformed by

$$\tilde{u} = u_e(x,y)f(x,y,z) \ , \quad \tilde{v} = u_e(x,y)g(x,y,z) \ ,$$
$$\tilde{w} = u_e(x,y)h(x,y,z) \ , \qquad (5.2)$$

If $\bar{x}(x,y)$, $\bar{y}(x,y)$, $\bar{z}(x,y)$ are the basic cartesian coordinates of Figure 5.1, the metric coefficients h_1, h_2 are defined by

$$h_1^2 = \bar{x}_x^2+\bar{y}_x^2+\bar{z}_x^2 \ , \quad h_2^2 = \bar{x}_y^2+\bar{y}_y^2+\bar{z}_y^2 \ . \qquad (5.3)$$

If s is the arc on a streamline, it follows

$$u_e = \partial\phi/\partial s \ , \quad h_1 = 1/u_e \qquad (5.4)$$

and we define δ as in the 2-D case by

$$\delta^2 = 2\int_0^s u_e(s^*)ds^*/u_e^2 = 2h_1^2\, x \ . \qquad (5.5)$$

The transformed 3-D bdl equations, continuity equation (CE), x-momentum equation (XME) and y-momentum equation (YME), are

$$CE \equiv h_z+K_1f_z+K_2g_z+K_3f_x+K_4g_y+K_5f+K_6g = 0 \ , \qquad (5.6a)$$

$$XME \equiv f_{zz}+I_{x,1}f_z+I_{x,2}ff_z+ I_{x,3}gf_z+ I_{x,4}hf_z+I_{x,5}ff_x+$$
$$+I_{x,6}gf_y+ I_{x,7}f^2+I_{x,8}fg+I_{x,9}g^2+I_{x,10}=0, \qquad (5.6b)$$

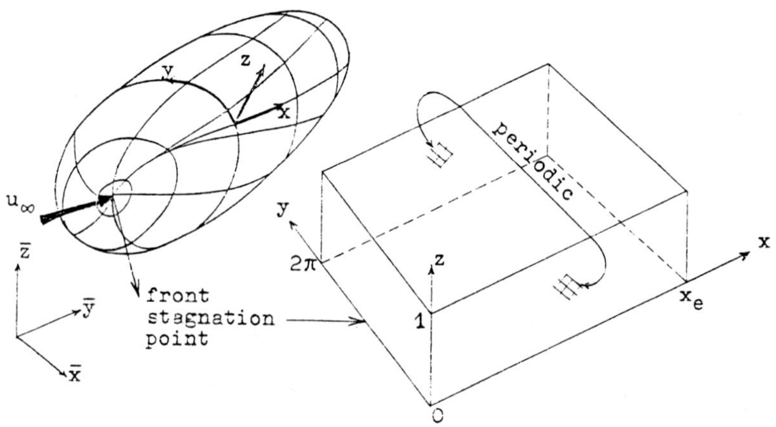

Figure 5.3 Coordinate systems and domain of solution

$$YME \equiv g_{zz}+I_{y,1}g_z+I_{y,2}fg_z+I_{y,3}gg_z+I_{y,4}hg_z+I_{y,5}fg_x+$$

$$+ I_{y,6}gg_y+I_{y,7}f^2+I_{y,8}fg+I_{y,9}g^2+I_{y,10}=0. \quad (5.6c)$$

The coefficients K_i, $I_{x,i}$ and $I_{y,i}$ can all be expressed by the metric coefficients h_1, h_2 and are given in Schönauer et al. (1977).

The surface of the body is transformed onto the x,y-domain shown in Figure 5.3 where lines parallel to the x- and y-axis represent streamlines and potential lines. At y=o=2π the domain is "closed" (periodic). At the outer edge (z=1) of the bdl the inviscid flow has the direction of the streamlines. Therefore the boundary condition are

$$\begin{aligned} z=o &: \text{BCW} \quad f=g=h=o \ , \\ z=1 &: \text{BCE} \quad f-1=g=o \ \text{ asymptotically} \ . \end{aligned} \quad (5.7)$$

5.2 Streamline-potential line coordinates for a general surface

Before we discuss the initial conditions at x=o we must discuss the generation of the streamline-coordinates. In practical applications the surface and the inviscid velocity potential ϕ are given only pointwise. From ϕ we easily get $x=\phi-\phi_o$. So we have, e.g. from a panel method, a set of points $P(\overline{x},\overline{y},\overline{z},\phi)$. From this data set we calculate the streamlines as follows: We interpolate locally the surface by a quadric

$$F \equiv F(\overline{x},\overline{y},\overline{z}) \equiv a_1\overline{x}+a_2\overline{y}+a_3\overline{z}+a_4\overline{x}^2+a_5\overline{y}^2+a_6\overline{z}^2+$$

$$+a_7\overline{xy}+a_8\overline{xz}+a_9\overline{yz} = 1 \ . \quad (5.8)$$

On the surface one of the coordinates $\overline{x},\overline{y},\overline{z}$ is a function of the two others. So the potential (only on the surface) is approximated by

$$\phi = \overline{\phi}(\overline{x},\overline{y},\overline{z}(\overline{x},\overline{y})) = \overline{\phi}(\overline{x},\overline{y}) \text{ or } \overline{\phi}(\overline{x},\overline{z}) \text{ or } \overline{\phi}(\overline{y},\overline{z}). \quad (5.9)$$

Then the tangent vector $\underline{t}_\psi$ to a streamline is given by

$$\underline{t}_\psi = \text{grad}^2 F \ \text{grad}\ \overline{\phi} - (\text{grad}\ \overline{\phi}\ \text{grad}\ F)\text{grad}\ F. \quad (5.10)$$

With the aid of (5.10) an integration method to generate the streamlines may be generated. But it is not possible to compute the streamlines <u>out</u> of the stagnation point. So we choose in the middle of the body starting points from which we integrate streamlines <u>into</u> the front and rear stagnation points. In regions where the streamlines diverge we choose again starting points at about 1/4 and 3/4 of the body. In each point of the calculation procedure the coefficients of the <u>local</u> approximations (5.8),(5.9) are determined by the least square method. The y-value on the potential lines is arbitrary, but finally normalized to $o \leq y \leq 2\pi$. So we get on the streamlines another data set of points $Q(x,y,\bar{x},\bar{y},\bar{z})$, from which we may generate local approximations

$$\bar{x}(x,y) = c_o+c_1x+c_2y+c_3x^2+c_4y^2+c_5xy \qquad (5.11)$$

and in the same way for $\bar{y}(x,y)$, $\bar{z}(x,y)$, whose coefficients are again determined by the least square method. From (5.11) we finally get the metric coefficients (5.3). But in the coefficients of the bdl equations (5.6) also derivatives of the metric coefficients are required which means that also second derivatives of (5.11) are to be computed.

If these second derivatives are not smooth enough, the selfadaptive solution method reacts as if the surface would be wavy. It turned out that the computation of the metric coefficients was a much more difficult problem than the solution of the bdl equations themselves. But on the other hand the solution of the bdl equation is nearly "visible" from the similarity coefficients if these coefficients are known.

At the stagnation point some of the bdl coefficients determined from (5.11) become singular. Therefore at the stagnation point a special expansion is used

$$\bar{x} = \bar{x}_s+f_1(y)\sqrt{x}+f_2(y)x \ , \ f_i = a_{i,o}+a_{i,1}\sin y+ +a_{i,2}\cos y+\ldots+a_{i,10}\cos 5y \ . \qquad (5.12)$$

Then at the stagnation point the bdl equations (5.6) reduce to an ordinary system of differential equations in z whose solution furnishes the initial condition at x=o.

If one goes round the body on a potential line one meets two y-values where for the metric coefficient holds $h_2=o$. In a symmetric flow these points are in the plane of symmetry, in a general nonsymmetric flow the streamlines leading through these points are "dividing streamlines" over which no information crosses. It can be shown that for $h_2=o$ the system of bdl equations (5.6) reduces to a system of two equations (2-D bdl) in x,z for the variables f,h, with g=o (no cross velocity). So numerically one has to care for the right (limiting) values of the bdl coefficients if h_2 has small values. This is the case in the region of converging streamlines in Figure 5.3. But these dividing streamlines are not used as "initial lines" for the calculation, they are not "known" to the program.

5.3 Difference method

Now the coefficients in the bdl equations (5.6) are assumed to be known and the principal ideas of solution of paragraph 3 are applied. The system of equations and boundary conditions is

$$Pu \equiv \begin{pmatrix} BCW \\ XME \\ YME \\ CE \\ BCE \end{pmatrix} = 0 , \quad \underline{u} = \begin{pmatrix} f \\ g \\ h \end{pmatrix} . \tag{5.13}$$

We apply a Newton-Raphson method and get a system of differential equations for the correction $\underline{\Delta u}$ which is solved by the difference method. In a point x_i, y_j, z_k of the difference grid we use for the x-derivatives nonequidistant backward difference formulae of order p (see Figure 2.1), for the y-derivatives centralized nonequidistant formulae of order r and in the z-direction (b.v.p.) central equidistant formulae of order q (see Figure 2.2). In the y-direction we have also a "boundary value problem" because of the closed potential lines although there are only frist derivatives with respect to y. The resulting difference star for the momentum equations is depicted in Figure 5.4. For the continuity equation in the z-direction formulae of Figure 3.3 are used, i.e. the equation is fulfilled at midpoints k-1/2. Instead also central formulae could be used if the unknowns would be ordered in the form $\underline{u}=(h,g,f)$.

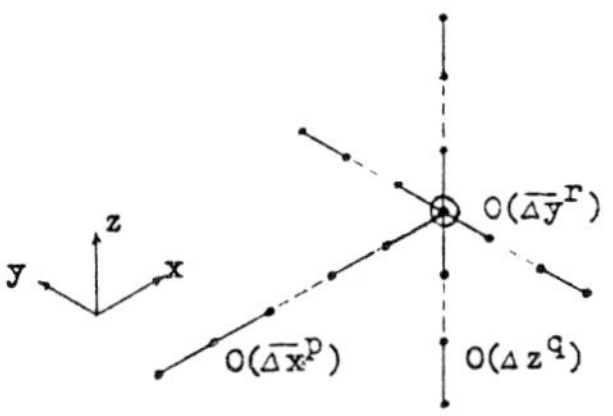

Figure 5.4 Difference star for momentum equation

The resulting error equation in analogy to (3.15) is

$$\underline{\Delta u} = \underline{\Delta u}_{Pu} + \Delta u_{D_x} + \underline{\Delta u}_{D_y} + \underline{\Delta u}_{D_z} = Q_d^{-1}[(P\underline{u})_d + D_x + D_y + D_z], \tag{5.14}$$

where D_x, D_y, D_z are the terms resulting from the (estimated) discretization errors. At the stagnation point the z-grid is determined for prescribed accuracy as in the 2-D bdl. Then D_z is the key error to which all other errors must be adapted. This is done in the x-direction on all streamlines separately by choosing the corresponding step size Δx and selecting from these the smallest one. The order p is independently determined for each solution streamline. In the y-direction the number M of streamlines and therefore the step size $\Delta y=2\pi/M$ is prescribed, rather by considerations to get enogh information on the surface than by accuracy requirements. The program chooses the order r=2, but if D_y is too large in comparison to D_z the order is increased to 4,6,... The iteration is stopped when the defect $(P\underline{u})_d$ is slightly smaller (always in the sense of the maximum norm) than D_z, but this has to be explained further. By the Newton-Raphson method and the difference star of Figure 5.4 all corrections in a plane x=const (y-z-plane, see Figure 5.3) are simultaneously coupled. Therefore at first they are uncoupled in the y-direction by extrapolating with order p the y-derivatives from foregoing profiles and the iteration is stopped according to the above mentioned criterium. So for each y_j on a potential line $x=x_i$ the solution profiles are computed separately. Then, if all profiles for all y_j have been calculated, one final iteration step at each y_j is made with y-derivatives computed from the actual profiles at y_j in order to guarantee stability.

If the iteration does not converge this is interpreted as separation. In 3-D flow there are two types of separation: the "take off" of the boundary layer where from two sides the wall streamlines converge, and the "regular" separation with flow reversal. The situation is illustred in Figure 5.5. At point A for the first time separation occurs (take off). Separated mesh points are eliminated from the grid and at the first or last nonseparated points onesided difference formulae in the y-direction are used. At point B another separation region begins. The domain of solution is split up in the domain C between the separated regions and the domain outside the separated regions. Several such enclosed separation regions may occur. At the lines D finally regular separation occurs.

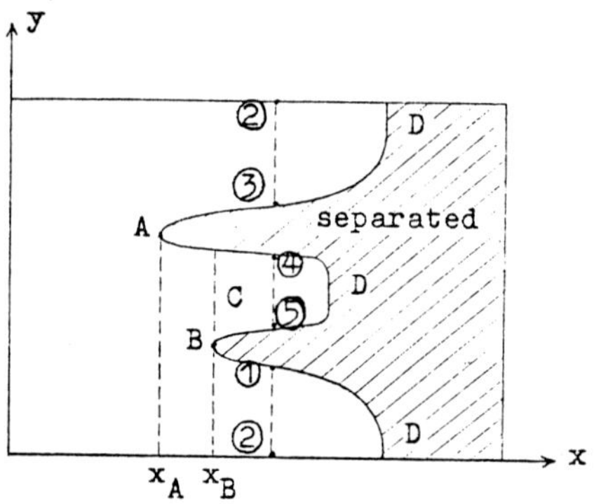

Figure 5.5 Separated regions

The part of the program for the solution of the bdl

equations has been developed using a flow configuration where the metric coefficients could be computed analytically: a flat plate bounded by streamlines and potential lines with a circular cylinder mounted on the plate. When we then used numerically given surfaces as described in paragraph 5.2 we met never expected difficulties and we are still ameliorating our results.The selfadaptation process detects the least nonsmoothness of the metric coefficients which would not be visible with a traditional difference method (and also not the errors caused by the nonsmoothness).

Here one result is presented of the actual program version for the general ellipsoid of Figure 5.6. The inviscid flow data was delivared by Dr.S.R.Ahmed of DFVLR Braunschweig from a panel method. There results a completely non-symmetric bdl. In Figure 5.7 the inviscid streamlines of the bdl calculation are shown together with the points where separation occured (crosses). The parameters of the (preliminary) computation are: relative accuracy of u/u_e is 5% (but the rough bdl coefficients generate larger errors); order in x-direction is limited by $p \leq 2$ (because of the rough-ness of the coefficients); step size Δx is selfadapted; order in y- and z-direction is prescribed: r=4, p=4; prescribed number of mesh points in y-direction: 30 points, and in z-direction: 10 points.

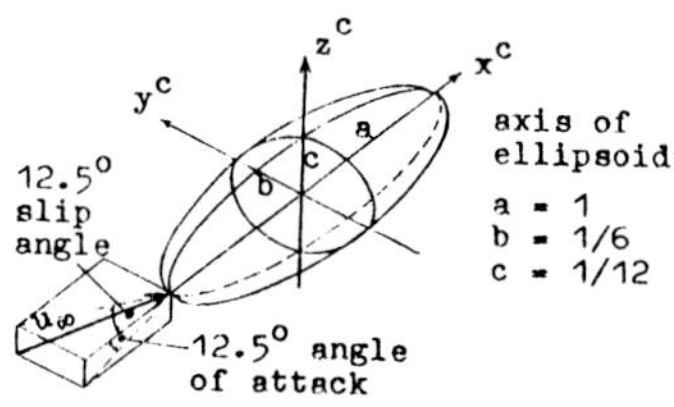

Figure 5.6 General ellipsoid

In Figure 5.7 the crosses mark the points where the iteration of the nonlinear equations diverges. Therefore these points are beyond the separation line, at maximum by the actual value of Δx.

The full selfadaptation process mentioned above will be applicable only if smooth enough bdl coefficients have been generated. If one has enough points in the y-direction, i.e. enough streamlines, one can find a nonseparated part of the flow between the two separation lines (take-off lines). In order to save storage space in the computer, the program will be split up into 4 separate steps: computation of smoothed data, computation of the streamlines, computation of the coefficients c_i of (5.11), solution of the bdl equations.

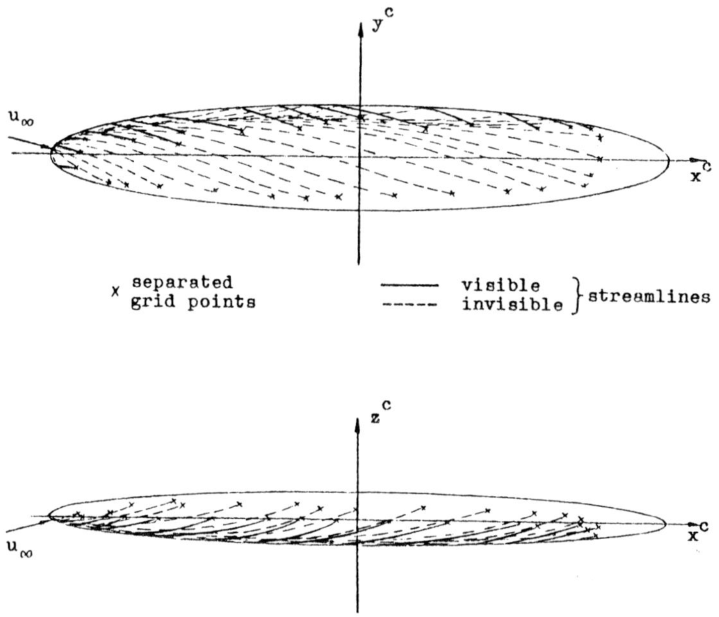

Figure 5.7 Streamlines of the numerical solution with the separated points marked by a cross

6. STANDARD SOFTWARE FOR PARABOLIC EQUATIONS

The procedure of selfadaptive solution for parabolic equations with the generation of difference formulae of arbitrary order and the generation of the corresponding matrices of the difference method with a variable number of diagonals (depending on the order) leads to a vast program with many sources for programming errors. Therefore a general purpose program package SL (selbstadaptive Lösung) is being developed for the numerical solution of a system of parabolic equations

$$F(t,x,u,u_t,u_x,u_{xx}) = 0 \tag{6.1}$$

where F and u are vectors with n components if there are n unknowns and n equations. The domain of solution is $t_0 \leq t \leq t_{end}$, $x_0 \leq x \leq x_{end}$. Here t and x are the initial and boundary coordinates. The part of the program package concerned with (6.1) is SL1PX, where 1 means one space boundary value problem coordinate and X denotes the version. There are versions with order and step size in the x-direction prescribed or selfadapted, with step size in the x-direction equidistant or nonequidistant and with the initial profile at $t=t_o$ prescribed or computed from an ordinary differential equation in x. All combinations of these possibilities are admitted, leading to 8 versions. We use the version SL1P4 (selfadapted, equidistant, computed initial profile) to solve the incompressible 2-D bdl equations (3.6). The ξ,η-coordinates of (3.6) correspond to the t,x-coordinates of (6.1).

The boundary conditions for x_o and x_{end} have the general form

$$G(t,x,u,u_t,u_x,u_{xx}) = 0 \tag{6.2}$$

which admits also a differential equation as boundary condition. The user of SL1P4 has to write a program for F and G and for the matrices

$$\frac{\partial F}{\partial u}, \frac{\partial F}{\partial u_t}, \frac{\partial F}{\partial u_x}, \frac{\partial F}{\partial u_{xx}}, \frac{\partial G}{\partial u}, \frac{\partial G}{\partial u_t}, \frac{\partial G}{\partial u_x}, \frac{\partial G}{\partial u_{xx}}, \tag{6.3}$$

which are needed for the Newton-Raphson procedure and for the calculation of the errors. Here e.g. $\partial F/\partial u$ means for a system of two equations

$$\frac{\partial F}{\partial u} = \begin{pmatrix} \frac{\partial F_1}{\partial u_1} & \frac{\partial F_1}{\partial u_2} \\ \frac{\partial F_2}{\partial u_1} & \frac{\partial F_2}{\partial u_2} \end{pmatrix} . \tag{6.4}$$

The programs for (6.1)-(6.3) must furnish also the values of the ordinary differential equation of the initial profile for $t=t_o$.

We apply the program SL1P4 to the bdl equations (3.6) for the flow around a circular cylinder. The inviscid potential flow in the nomenclature of paragraph 3 is

$$u_e(\xi) = 2 \sin \xi \tag{6.5}$$

if the radius R of the cylinder is taken as the reference length. The value $C=y_e/\delta$ of the mesh stretching function (3.5) which gives the adequate asymptotic behaviour of Figure 3.6 is not known a priori. Therefore after the computation of the initial profile the asymptotic behaviour is checked and adapted according to Figure 3.6 to the accuracy of the solution.

The solution has been computed up to $\xi=1.8$, because at $\xi=1.823$ there is the separation point of the flow (6.5) which is a singularity of the bdl equations and such a singularity is not suited for accuracy checks. In Figure 6.1 there is depicted the <u>global error</u> of

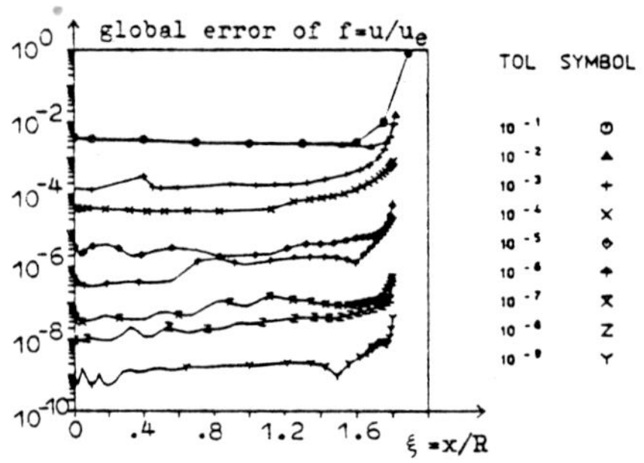

Figure 6.1 global error of $f=u/u_e$ for different values of tol

$f=u/u_e$ for different values of a prescribed tolerance which is

$$tol = \max \left\{ \frac{\max\limits_k |\Delta f_k|}{\max\limits_{i,k} |f_{i,k}|}, \frac{\max\limits_k |\Delta g_k|}{\max |g_{i,k}|} \right\} . \qquad (6.6)$$

The high accuracy values have only been computed to demonstrate the behaviour of the program. It is clearly visible that the error increases when the separation point is approached.

Table 6.1 Results of selfadaptation

tol	$N\eta$	q	$C=\frac{y_e}{\delta}$	$N\xi$	CPU-time U1108, sec
10^{-1}	6	2	1.29	8+ 2	3.31
10^{-2}	8	2	3.00	9+ 4	4.37
10^{-3}	12	4	3.47	18+ 5	7.99
10^{-4}	16	4	4.21	28+10	13.45
10^{-5}	29	4	4.41	47+ 5	29.79
10^{-6}	25	6	5.58	70+ 4	41.72
10^{-7}	38	6	6.26	113+ 4	97.91
10^{-8}	31	10	6.24	208+20	224.71
10^{-9}	43	10	7.62	344+11	499.69

In Table 6.1 the result of the selfadaptation process is demonstrated. For different values of tol (6.6) are given the number $N\eta$ of grid points and the order q in η-direction determined according to Figure 3.4, the value $C=y_e/\delta$ determined according to Figure 3.6, the number of solution stations ξ_i plus the number of dropped solutions (where the accuracy requirement failed), and the CPU-time of a UNIVAC 1108. Below $tol=10^{-7}$ round-off errors become more and more important. The order p in the ξ-direction always reached quickly the maximum value p=5 admitted by the program.

The program SL1P4 could easily be used in the same way for the solution of the compressible or turbulent bdl equations with complex relations or transport equations for the turbulence quantities. At present similar program packages are under development for general parabolic and elliptic equations with two or three space coordinates.

REFERENCES

1. Blottner,F.G. (1970) Finite Difference Methods of Solution of the Boundary Layer Equations, AIAA-Journal 8,p.193-205.
2. Blottner,F.G. (1975) Computational Techniques for boundary layers, AGARD-LS-73,p.3-1 to 3-51.
3. Däubler,H.-G.,Glotz,G., Grüning,J., Schönauer,W. (1976) Differenzenverfahren mit Selbststeuerung der Fehlerordnung zur Lösung zweidimensionaler Grenzschichtgleichungen, ZAMM 56,p.T171-T174.
4. Däubler,H.-G.,Schönauer,W., Glotz,G., Grüning,J. (1976a) Effektive Differenzenverfahren zur Lösung der laminaren Hyperschall-Grenzschichtgleichungen für chemisches Nichtgleichgewicht, Interner Bericht Nr. 6/76 des Rechenzentrums der Universität Karlsruhe.
5. Däubler,H.-G.(1977) Der Einfluß von Fehlern bei den Anpaßparametern der Thermostatik, Thermodynamik und Reaktionskinetik auf die Lösung der Gleichungen für zweidimensionale laminare Hyperschallgrenzschicht, Interner Bericht Nr. 10/77 des Rechenzentrums der Universität Karlsruhe, auch Dissertation Karlsruhe 1977.
6. Däubler,H.-G., Glotz,G., Schönauer,W. (1977a) Numerische Lösung der instationären, zweidimensionalen Hyperschallgrenzschichtgleichungen mit einem selbststeuernden Differenzenverfahren, ZAMM 57,p.T178-T180.
7. Däubler,H.-G., Glotz,G., Schönauer,W. (1977b) Effektive Differenzenverfahren zur Lösung der instationären zweidimensionalen Hyperschallgrenzschichtgleichungen für chemisches Nichtgleichgewicht, Interner Bericht Nr.9/77 des Rechenzentrums der Universität Karlsruhe.
8. Glotz,G., Schönauer,W., Däubler,H.-G., Grüning,J. (1975) Aufstellung eines Rechenprogramms zur Lösung der laminaren Hyperschall-Grenzschichtgleichungen bei chemischem Nichtgleichgewicht, Interner Bericht Nr. 5/75 des Rechenzentrums der Universität Karlsruhe.

9. Glotz,G., Schönauer,W. (1977) Einfluß des Chemiemodells und der Randbedingungen auf die Hyperschallgrenzschicht bei Luft, Interner Bericht Nr. 12/77 des Rechenzentrums der Universität Karlsruhe.

10. Glotz,G. (1979) Der Einfluß von Idealisierungen der Luft und der Randbedingungen an der Wand auf die Lösung der laminaren Hyperschallgrenzschichtgleichungen, Dissertation Karlsruhe 1979.

11. Grüning,J., Schönauer,W. (1975) Untersuchungen zur Entwicklung effektiver Lösungsverfahren für Grenzschichtgleichungen, ZAMM 55, p.T131-T133.

12. Raith,K. (1976) Numerische Untersuchung der Stabilität einer Familie von Differenzenverfahren zur Lösung der Grenzschichtgleichungen, Interner Bericht Nr. 7/76 des Rechenzentrums der Universität Karlsruhe.

13. Schönauer,W., Straub,D., Spreuer,H., Adams,E. (1974) Die numerische Lösung der stationären laminaren Hyperschall-Grenzschichtgleichungen bei chemischem Nichtgleichgewicht, DLR-FB 74-71.

14. Schönauer,W., Däubler,H.-G., Glotz,G., Grüning,J. (1976) Selbststeuernde Differenzenverfahren zur effektiven Lösung von stark rechenintensiven Problemen der Grenzschichttheorie, DGLR 76-185, in DGLR-Jahrbuch 1976, Band I, Nr.13.

15. Schönauer,W., Grüning,J., Glotz,G., Däubler,H.-G. (1977) Selfadaptive Solution of the Incompressible 3-D Boundary Layer Equations with Controlled Error, Proceedings of the Second GAMM-Conference on Numerical Methods in Fluid Mechanics, Ed. E.H.Hirschel and W.Geller, DFVLR, Köln.

16. Schönauer,W., Grüning,J., Däubler,H.-G., Glotz,G. (1978) Streamline Coordinates and Similarity Transformation for the 3-D Boundary Layer, ZAMM 58, p.T320-T321.

17. Schönauer,W., Däubler,H.-G., Glotz,G. (1979) The Problem of Accuracy in Numerics and Thermodynamics at the Solution of the Hypersonic Boundary Layer Equations Including Chemical Reactions, Recent Developments in Theoretical and Experimental Fluid Mechanics, Editors W.Müller, K.G. Roesner, B.Schmidt, Springer-Verlag 1979, p.257-266.

CHAPTER 4

GALERKIN AND COLLOCATION FINITE ELEMENT METHODS FOR NEWTONIAN AND VISCOELASTIC FLUIDS

Ping-Wu Chang, Thomas W. Patten, Fred T. Adams, Bruce A. Finlayson

Department of Chemical Engineering, University of Washington, Seattle, Washington, U.S.A.

INTRODUCTION

Galerkin and collocation finite element methods are applied to predict the flow of Newtonian and non-Newtonian fluids, including viscoelastic fluids. The Galerkin method uses a nine-node quadratic element for velocity and a bilinear element for pressure and the frontal solution to solve the matrix equations. The collocation method uses Hermite cubics and a block lower-upper decomposition to solve the equations. The methods are compared for efficiency.

The viscoelastic fluid is modeled using a non-linear convective Maxwell model with a viscosity function and a time constant function which can depend upon shear rate.

EQUATIONS

The dimensionless equations are the momentum equation

$$\mathrm{Re}\ \underset{\sim}{u} \cdot \nabla \underset{\sim}{u} = -\nabla p - \nabla \cdot \underset{\approx}{\tau} \tag{1}$$

continuity equation

$$\nabla \cdot \underset{\sim}{u} = 0 \tag{2}$$

and a constitutive equation for the fluid:

Newtonian $$\underset{\approx}{\tau} = -2\underset{\approx}{e} \tag{3}$$

Purely viscous $$\underset{\approx}{\tau} = -2g(II)\underset{\approx}{e} \tag{4}$$

Viscoelastic $$\underset{\approx}{\tau} + \mathrm{We}\ f(II)\frac{\Delta_{abc}\underset{\approx}{\tau}}{\Delta t} = -2g(II)\underset{\approx}{e} \tag{5}$$

$$\frac{\Delta_{ooo}\tau}{\Delta t} = \frac{\mathcal{D}\tau}{\mathcal{D}t} = v \cdot \nabla\tau + w \cdot \tau - \tau \cdot w \tag{6}$$

$$\frac{\Delta_{abc}\tau}{\Delta t} = \frac{\mathcal{D}\tau}{\mathcal{D}t} - a[e \cdot \tau + \tau \cdot e] + b\, e:\tau\, \delta + c\, \delta:\tau e \tag{7}$$

The derivative $\mathcal{D}\tau/\mathcal{D}t$ is the Jaumann derivative. The vorticity dyadic and deformation dyadic are

$$w = \frac{1}{2}(\nabla v - \nabla v^T),\quad e = \frac{1}{2}(\nabla v + \nabla v^T) \tag{8}$$

and the shear rate is $II = 2e : e$. We use the Gibbs notation (Bird, et al. 1977) and make variables non-dimensional using average velocity, $\langle u\rangle$, radius, R, and for pressure and stress use $\mu\langle u\rangle/R$. The function $g(II)$ allows the viscosity to be shear-rate dependent. The Weissenberg number $We = \lambda\langle u\rangle/R$ (also called De in some figures) measures the elasticity of the fluid, expressed through its time constant, λ. The model with a=1, b=c=0 is then a White-Metzner (1963) nonlinear Maxwell model, which has apparent viscosity which is independent of shear rate. We also do calculations with a=1, b=2/3, which correponds to a model from the first term in the series for the four-constant Spriggs model (1965).

Since the stress cannot be explicitly evaluated for the viscoelastic fluid, it is necessary to solve equation (5) simultaneously with (1). We solve the equation using a Newton-Raphson iteration on the convective terms and a successive substitution or Picard iteration on the viscosity function and elastic terms. At the s-th iteration

$$Re[u^s\cdot\nabla u^{s+1} + u^{s+1}\cdot\nabla u^s - u^s\cdot\nabla u^s] = -\nabla p^{s+1} + \nabla\cdot[2g(II^s)e^{s+1}] - \nabla\cdot\tau^{e,s} \tag{9}$$

where the τ^e is the elastic part of the stress equation. In the collocation method, the equations are solved as they stand. For the Galerkin method, we integrate the viscous terms by parts.

$$0 = \int_V g(II^s)\,\nabla\delta u : (\nabla u^{s+1} + \nabla u^{T,s+1})dV - \int_V \delta u \cdot \nabla p^{s+1} dV$$

$$- \int \delta p\, \nabla\cdot u^{s+1} dV - \int \nabla\delta u^T : \tau^{e,s} dV - \int_S \delta u \cdot f_s dS$$

$$+ Re \int_V \delta u \cdot [u^s \cdot \nabla u^{s+1} + u^{s+1} \cdot \nabla u^s - u^s \cdot \nabla u^s]dV \tag{10}$$

where $\underset{\sim}{f}_s$ are the applied forces on the surface S of volume V. The form of this equation corresponds to a variational principle (Finlayson, 1972, p. 272) when Re=We=0. It leads to the natural boundary conditions

$$\int_S \delta u_n [2g(II^s)\underset{\sim}{n}\cdot\underset{\approx}{e}^{s+1}\cdot\underset{\sim}{n}-\underset{\sim}{n}\cdot\underset{\approx}{\tau}^{e,s}\cdot\underset{\sim}{n}-p^{s+1}-\underset{\sim}{n}\cdot\underset{\sim}{f}_s]dS=0 \tag{11}$$

$$\int_S \delta u_t [2g(II^s)\underset{\sim}{t}\cdot\underset{\approx}{e}^{s+1}\cdot\underset{\sim}{n}-\underset{\sim}{t}\cdot\underset{\approx}{\tau}^{e,s}\cdot\underset{\sim}{n}-\underset{\sim}{t}\cdot\underset{\sim}{f}_s]dS = 0 \tag{12}$$

where $\underset{\sim}{n}$ is the outward pointing normal and $\underset{\sim}{t}$ is the tangent to the surface.

Most polymer melt flows are for a low Reynolds number (10^{-6} to 10^{-3}), whereas polymer solutions may have Reynolds numbers of several hundred. We consider here Reynolds numbers up to about 25, although we have used higher values. Weissenberg numbers for situations we would like to solve are as high as 25; the iteration scheme is limited to We < 0.2.

Method of Solution

The Galerkin method program (FLUID) is an adaptation of the program NASS developed by Dr. P. Hood at the University College of Swansea, Wales, under the direction of Professor O. C. Zienkiewicz and Dr. C. Taylor. Since the details are readily available (Taylor and Hood 1973, Chang, *et al*. 1979) we only sketch them. We changed the shape functions for velocity from eight-node serendipity quadratic polynomials to nine-node Lagrangian polynomials. Huyakorn, *et al*. (1978) found, as we do, that the nine-node elements give much better pressure profiles at the inlet to ducts. The Lagrangian element has a center node while the serendipity element does not. We added the capability to allow free surfaces, changed to the divergence formulation, and added the viscoelastic terms. The pressure is represented as a four-node bilinear function. The elements were isoparametric and an irregular element was transformed into a rectangle to calculate integrals. The frontal routine was used to perform the LU decomposition of the final matrix.

The Hermite interpolation used with the collocation method is described in detail elsewhere (Chang and Finlayson 1977, 1978, Chang, *et al*. 1979). The unknown is expanded in a cubic polynomial in both z and y directions. The differential equations (1) and (2) are satisfied at the collocation points interior to each element and the boundary conditions are satisfied on the boundary. At a given boundary collocation point the velocity or stress was specified. The boundary conditions for pressure are less obvious. If the pressure is known (such as at an inlet or outlet region) it is specified. If pressure cannot be specified, a pressure condition is obtained by satisfying at a collocation point the component of the momentum equation in the normal direction.

The global approximate solution for u,v and p is in C^1; that is, it has continuous x,y and xy derivatives. The effect of the boundary conditions is to increase the bandwidth of the matrix (see Chang and Finlayson 1978) but an LU block decomposition is suitable, with partial pivoting within the block. Only planar geometry is solved by collocation.

We can compare the efficiency of the combined methods. The Galerkin method uses quadratic basis functions for velocity, linear ones for pressure, giving continuous solutions, and a frontal LU decomposition (without pivoting) is used. The collocation method uses cubic polynomials for velocities and pressure, giving solutions which are continuous and whose first derivatives are continuous. The collocation method also uses a block diagonal decomposition of the matrix with partial pivoting. Thus each method makes different choices of criterion (Galerkin vs. collocation), basis function (quadratic and linear vs. Hermite cubic), and matrix decomposition (frontal vs. block diagonal). We compare only the combination of these choices. Comparisons of Galerkin vs. collocation when both use Hermite cubics are in the cited papers. One of the differences in the methods is the number of unknowns. For a rectangular NEZ x NEY grid, we have for velocity and pressure:

$$\text{Galerkin:} \quad NT = 9NEZ \cdot NEY + 5(NEZ + NEY) + 3 \tag{13}$$

$$\text{Collocation:} \quad NT = 12NEZ \cdot NEY + 12(NEZ + NEY) + 12 \tag{14}$$

For the same size grid the collocation method has more unknowns, but perhaps fewer elements can achieve the same accuracy due to the higher degree of polynomial. Results given below relate to this question.

The computational work (in CPU seconds) obeys the following empirical formulas (for a CDC 6400 Cyber 73 computer).

$$\begin{aligned} &\text{Galerkin} \quad CPU = 0.0373\ NT^{1.15} \\ &\text{Collocation} \quad CPU = 0.0109\ NT^{1.57} \end{aligned} \tag{15}$$

For the same CPU the Galerkin method is about 5/3 as expensive as the collocation program because of a larger memory and more frequent transfer of information to disc. The above formulas are used to compute relative work and the relative cost is computed by multiplying the Galerkin work by 5/3. These are approximations only.

STICK SLIP PROBLEMS

In the stick slip problem fluid is flowing between two parallel plates with its fully developed velocity profile. Down stream the plates end and the fluid is free to accelerate near the boundary, and decelerate at the centerline until the velocity profile is flat at its average value. The free surfaces, on which slip is allowed, are maintained in position as shown in Figure 1.

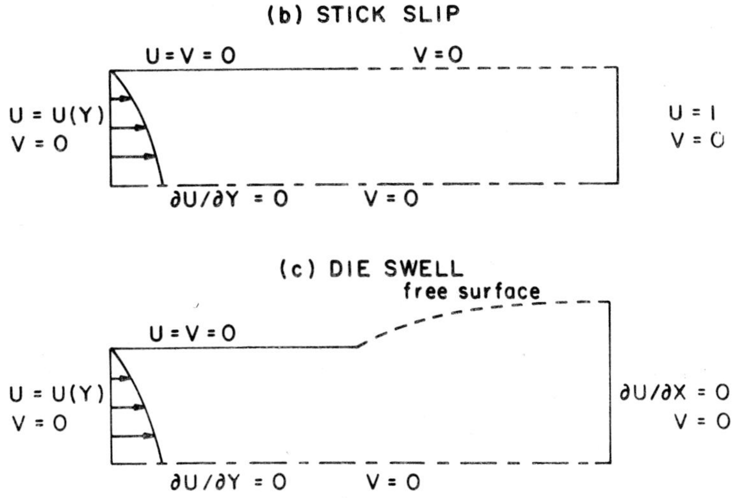

Figure 1. Flow geometries studied

This is a useful prototype problem because it exhibits all the features of the die swell problem, except the swell, and has been solved analytically for a Newtonian fluid at Re = 0. For this special case, the velocity gradient, $\partial u/\partial z$, and pressure, p, have a singularity at the die lip.

We report calculations using several grids and the 6 x 18 grid is shown in Figure 2. Galerkin calculations using quadratic velocities and linear pressure on triangles have been reported by Nickel, *et al*. (1974) and all our calculations agree with theirs.

Results for the axial velocity on the free surface are shown in Figure 3 for a Newtonian fluid at Re = 0. Richardson (1970) has solved this problem analytically and near the die lip

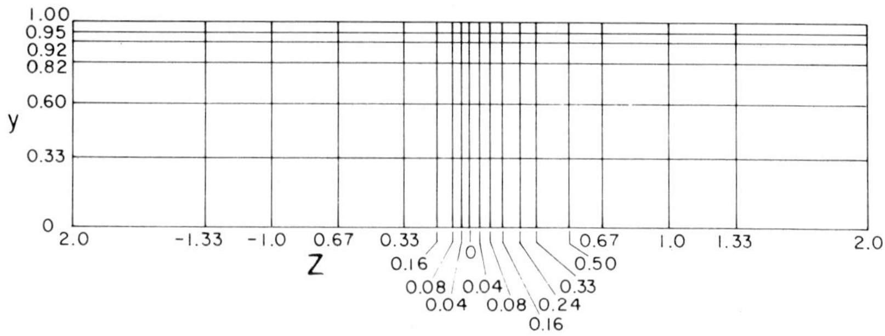

Figure 2. Element location for stick slip problem

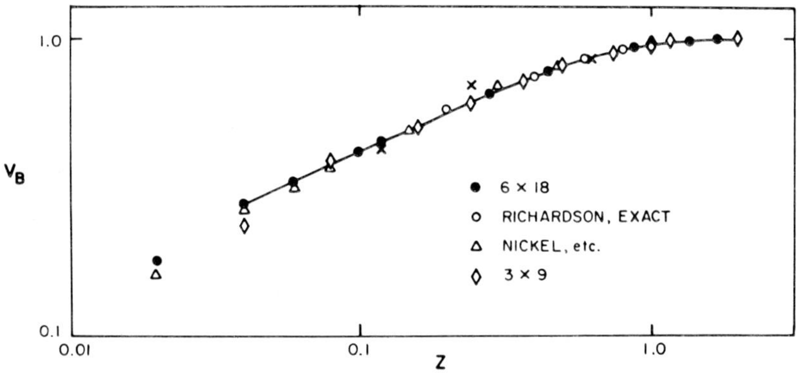

Figure 3. Stick slip flow of Newtonian fluid, Re=0, Velocity on free surface

(z=0) $v_B = Cz^{1/2}$. All the solutions demonstrate the correct asymptotic behavior as $z \to 0$ except for the very first element. The power a in the formula $v_B = cz^a$ is deduced from the results in the second element from the die lip and are given in Table I. The Galerkin method gives a slightly better value of a for less cost.

The pressure along the centerline is shown in Figure 4. All the calculations agree well except the collocation results for pressure, which oscillate. The nodal values of pressure are quite accurate, but the pressures at the collocation points give the oscillating solution. The reason for this is not known, but may represent an overspecified system of equations. The accuracy of the different solutions can be ascertained by examining the pressure at the centerline and die lip, p(0,0), as listed in Table I. Another quantity of interest in rheology is $p(0,0)/p_{w,s}$, where $p_{w,s}$ is the wall pressure extrapolated from upstream to the die exit. Values of $p_{w,s}$ and the ratio are also shown in Table I. Both methods give about the same accuracy, so that the Galerkin method is preferred because of its lower cost.

Table I - Stick Slip Problem

	Galerkin			Collocation		
	2x4 L	3x9 L	6x18 S	2x6	3x10	Exact
a in $u_B=Cz^a$	0.30	0.43	0.40	0.29	0.37	0.5
p(0,0)	0.960	1.11	0.982	0.818	0.962	1.04
$p_{w,s}$	1.61	0.842	0.872	0.836	0.844	0.81
$p(0,0)/p_{w,s}$	0.59	1.32	1.22	0.98	1.14	1.28
NT	105	306	409	252	528	
CPU	5	27	38	64	205	
Cost	8	45	63	64	205	

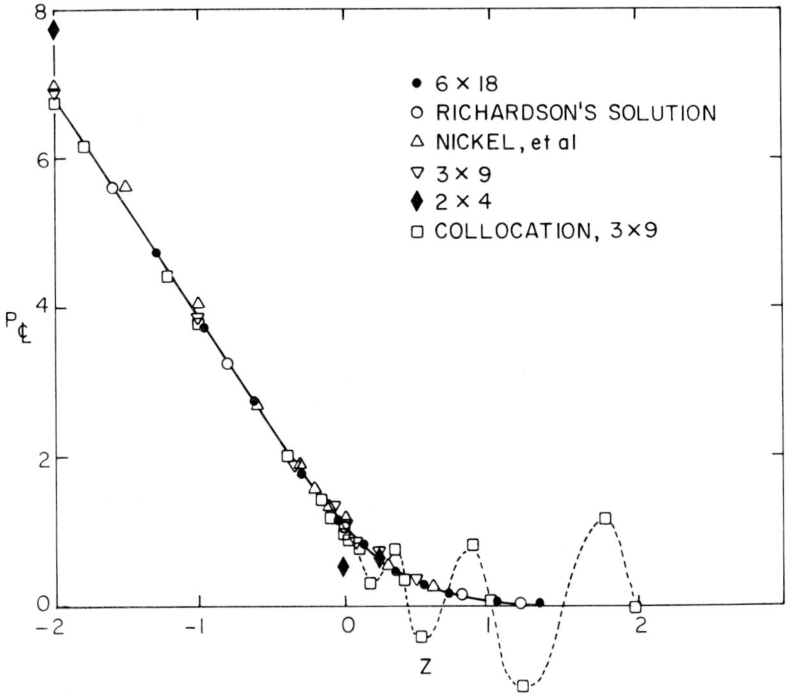

Figure 4. Pressure on centerline for Newtonian fluid in stick slip flow, Re=0

The total normal stress along the free stream (y=1, z>0) is shown in Figure 5. The total normal stress is the sum of the pressure and viscous normal stress.

$$t_{yy} = - p - \tau_{yy} \tag{17}$$

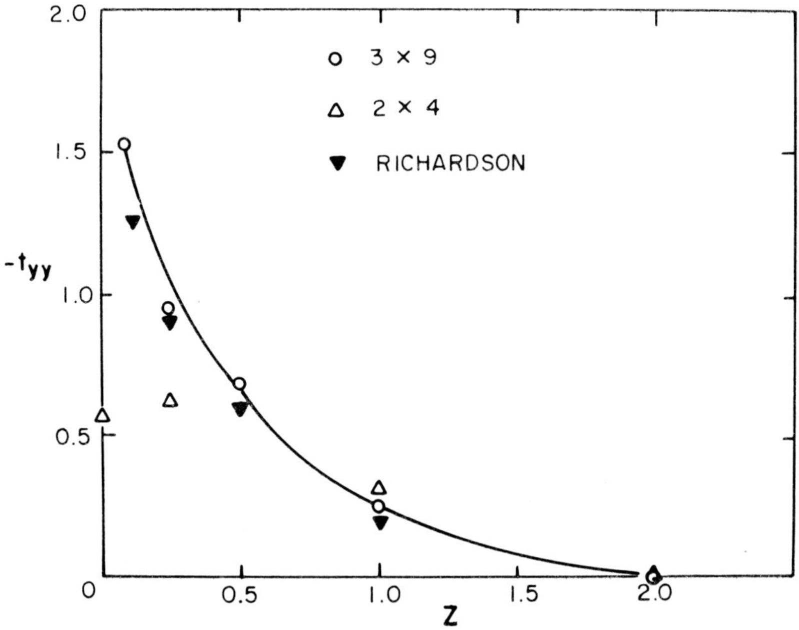

Figure 5. Normal Force on Boundary for Newtonian fluid in stick slip flow, Re=0.

The pressure and viscous stress are singular at y=1, z=0, but the sum is not. Figure 5 shows that even a 3x9 mesh is successful in modeling the normal force except very near the die lip.

The stick slip problem can also be solved for viscoelastic fluids. One such solution is illustrated in Figure 6. The purely viscous fluid uses

$$g(II) = \frac{1}{(1+II^2)^{0.25}} , \quad f(II) = We = 0 \tag{18}$$

whereas the viscoelastic fluid uses the same g(II) and

$$f(II) = g(II), \; We = 0.05, \; a = 1, \; b = c = 0 \tag{19}$$

At the inlet and wall the apparent viscosity is about half that of the Newtonian fluid. Using (18) or (19) in place of the Newtonian fluid causes the velocity along the free surface to accelerate more rapidly. Figure 7 illustrates the effect on

t_{yy}: the normal stresses show large deviations from the Newtonian results due to the variation of viscosity with shear rate. For Weissenberg numbers this small elasticity plays little role, since using (19) gives similar results to those obtained with (18).

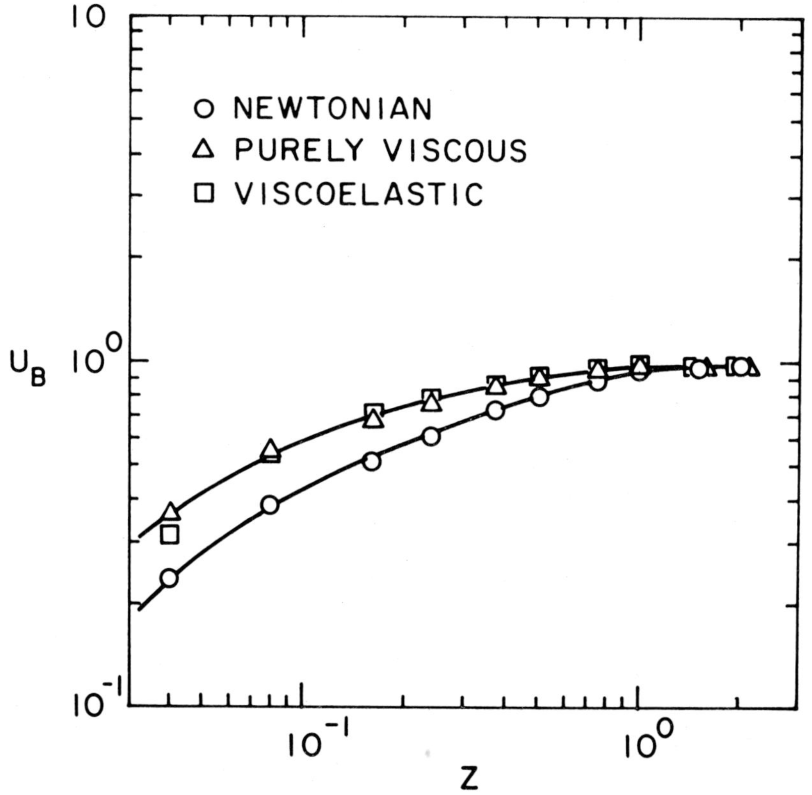

Figure 6. Stick slip flow of viscoelastic fluid, Re=0, Galerkin, 3x9 Grid

In a two-dimensional flow even a Newtonian fluid exerts a normal traction

$$\tau_{yy} = -\mu \frac{\partial u}{\partial z} \tag{20}$$

and for the non-Newtonian fluids (18-19) the μ is reduced by about 1/2. Thus the normal force is reduced, as shown in Figure 7.

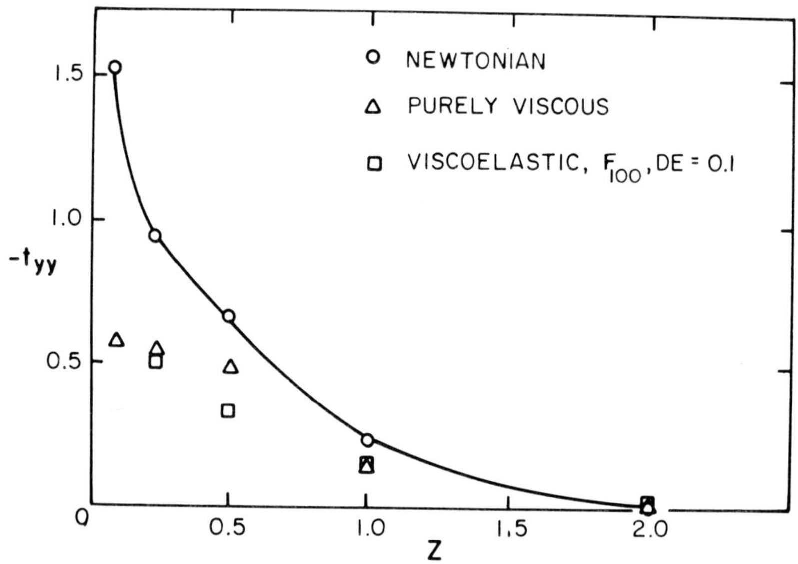

Figure 7. Total normal force on upper surface for stick slip flow, Galerkin, 3x9 Grid

The accuracy of the calculations can be partially checked because at zero Reynolds number an integral momentum balance gives

$$I = \int_0^1 T_{zz} dy = 0 \tag{21}$$

Values of I are listed in Table II and the collocation method is far and away the most accurate as far as this integral balance of momentum is concerned. Unfortunately, it is also the most expensive.

DIE SWELL

Location of Free Surface

We solve for the location of the free surface in an iterative fashion using the equation for the free surface to be a streamline.

$$\frac{dR^{s+1}}{dz} = \frac{v^s(R^s(z),z)}{u^s(R^s(z),z)} \tag{22}$$

We then solve the problem on a fixed geometry for the next

Table II - Comparison of Collocation and Galerkin

Method	We	Geometry	Grid	No. of Unknowns	CPU Work	Relative Cost	I(Eq.21)
Galerkin	0.	Cyl.	3x9	306	126	210	0.29*
	0.	Slit	3x9	306	126	210	-0.21
	0.05	Slit	3x9	306	288	480	-0.068
Collocation	0.	Slit	2x6	252	144	144	2.3×10^{-5}
	0.05	Slit	2x6	252	1400	1400	-4.6×10^{-3}

*For cylindrical geometry

$$I = \int_0^1 T_{zz} y\,dy \Big/ \int_0^1 y\,dy$$

iteration. The location of the free surface is thus defined by the function R(z). In the Galerkin method we stretch all elements by R(z) and relocate the element corners. The isoparametric elements are then used to transform the quadrilaterals element by element. In the collocation method we transform the domain (z,y) to the rectangular domain ($\xi = z/R(z),y$). The transformed differential equation is then solved at the collocation points on a regular grid. Transformation of the elements is also possible (Pinder, *et al*. 1978).

One important difference in the Galerkin and collocation methods is that the Galerkin method uses natural boundary conditions while the collocation method satisfies the boundary conditions at collocation points on the boundary. Eqs. 11-12 give the Galerkin natural boundary conditions and only in the limit of infinitely many elements are the integrands zero everywhere on the boundary. Indeed, by using natural boundary conditions we are essentially satisfying an integral momentum balance on the element, one of whose sides is on a boundary. It may happen that errors interior to the fluid offset or magnify errors on the boundary, and the integrand may not be zero anywhere on the boundary. In the collocation method we set the boundary condition to zero at collocation points on the boundary. The boundary condition is not necessarily satisfied at points between the collocation points, except in the limit of infinitely many collocation points. However, the boundary conditions are definitely satisfied at the collocation points on the boundary.

Die Swell for Newtonian Fluids

Die swell from a slit die is shown in Figure 8. The experiments of Whipple (1974) are for a 5% Separan AP in a 50-50 water-glycerine mixture, which exhibits some viscoelasticity. The Galerkin results give an ultimate die swell which decreases with mesh refinement attaining the value 1.20 for the most refined mesh used here. The collocation method gives a die swell which increases with mesh refinement, attaining the value 1.17 for the finest mesh. The effect of using natural boundary conditions in the Galerkin method is to make the die swell larger than it should be, and the collocation solutions have converged so that we choose the value of 1.17 as the correct one.

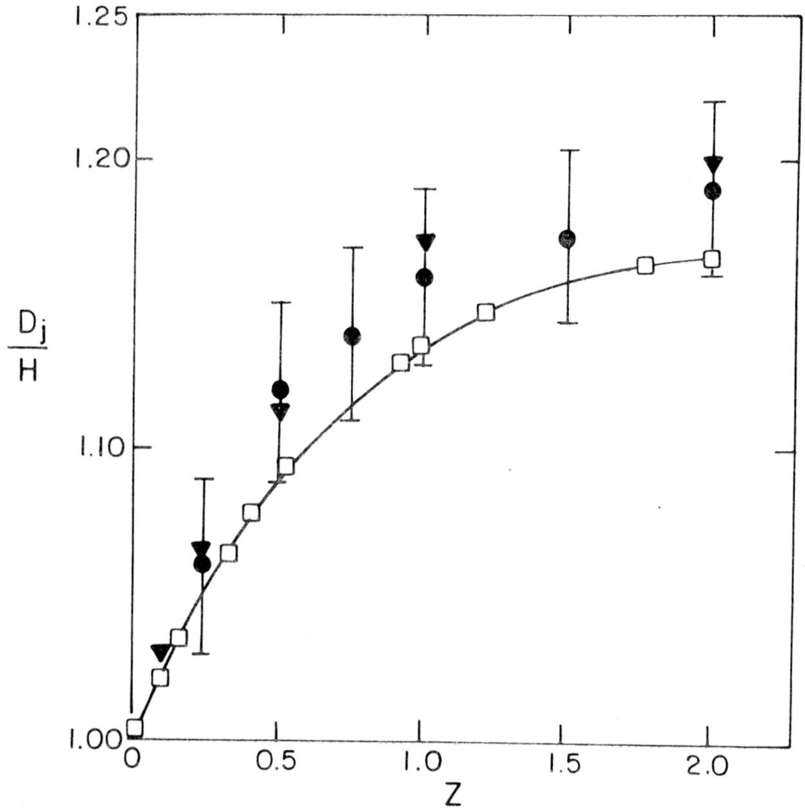

Figure 8. Comparison to Experiment of Die Swell for Slit Die, Re=0. Experimental data (●) due to Whipple (1974) for 5% Separan AP in 50-50 water-glycerin mixture at 0.10 cm/sec, H-0.32 cm, Re=7 x 10^{-4}. ▼ Galerkin 3 x 9, □ Collocation 3 x 10

Die Swell for Viscoelastic Fluids

Die swell ratios were calculated for a "one-term" Spriggs model a=1, b=2/3. The values of the integrated normal force at the die lip and the average velocity in the expanding jet show the collocation results are more accurate, and the accuracy of both Galerkin and collocation methods is improved by mesh refinement (more elements). The pressure on the centerline and boundary, and the normal stress are similar to the Newtonian case.

Collocation solutions of die swell for We=0, 0.05, 0.1, 0.2 are shown in Figure 9. The other results suggest this grid gives reasonable results, and it is relatively inexpensive (120 unknowns, CPU work = 20). At the highest Weissenberg number the

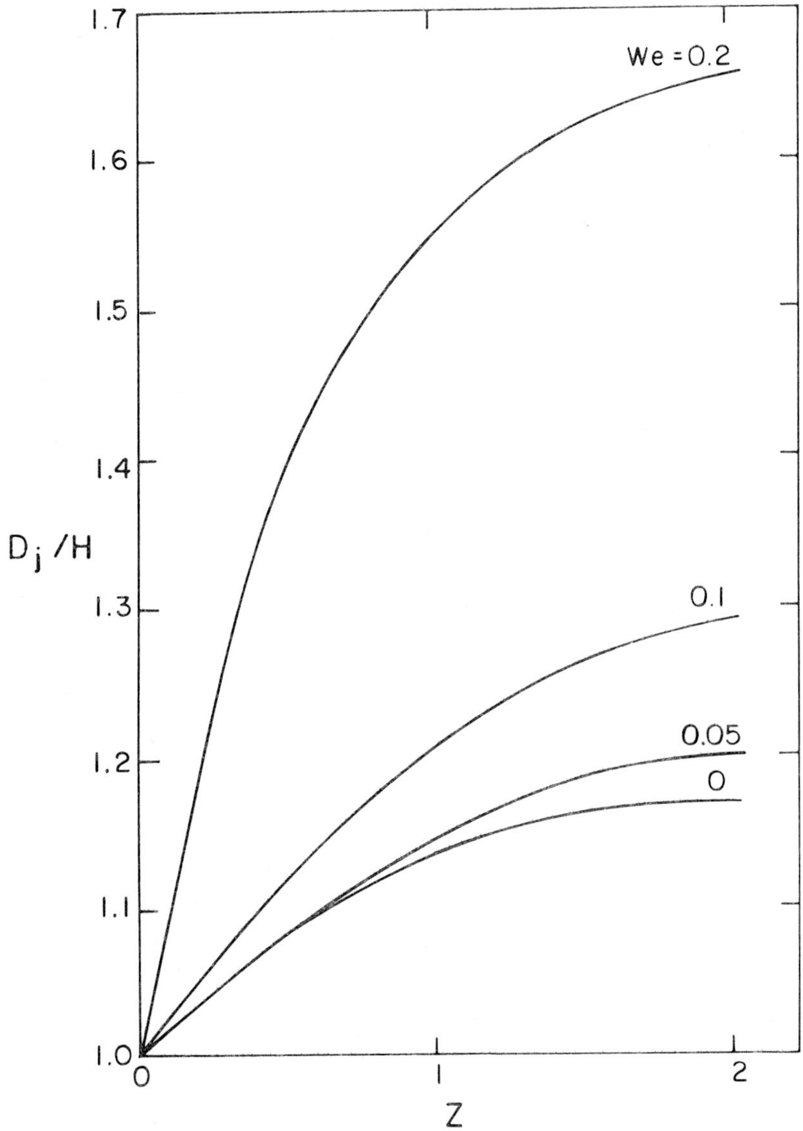

Figure 9. Die Swell for Viscoelastic Fluid, a=1, b=2/3, Collocation, 1 x 4 grid, slit flow

swell is 70% which is the largest swell predicted by numerical calculations. Experimental results give larger values, but simulations of these cases will require a different iteration scheme. We cannot compare these results to experimental data because the viscosity function corresponding to the fluid represented in Figure 9 is constant with shear rate, in contrast to realistic behavior. We can, however, see if the predicted die swells fit the empirical correlation derived by Tanner (1970), which is in the present notation.

$$We^2 = C\left[\left(\frac{D_j}{H}\right)^6 - (1.17)^6\right] \tag{23}$$

The calculated results from Figure 9 are consistent with this relation. This suggests the calculations are reasonable and provides impetus to calculations with higher Weissenberg numbers.

Galerkin solutions of die swell for We=0, 0.05 and 0.1 are shown in Figure 10. Unfortunately, the trend with increasing Weissenberg number is wrong: less swell occurs for the higher We. To understand these results we look at the total forces exerted on the free surface for the different cases. The boundary condition is that these forces are zero. Table III illustrates that these forces are not zero, and the discrepancy is increased for viscoelastic fluids. The negative normal forces shown in Table III tend to make the die swell smaller as do negative tangential forces. As the Weissenberg number increases the normal forces work towards a smaller die swell and the smaller but positive tangential forces cause a larger die swell. The net effect is decreased swell as We increases. The failure of the natural boundary conditions to be satisfied is probably also related to the fact that the Galerkin method predicts larger die swells than the collocation method, which satisfies the boundary conditions point wise. In any case, it is not possible to use the Galerkin method with natural boundary conditions in this way for viscoelastic fluid calculations. In this case the collocation method is plainly superior.

NORMAL STRESSES DEDUCED FROM EXIT PRESSURES

Normal stress measurements of polymer melts are frequently made using a die swell experiment. The normal force is measured on the wall at different positions along the die, and extrapolated to the die exit. The result is called the exit pressure, $p_{w,c}$ or $p_{w,s}$ for capillary or slit geometry. Davies, *et al.* (1973) and more recently Boger and Huilgol (1977) derive the following relations for the first and second normal stress difference.

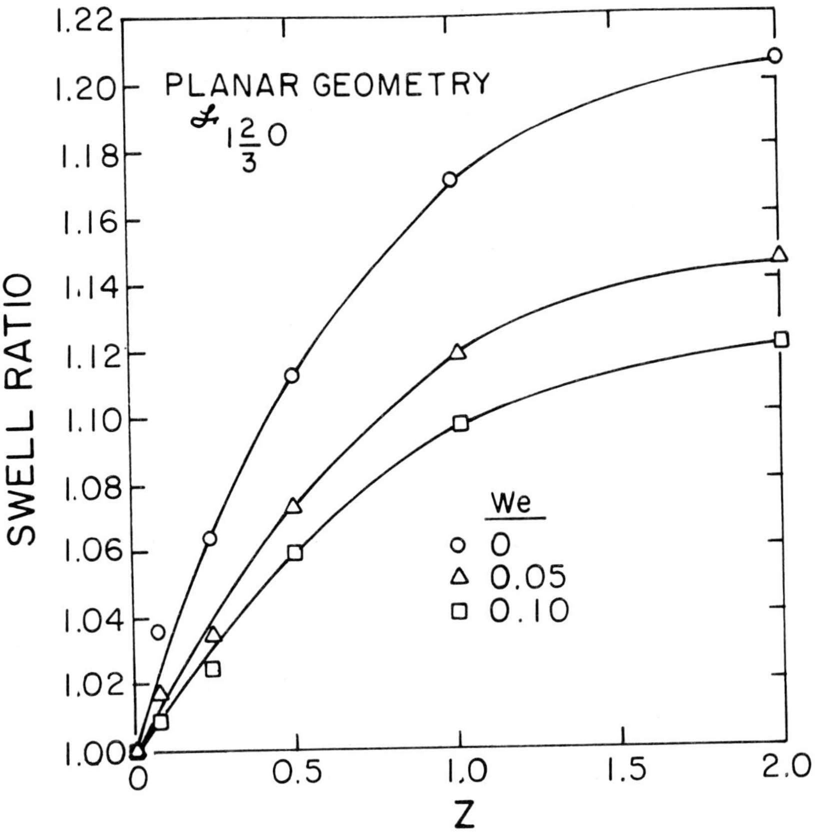

Figure 10. Die Swell for Viscoelastic Fluid, a=1, b=2/3, Galerkin, 3 x 9 grid, slit flow

$$(N_1 + N_2/2)_w = p_{w,c} + \tau_w \frac{\partial p_{w,c}}{\partial \tau_w} /2 \tag{24}$$

$$(N_1)_w = p_{w,s} + \tau_w \frac{\partial p_{w,s}}{\partial \tau_w} \tag{25}$$

The subscript w refers to the wall conditions and all terms are evaluated for fully developed flow. Different equations are derived by Han (1974).

$$N_1 = p_{w,c} + \tau_w \frac{dp_{w,c}}{d\tau_w} \tag{26}$$

$$N_2 = -\tau_w \frac{dp_{w,c}}{d\tau_w} \tag{27}$$

Table III - Forces on Free Surface

3x9 Galerkin Solution

	We=0		We=0.025		We=0.05	
Z	Normal*	Tangential**	Normal	Tangential	Normal	Tangential
0.04	-1.97	0.49	-2.92	0.43	-3.79	0.26
0.16	-0.06	0.03	-0.15	-0.06	-0.15	-0.10
0.37	+0.01	0.00	0.03	-0.01	0.05	-0.01
0.75	-0.01	0.00	0.02	0.01	-0.01	-0.02
1.0	0.00	-0.01	0.00	-0.01	0.00	-0.03

*Normal Force = $\underset{\sim}{n}\cdot\underset{\approx}{T}\cdot\underset{\sim}{n}$, positive in $\underset{\sim}{n}$ direction, outward pointing normal.

**Tangential Force = $\underset{\sim}{t}\cdot\underset{\approx}{T}\cdot\underset{\sim}{n}$, positive in downstream direction.

Han assumes that $p_{w,c} = p_{w,s}$ and Boger and Huilgol (1977) show that this is equivalent to assuming $p(0,0) = 2p_{w,c}$. All equations are derived assuming fully developed flow up to the die exit, which the results above suggest is incorrect. Thus the question can be raised as to whether normal stresses deduced in this way are in fact due to the two-dimensional flow rather than the elastic terms associated with viscoelastic polymers. We show here that the normal stresses deduced from Eq. (24-27) are small for both Newtonian fluids and a power law fluid modeling the purely viscous part of high density polyethylene.

First consider a Newtonian fluid. Results for cylindrical geometry are in Figure 11. The usual experimental practice is to measure the wall pressure at different axial locations and extrapolate to the die exit. The data (Han, et al. (1970)) has the last measurement approximately two radii upstream from the die exit (i.e., the left most point of Figure 11). We thus extrapolate this fully developed region to the die exit to obtain $p'_{w,c} = 1.84$ and $p'_{w,s} = 0.88$. (For Newtonian fluid in this region the wall normal stress is the pressure.) Also

$$\frac{dp_{w,c}}{d\tau_w} = \frac{1.84}{4} = 0.46, \quad \frac{dp_{w,s}}{d\tau_w} = \frac{0.88}{3} = 0.29 \tag{28}$$

We thus obtain from Eq. (24-25)

$$(N_1 + N_2/2)_w = 2.76, \; N_{1w} = 1.76 \tag{29}$$

For a typical case $\mu = 4500$ poise, $4u_o/R = 31.2\ \text{sec}^{-1}$, and

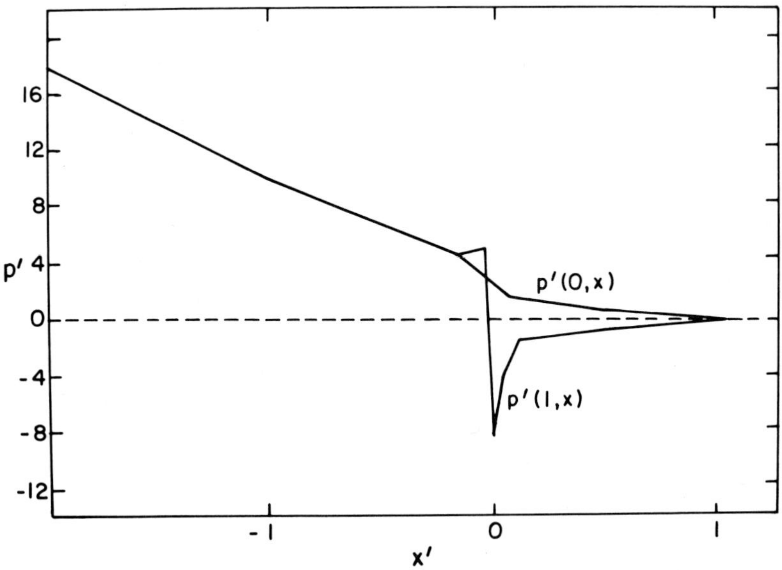

Figure 11. Pressure in die swell of Newtonian fluid in cylindrical geometry

$p_s = \tau_s = \mu\langle v\rangle/R = 35100$ dyne/cm^2 = 0.509 psia. This yields $N_1 = 0.90$ psi, $N_2 = 1.0$ psi from Eq. (29). Han's equations (26,27) give $N_1 = 1.9$ psi, $N_2 = -0.94$ psi. The exit pressures are $p_{w,c} = 0.94$ psi, $p_{w,s} = 0.48$ psia. Such values (~7 x 10^4 dyne/cm^2) are too small to be measured in a capillary or slit die. Thus even though a Newtonian fluid is not fully developed at the die exit, and normal stresses are generated due to the two-dimensional character of the flow, the normal stresses are small. This is in agreement with the data for the Newtonian fluid Indopol H1900 (Han _et al_. (1970))which shows negligible exit pressure, from -0.35 to +0.35 psia for a capillary, on a scale of 0 to 700 psia.

We next report calculations for a purely viscous fluid which is chosen to match the stress-shear rate curve for high density polyethylene at 180°C (Han, 1976, p. 96). The results are reported in Table IV. The pressure drop per length agrees with

Table IV. Die Swell Calculations for Power-Law Fluid
$n = 0.2886$ $m = 2.31 \times 10^5$ cgs, $R = 0.0625$ in.
$\mu_s = 1.189 \times 10^6$ poise (viscosity at $\gamma = 0.1$ sec^{-1})

Geometry	Avg. Velocity (in/sec)	$\frac{\Delta p'}{L'}$	p' z'=-2	p' z'=-1	p(0,0) psig	$p_{w,c}$ & $p_{w,s}$* psig	Die Swell
Cylinder	2	0.111	0.248	0.137	11.0	6.8	1.01
Cylinder	4	0.143	0.309	0.166	14.1	6.8	1.01
Cylinder	6	0.161	0.347	0.186	15.7	6.8	1.01
Slit	2	0.055	0.133	0.068	5.52	0.92	1.00
Slit	4.5	0.070	0.167	0.085	7.17	0.92	0.99
Slit	7	0.079	0.190	0.097	8.28	0.92	0.99

*Obtained by extrapolating p' on the wall from upstream to the die exit, averaged for all 3 cases.

the data of Han (1976, p. 92). We can analyze the calculated results to see what normal stresses are predicted by the calculations. We do this using both equations (26,27) and (24,25). The results in Table V for equations (26,27) show that the first normal stress is 4.7 x 10^5 dynes/cm^2, which is only 4% of the normal stress measured for the viscoelastic fluid (6-10 x 10^6 dyne/cm^2). Using equations (24,25) gives a first normal stress of only 1% of that measured for the viscoelastic fluid. Thus we can conclude that even though a power law fluid gives normal stresses in a die swell situation, those normal stresses are small compared to those expected for a fluid having the same viscous properties but being an elastic fluid as well.

Table V. Normal Stresses Predicted from Calculations for Power Law Fluid in Die Swell

First Normal Stress Difference	(dyne/cm^2)
Eq. (26,27)	4.7×10^5
Eq. (24,25)	6.3×10^4
Second Normal Stress Difference	(dyne/cm^2)
Eq. (26,27)	0.
Eq. (24,25)	8.1×10^5

Both first and second normal stress differences are constant for the range of conditions in Table IV.

We would like to do similar calculations for viscoelastic fluids. Unfortunately, we do not yet have complete calculations for both geometries and for fluids with a large enough elastic contribution to be important.

SUMMARY AND CONCLUSIONS

Both Galerkin and collocation finite element methods can solve fluid flow problems with singularities, such as the stick slip problem and the die swell problem. These calculations agree with experimental data or calculations reported previously when available. The finite element method permits mesh refinement near the singularities and this reduces the calculation cost compared to a fine mesh covering the entire domain.

The Picard-type iteration on the viscoelastic terms is suitable only for Weissenberg numbers lower than 0.1. The viscoelastic terms caused the strength of the singularities to change but otherwise the solutions were similar to those for Newtonian fluids.

For engineering accuracy, the Galerkin method with Lagrangian quadratic polynomials and a frontal LU decomposition gives solutions at a lower cost than collocation with Hermite cubics and a block diagonal decomposition. The collocation method gave very much more accurate average velocities and integrated forces.

Die swells as large as 70% are predicted for certain viscoelastic fluids and the predicted die swells agree reasonably well with an empirical correlation of experimental data. The collocation finite element method can be used for viscoelastic die swell while the Galerkin method with natural boundary conditions cannot, as applied here. For die swell with Newtonian fluids both collocation and Galerkin methods can be used and give comparable accuracy for comparable cost.

REFERENCES

Bird, R. B., Armstrong, R. C. and Hassager, O. (1977) Dynamics of Polymeric Liquids: Volume I. Fluid Mechanics, Wiley.

Boger, D. V. and Huilgol, R. R. (1977) Trans. Soc. Rheol. 21 447.

Chang, P. W. and Finlayson, B. A. (1977) pp. 79-86 in Advances in Computer Methods for Partial Differential Equations - II, (ed.) R. Vichnevetsky, IMACS.

Chang, P. W. and Finlayson, B. A. (1978) Math. Comp. Simulation 20 83.

Chang, P. W., Patten, T. W. and Finlayson, B. A. (1979) to be published in Comp. Fluids.

Davies, J. M., Hutton, J. F., and Walters, K. (1973) J. Phys. D: Appl. Phys. 6 2259.

Finlayson, B. A. (1972) The Method of Weighted Residuals and Variational Principles, Academic Press.

Han, C. D. (1974) Trans. Soc. Rheol. 18 163.

Han, C. D., "Rheology in Polymer Processing," Academic Press, 1976.

Han, C. D., Charles, M. and Philippoff, W. (1970) Trans. Soc. Rheol. 14 393.

Huyakorn, P. S., Taylor, C., Lee, R. L. and Gresho, P. M. (1978) Comp. Fluids 6 25.

Nickel, R. E., Tanner, R. I. and Caswell, B. (1974) J. Fluid Mech. 65 189.

Pinder, G. F., Frind, E. O., and Celia, M. A. (1978) "Groundwater Flow Simulation Using Collocation Finite Elements," Proc. Second Ind. Conf. Finite Elements in Water Resources, London, Eng., July 10-14, 1978.

Spriggs, T. W. (1965) Chem. Eng. Sci. 20 931.

Tanner, R. I. (1970) J. Poly. Sci. A-2, 8 2067.

Taylor, C. and Hood, P. (1973) Comp. Fluids 1 73.

Whipple, B. A. (1974) Ph.D. Dissertation, Washington Univ.

White, J. L. and Metzner, A. B. (1963) J. Polymer Sci. 7 1867.

CHAPTER 5

NUMERICAL COMPUTATION OF MULTI-PHASE FLUID FLOW AND HEAT TRANSFER.

D. Brian Spalding

Reilly Professor of Combustion, Purdue University, USA

Professor of Heat Transfer, Imperial College, London, U.K.

ABSTRACT

The paper describes the IPSA (≡ inter-phase-slip algorithm numerical procedure for solving the equations of heat, mass and momentum transfer governing multi-phase flow phenomena. The procedure is an iterative one, operating on finite-difference equations connecting the values of variables pertaining to points arrayed on a "staggered" topologically-cartesian grid.

The IPSA procedure can handle one-, two- and three-dimensional distributions of the variables in space; both steady and transient processes can be handled; the fluids may be compressible or incompressible; buoyancy, phase-change and chemical-reaction effects may be present; and a special version of IPSA is available which facilitates the handling of gravity-stratified flows.

1. PRACTICAL IMPORTANCE OF THE SUBJECT

1.1 Nature of multi-phase flows

A multi-phase flow, in the context of the present paper, is one which requires for its characterisation more than one set of velocities, temperatures, masses per unit volume, etc., at each location in the calculation domain. The material associated with each set of characterising variables is what is meant by a "phase", in this connexion.

Often the phases are distinct in a way which a thermodynamicist would recognise; for example, one phase may be steam (vapor-phase H_2O) and the other water (liquid-phase H_2O); and the steam may have velocity components, at each location in space, which differ from those of the water. Thus, in a boiling kettle, the steam rises while the water has a slight downward motion. This phenomenon can be regarded as a <u>two</u>-phase flow.

Sometimes, however, more distinctions are needed than the thermodynamicist would be inclined to recognise. For example, when water is injected upwards into the air, to make a decorative fountain, there can be found, at any location, both upward- and downward-moving water. In this case it is necessary, if the process is to be described adequately, to treat upward-moving water and downward-moving water as distinct phases; and the process is a three-phase one, the third phase being the air.

An example in which it is necessary to consider many phases is that of a burning liquid-fuel spray. The gas in which the liquid burns is one phase; and the other phases are all of liquid fuel, distinguished however at least according to droplet size. The distinction is needed, for realistic description, because the smaller droplets are carried along with the gas stream, while the motion of the larger ones can exhibit different velocities and directions.

Multi-phase flows are of frequent occurrence in nature and in man-made devices. Some illustrations will now be presented, with the purpose of demonstrating the desirability of having a reliable and general means of predicting the phenomena.

1.2 Examples from nature

Rainfall, snowdrifts, sandstorms and silted rivers are all examples of multi-phase flow or of its results. Droplets of water, flakes of snow, or particles of sand, are carried along by the wind, if it is travelling fast enough; and, when it is not, their greater density causes them to fall downwards, leaving the air above them.

It follows that the quantitative understanding of weather phenomena, and the proper design of snow fences, drainage systems and sand-storm-resisting structures, must be based on an analysis of multi-phase flows.

The same is true when water is the continuous phase and particles of sand, or other materials of different density, are carried by it in high-velocity regions and dropped from low-velocity ones. The gradual deformation of the courses of rivers, resulting from erosion and deposition, is a multi-phase-flow process.

1.3 Examples from engineering

The list of multi-phase-flow phenomena which appear in man-made equipment is immensely long. A brief extract is all that can be provided here, as follows:...

- Steam-water flows in boilers, condensers, turbines, cavitating pumps, pipes, valves.
- Water-air flows in cooling towers, air-conditioning equipment, jet-engine intakes, fire-fighting devices.

- Dusty-gas flows in coal mines, cyclone separators, electrostatic precipitators, fluidised-bed combustors, sand-blast cleaners, magnetohydrodynamic devices, rocket-exhaust plumes.
- Spray-suspension flows in Diesel engines, furnaces, rocket motors, gas-turbine combustors, powdered-milk driers, painting devices, fumigation processes, distillation columns.
- Liquid-solid flows in coal-transportation pipe lines, sewage plant, heat exchangers.
- Liquid-liquid flows in settling tanks, emulsifiers, petroleum-winning processes, electrolytic smelters.
- Complex phenomena such as occur in blast furnaces for iron-ore reduction, under-ground gasification, sea-water distillation, electrophoresis, glass manufacture, nuclear-rector cooling.

Were multi-phase phenomena to be important in only a fraction of these situations, the case for developing a reliable predictive procedure would be overwhelming. It is therefore not surprising that multi-phase flows are now increasingly engaging the attention of practitioners of numerical fluid mechanics; the only surprise is that the development has started so late.

1.4 Purpose of present paper

There exists a growing literature concerned with the subject; and much of it is, as is common in the early stages of development of a new subject, misleading in content or emphasis. To list publications uncritically would do no service to the reader; and the task of sifting better from worse is too large (and dangerous!) to contemplate here.

For these reasons, the present paper will be confined to the presentation of the method of analysis developed by the present author and his colleagues. Specifically, the IPSA algorithm will be described, and illustrated. It is discussed in finite-difference rather than finite-element language ; but those conversant with both should find the translation easy.

2. THE MATHEMATICAL PROBLEM

2.1 The notion of inter-penetrating continua

In section 1.1 it was implied that more than one phase can exist at the same location at the same time. Of course, this is only a convenient fiction; and it rests on the ideas of time- and space-averaging.

These ideas are easier to grasp imaginatively than to set down rigorously (although rigour can be provided if demanded). Here the reader will be asked to accept that it is meaningful to conceive that:-

(i) any small volume of the space in question, at any particular time, can be regarded as containing a volume fraction r_i of the i-th phase; so that, if there are n phases altogether:

$$\sum_{i=1}^{i=n} r_i = 1 \qquad .(2.1\text{-}1)$$

(ii) The mass flow rate of any phase across any elementary surface within the domain at any time can be expressed in terms of the quantity $r_i\rho_i\vec{V}_i$, where ρ_i is the density of the i-th phase, and $\vec{V}_i$ is the local value of the velocity vector of that phase.

(iii) When the contents of finite volumes and the flow rates across finite areas are to be computed over finite time intervals, a suitable averaging over space and time will be carried out.

Observance of these rules amounts to treating each phase as a continuum in the space under consideration. The phases "share" this space; and they may, as they move within it, interpenetrate.

2.2 The partial differential equation

(a) The mass-conservation equation The volume fractions, densities and velocities of the i-th phase must, in order to satisfy the mass-conservation principle, obey the differential equation:

$$\frac{\partial}{\partial t}(\rho_i r_i) + \text{div}\,(\rho_i r_i \vec{V}_i) = M_i \qquad ,(2.2\text{-}1)$$

wherein M_i is the "source" of mass per unit volume.

Mass sources are often non-zero, as when liquid H_2O (water) disappears and gaseous H_2O (steam) appears in its place, as a consequence of heat transfer.

However, summation over all the phases leads to the "overall" mass-conservation equation:

$$\sum_{i=1}^{i=n}\left(\frac{\partial}{\partial t}(\rho_i r_i) + \text{div}\,(\rho_i r_i \vec{V}_i)\right) = 0 \qquad .(2.2\text{-}2)$$

This has a zero on the right-hand side, signifying that there is no net creation or destruction of mass.

(b) The momentum-conservation equation Let $u_{i\ell}$ be the velocity of phase i in direction ℓ, i.e. the ℓ-direction momentum per unit mass of the i-th phase.

Then the momentum-conservation equation becomes:

$$\frac{\partial}{\partial t}(\rho_i r_i u_{i\ell}) + \text{div}\,(\rho_i r_i \vec{V}_i u_{i\ell}) =$$

$$= r_i\left(-\vec{\ell}\cdot \text{grad}\, p + B_{i\ell}\right) + F_{i\ell} + I_{i\ell} \qquad .(2.2\text{-}3)$$

Here p stands for the pressure, which is regarded as being "shared" between the phases; ℓ is a unit vector in the ℓ-direction; $B_{i\ell}$ is the ℓ-direction body force per unit volume of phase i; $F_{i\ell}$ is the friction force exerted on phase i by stresses within that phase, i.e. by viscous action; and $I_{i\ell}$ is the momentum transfer to phase i resulting from interactions with other phases occupying the same space. The latter terms vanish when the "overall" ℓ- direction momentum equation is formed; thus:

$$\sum_{i=1}^{i=u} \frac{\partial}{\partial t}(\rho_i r_i u_{i\ell}) + \text{div}\,(\rho_i r_i \vec{V}_i u_{i\ell}) =$$

$$= -\vec{\ell}\cdot \text{grad}\, p + \sum_{i=1}^{i=n}(r_i B_{i\ell} + F_{i\ell}) \qquad .(2.2\text{-}4)$$

The formulation of the pressure term has occasioned some uncertainty among writers on this subject; and it is sometimes thought that different pressures "ought" to be provided for each phase. There are some cases in which this is indeed desirable (see section 5 below); but the shared-pressure presumption is correct for most practical circumstances. It will be noted that the r_i's have vanished from the pressure-gradient term, by reason of equation (2.1-1).

(c) The energy equation Let $\tilde{h}_i$ stand for the stagnation enthalpy of phase i per unit mass, by which is meant the thermodynamic enthalpy plus the kinetic energy of the phase ($\vec{V}_i \cdot \vec{V}_i/2$) plus any potential energy associated with the position of the fluid in a force field (e.g. that of gravitation). Then the First Law of Thermodymics implies:

$$\frac{\partial}{\partial t}\left(r_i(\rho_i \tilde{h}_i - p)\right) + \text{div}\,(\rho_i r_i \vec{V}_i \tilde{h}_i) =$$

$$= r_i Q_i + H_i + J_i \qquad .(2.2\text{-}5)$$

Here Q_i represents the heat transfer to phase i per unit volume, resulting perhaps from the net absorption of radiation; H_i represents heat transfer from neighbouring elements of the same phase e.g. by thermal conduction and viscous action; and J_i represents the effect of interactions with other phases occupying the same space.

Once again, the "overall" energy equation loses the

interaction terms:

$$\sum_{i-1}^{i=n}\left(\frac{\partial}{\partial t}\left(r_i(\rho_i\tilde{h}_i-p)\right) + \text{div}\,(\rho_i r_i\vec{V}_i\tilde{h}_i)\right) =$$

$$= \sum_{i=1}^{i-n} (r_i Q_i + H_i) \qquad .(2.2\text{-}6)$$

It should be remarked that there are advantages in replacing the enthalpy equation by the entropy equation. However, since these advantages are not specific to multi-phase flows, the matter will not be further discussed here.

(d) The species conservation equation Let a chemical species k be present in the various phases i in various mass fractions m_{ik}. Then the partial differential equation which expresses the conservation of species k is:

$$\frac{\partial}{\partial t}(\rho_i r_i m_{ik}) + \text{div}\,(\rho_i r_i \vec{V}_i m_{ik}) =$$

$$= r_i C_{ik} + D_{ik} + K_{ik} \qquad .(2.2\text{-}7)$$

Here C_{ik} represents the rate of creation of k, per unit volume of i actually present, by chemical transformation; D_{ik} represents the effects on the k-content of phase i exerted by neighbouring elements of the same phase, e.g. by diffusion; and K_{ik} represents the result of interactions of various kinds with other phases occupying the same space.

The "overall" k-conservation equation takes the form, without K_{ik}'s:

$$\sum_{i=1}^{i=n}\left(\frac{\partial}{\partial t}(\rho_i r_i m_{ik}) + \text{div}\,(\rho_i r_i \vec{V}_i m_{ik})\right) =$$

$$= \sum_{i=1}^{i=n} (r_i C_{ik} + D_{ik}) \qquad .(2.2\text{-}8)$$

2.3 Auxiliary relations

The above set of differential equations ordinarily has to be solved in conjunction with observance of constraints on the values of the variables represented by algebraic relations. These auxiliary relations express physical laws of various kinds. A few examples will be indicated in the remainder of this section.

(a) Equations of state

$$\rho_i = \rho_i\,\{p,\ h_i,\ m_{ik}.\ .\ .\} \qquad ,(2.3\text{-}1)$$

where h_i is the stagnation enthalpy $\tilde{h}_i$, <u>less</u> the kinetic-energy and potential-energy terms, and m_{ik} . . . indicates that all the chemical species (k) present in phase i can contribute to the phase density ρ_i.

(b) Temperature ∿ enthalpy relation

$$T_i = T_i \{h_i, m_{ik} \ldots, p\} \quad ,(2.3\text{-}2)$$

where T_i stands for the temperature of phase i.

(c) Heat-transfer relations

$$Q_i = Q_i \{T_i \ldots\} \quad ,(2.3\text{-}3)$$

$$H_i = \text{div}\,(\lambda_i r_i \text{ grad } T_i) \quad ,(2.3\text{-}4)$$

$$J_i = \sum_{i=1}^{i=n} \alpha_{ij}(T_j - T_i) \quad ,(2.3\text{-}5)$$

where λ_i represents the thermal conductivity of phase i, α_{ij} represents an interphase volumetric heat-transfer coefficient, and j is, like i, an index of the phase number.

(d) Chemical-species relations

$$C_{ik} = C_{ik}\{T_i, p, m_{ik} \ldots\} \quad ,(2.3\text{-}6)$$

where m_{ik} . . . indicates that the concentrations of all the species present in phase i may, in principle, contribute to the reaction rate.

$$D_{ik} = \text{div}\,(\Gamma_{ik} r_i \text{ grad } m_{ik}) \quad ,(2.3\text{-}7)$$

where Γ_{ik} is the exchange coefficient (diffusion coefficient times density) of species k in phase i.

$$K_{ik} = \sum_{j=1}^{j=n} K_{ikj}\,(m_{ik}, m_{jk}, \text{etc}) \quad ,(2.3\text{-}8)$$

where K_{ikj} represents the result of interactions between the i'th and j'th phase.

(e) Boundary conditions

In order to represent a complete mathematical problem, the partial differential equations and auxiliary relations must be supplemented by initial and boundary conditions, expressing the particular situation to be investigated. The point is noted here, for completeness; but there is nothing further of general significance which requires to be stated.

2.4 Discussion of the mathematical problem

(a) Comparison with single-phase-flow problems The reader will be supposed to be familiar with the problems which arise when a single-phase flow is in question (n=1); the discussion will therefore be concentrated on three points in which the multi-phase problem differs from the single-phase problem. The first is obvious, the second rather less so, and the third may easily escape notice.

Point (i) A multi-phase problem necessitates the solution of more equations, and the specification of more auxiliary relations and boundary conditions, than a single-phase problem. However, the equations for multi-phase flows are of the same kind as those for single-phase flows; therefore the increase in the number of equations will not, of itself, increase the difficulty of obtaining solutions.

Point (ii) A difficulty with single-phase flows is that the pressure p is shared between three momentum equations; and a way has to be found of computing it. One popular way is that published by Patankar and Spalding in 1972, Reference (1), and called the SIMPLE algorithm; it obtains p from a special treatment of the three momentum equations and one continuity equation. However, for multi-phase flows, there are numerous continuity equations. Which one should be turned to? And how should the pressure be deduced? The answers to these questions are not obvious.

Point (iii) It is customary for single-phase flow to solve the equations iteratively; and, within one cycle of the iteration, a single dependent variable is handled at a particular time.

A straightforward application of this practice to multi-phase flows would entail, for example, an iteration cycle such as:

- Solve for T_i for i=1,
- Solve for T_i for i=2,
- Solve for T_i for i=3,
- Solve for . . .
- . . .
- . . .
- Solve for T_i for i=1,
- etc.

Continuation of this cyclic process would be expected to lead, finally, to a solution of all the equations for the temperatures of the phases.

In practice, however, the process is often found to converge with intolerable slowness, a remedy for which must be found.

(b) The method of solving for pressure What is required, to replace the SIMPLE algorithm (or to generalise it to multi-phase flows) is a procedure for satisfying the continuity equations, and deducing the pressure, which will work satisfactorily over the whole range of conditions likely to be encountered in practice. When all the r_i's but one vanish, the method should reduce to that for single-phase flow; and this should happen whichever of the r_i's remains finite. Thus, impartiality as between phases, and velocity directions for that matter, is a desideratum.

The method developed by the present writer is called IPSA; this stands for inter-phase slip algorithm. Its first form was described in Reference (2); and illustrations of its use have also been published (3, 4, 5). Since its first use in 1976, IPSA has been refined, in the interests of economy and reliability. The present paper constitutes the first publication of the refined version of IPSA. The method is described, in detail in section 4 below.

(c) The method of escaping the slow-convergence difficulty What is required, in order to ensure fast convergence in multi-phase-flow problems, is a degree of implicitness, which anticipates the effects of a change in the local property of one phase on the properties of the other phases at the same location.

This can be conveniently accomplished by the "partial-elimination algorithm" (PEA). This was referred to briefly in Reference (2); but a more extended treatment is provided in Sections 3.6 and 4. below, because of its importance.

(d) Further manipulations of the differential equations In a comprehensive treatise, the differential equations derived in section 2.2 would be subjected at this stage to two further treatments. First, because many two-phase flows are turbulent, a time-averaging process would be carried out, with the result that terms like

$$\rho_i r_i \vec{V}_i \phi_i \quad ,$$

for example, would take the form:

$$\overline{(\rho_i r_i \vec{V}_i)}\,\overline{\Phi}_i + \overline{(\rho_i r_i \vec{V}_i)'\overline{\Phi}_i'} \quad ,$$

wherein the bar denotes a time-average value and a prime denotes a fluctuation. ϕ_i stands for any extensive property of the i'th phase.

Secondly, versions of the equations would be written out in terms of cartesian, cylindrical-polar, and other coordinate systems; and the equations for one-, two- and three-dimensional systems would be distinguished. Further attention could then be given to special cases, in which specific terms became either dominant or unimportant, with interesting consequences.

Neither of these exercises will be undertaken here; for they are conventional; and they would merely distract attention from the matter in hand. Indeed, attention to the "turbulent-diffusion" term, $\overline{(\rho_i r_i V_i)'\phi_i'}$, would raise questions about how it should be related to local time-average variables, and to their gradients, which can be only speculatively answered for multi-phase flows.

Instead, the equations will be immediately put into finite-difference form, in order that attention can be paid to the method of solution, the presentation of which is the main subject of the paper.

3. FINITE-DIFFERENCE EQUATIONS

3.1 The grid system

The discussion of the solution procedure will be conducted in terms of a cartesian-grid system, for brevity; but the procedure can be applied to more general grid systems equally well. Fig. 3.1-1 illustrates an element of the grid, and the nomenclature to be employed. Because of its conventional character, familiarity with the use of such grids for single-phase flow will be presumed. The grid is of the "staggered" variety, with the velocity components u,v and w being stored at the mid-points of the corresponding links between the other nodes; once again, familiarity with this practice will be presumed. Reference (6) provides a discussion of this and other relevant practices.

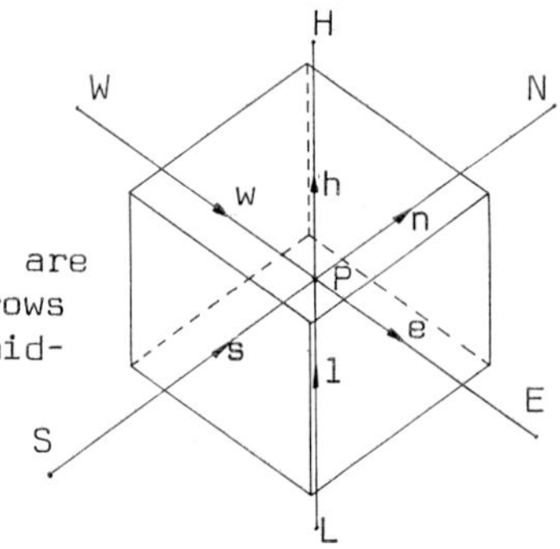

Figure 3.1-1

The "staggered" grid. Velocities are stored at locations marked by arrows (n,s,e,w,h,l). They are at the mid-points of cell-wall surfaces.

3.2 Origin of the finite-difference equations

In IPSA, as in SIMPLE, the finite-difference equations are formulated by integrating the partial differential equations over control volumes of finite size. For the variables r_i, h_i, m_{ik}, these control volumes are the ones centered on grid nodes such as P (in Fig. 3.1-1), and having faces which bisect the lines joining P to N, S, E, W etc., at right angles. For the velocity components u_i, v_i, w_i, the control volumes are like those just described; but they are displaced in the directions x, y or z respectively by distances which place the relevant velocity locations in their centres. This practice is conventional (6).

The result of integrating the equations can be most conveniently expressed in words, as follows. Let the general equation be expressed as:

$$\frac{\partial}{\partial t}(\rho_i r_i \phi_i) + \text{div}\,(\rho_i r_i \vec{V}_i \phi_i) = \text{div}\,(\Gamma_\phi r_i \text{ grad } \phi_i) + \Phi_i \quad ,(3.2\text{-}1)$$

wherein the first term on the right-hand side expresses the whole of that part of the source term which can without perverseness be so expressed, while Φ_i contains the remainder of the source term. Then integration over the control volume and over a finite time interval yields:

(Volume divided by time interval) ×
i. {(volume average of $\rho_i r_i \phi_i$ for end of time interval)
ii. -(volume average of $\rho_i r_i \phi_i$ for start of time interval)}
+(sum for all six faces) of {(face area) ×
×(area-and-time average for the face of the
iii outward - normal component of $(\rho_i r_i \vec{V}_i \phi_i)$}
=(sum for all six faces) of {(face area) ×
×(area-and-time average for the face of the
iv. outward-normal component of $\Gamma_\phi r_i$ grad ϕ_i)} +
v. the volume of Φ_i) .(3.2-2)

The task is now to express the quantities appearing in (3.2-2) in terms of the values of the variables pertaining to grid points. This is done by way of interpolation rules, of which a recommended set now follows. No justification of the recommendations will be provided, because they do not significantly differ from those which are commonplace for single-phase prediction procedures (6).

3.3 Interpolation rules

The reader's attention is drawn to the terms in equation (3.2-2) which are indicated by: i, ii, iii, etc. Each of these requires an interpolation rule.

Term i. The values of ρ_i, r_i, and ϕ_i prevailing at the grid point (P) in the centre of the control volume are presumed to prevail throughout the volume; so the grid-point and volume-average values are identical.

Term ii. As for term i.; but of course the values at the start of the interval are taken, viz. $(\rho_i r_i \phi_i)_{P-}$ where P- is the subscript denoting conditions at point P at the earlier time.

Term iii. The value of the area-average normal component of

$\vec{V}_i$ is taken as the velocity component stored at the velocity node at the centre of the area. The value of the $\rho_i r_i \phi_i$ product is taken as $(\rho_i r_i \phi_i)_P$ if the direction of the velocity is outward from the control volume; if the direction of the velocity is inward, it is $(\rho_i r_i \phi_i)_N$ for the n face, $(\rho_i r_i \phi_i)_S$ for the s face, $(\rho_i r_i \phi_i)_E$ for the e face, etc. This is the so-called "upwind-differencing practice." No minuses have been attached to P, N, S, E etc. in this recommendation; this implies that end-of-interval values are recommended. The practice is often referred to as being "fully implicit."

Term iv. The face-average value of the gradient term is to be taken as the difference between the grid-point values on either side of the face, divided by the distance between the nodes. The value of $\Gamma_\phi r_i$ is to be taken as the harmonic mean of the values pertaining to the same two grid points. Once again, the end-of-interval values are to be taken.

Term v. The volume-average value of Φ_i should be taken as a linear function of $\phi_{i,P}$, expressed as:
$\Phi_{i,P}^* + \Phi'_{i,P} \cdot \phi_{i,P}$, provided that $\Phi'_{i,P}$ is negative. This linearisation practice is an aid to convergence, because it anticipates, to some extent, the change which will occur in the value of the volume-source term.

Two further remarks are in order for the velocity equations, based upon the displaced control volumes. They are:

It is preferable to take arithmetic means for convection terms $(\rho_i r_i \vec{V}_i)$, where appropriate; and the pressure-gradient term should be obtained from the difference of the pressures on the two grid nodes which lie centrally in control-volume faces.

Nothing that has been stated in this section, it should be understood, represents a significant departure from single-phase practices. The only features specific to multi-phase flows are the r_i's, and the inter-action parts of the source terms.

3.4 Finite-difference equations for $\phi_{i,P}$

Application of these rules results in an equation for $\phi_{i,P}$ which can be usefully expressed in the form:

$$\phi_{i,P} = \frac{a_N\phi_{i,N}+a_S\phi_{i,S}+a_E\phi_{i,E}+a_W\phi_{i,W}+a_H\phi_{i,H}+a_L\phi_{i,L}+a_P\phi_{i,P-}+S_{i,P}}{b_N+b_S+b_E+b_W+b_H+b_L+b_P-S'_{i,P}} \quad .(3.4\text{-}1)$$

Discussion of a_N and b_N will serve to illustrate how the a's and b's are constructed. a_N consists of an inflow-convection contribution according to term iii. and a diffusion contribution according to term iv.

Thus:

$$a_N = \{((-\rho_{i,n} r_{i,N} v_{i,n})) + \Gamma_{i,N} r_{i,n}/(y_N - y_P)\}\, A_n \qquad ,(3.4\text{-}2)$$

wherein ((variable)) equals "variable" when the latter is positive, but otherwise equals zero. Further, A_n is the cell area; $(y_N - y_P)$ is the internode distance; and $\Gamma_{i,n} r_{i,n}$ is the harmonic mean of $\Gamma_{i,N} r_{i,N}$ and $\Gamma_{i,P} r_{i,P}$.

b_N, correspondingly, is given by:

$$b_N = \{((\rho_{i,P} r_{i,P} v_{i,n})) + \Gamma_{i,n} r_{i,n}/(y_N - y_P)\} A_n \qquad ,(3.4\text{-}3)$$

a definition which implies that the convection part of b_N is zero when that of a_N is finite, and vice versa. Thus, in contrast to a_N, b_N consists of an <u>out</u>flow contribution and a diffusion contribution.

The case of a_P and b_P is different in detail, but not in spirit; for the "velocity through time" is always positive; and there is no "diffusion in time." The appropriate formulae are:

$$a_P = \rho_{i,P-} r_{i,P-} V_P/\delta t \qquad ,(3.4\text{-}4)$$

and:

$$b_P = \rho_{i,P} r_{i,P} V_P/\delta t \qquad .(3.4\text{-}5)$$

Here V_P stands for the volume of the control volume around point P, and δt stands for the time interval.

The quantities $S_{i,P}$ and $S'_{i,P}$, appearing in equation (3.4-1), are defined by:

$$S_{i,P} = \Phi^*_{i,P} \quad V_P/\delta t \qquad ,(3.4\text{-}6)$$

and

$$S'_{i,P} = \Phi'_{i,P}\, V_P/\delta t \qquad .(3.4\text{-}7)$$

It is useful, at this stage, to recognise that the source term $\Phi_{i,P}$ is often constituted of terms having similar significances to the a's and b's. Thus, if there exists a mass transfer $\dot{m}_j$ from phase j into phase i, and also an interchange of the entity which affects ϕ_i <u>without</u> mass transfer, equal to c_j times $\phi_{j,P}$ minus $\phi_{i,P}$, Φ_i should be written as:

$$\Phi_i = \sum_{j=1}^{j=n} \left(((\dot{m}_j))\ \phi_{j,P} - ((-\dot{m}_j))\ \phi_{i,P} + c_j(\phi_{j,P} - \phi_{i,P}) \right) \qquad ;(3.4\text{-}8)$$

or

$$\Phi_i = \sum_{j=1}^{j=n} \left(((\dot{m}_j)) + c_j \right) \phi_{j,P} - \phi_{i,P} \left(\sum_{j=1}^{j=n} c_j + \sum_{j=1}^{j=n} ((\dot{m}_j)) \right)$$

$$\equiv \sum_{j=1}^{j=n} \alpha_j \phi_{j,P} - \phi_{i,P} \sum_{j=1}^{j=n} c_j \qquad .(3.4\text{-}9)$$

With this substitution, and further rearrangement ensuring that $\phi_{i,P}$ appears only on the left-hand side, the finite-difference equation takes the form:

$$\phi_{i,P} = \frac{\left(a_N \phi_{i,N} + a_S \phi_{i,S} + a_E \phi_{i,E} + a_W \phi_{i,W} + a_H \phi_{i,H} + a_L \phi_{i,L} + a_P \phi_{i,P-} + \sum_{\substack{j=1 \\ j \neq i}}^{j=n} \alpha_j \phi_{j,P} \right)}{b_N + b_S + b_E + b_W + b_H + b_L + b_P - \sum_{j=1}^{j=n} c_j} \qquad .(3.4\text{-}10)$$

It is worth noting that the definitions imply, in this case, that $\phi_{i,P}$ is a weighted average of $\phi_{i,N} \ldots \phi_{i,P-}$ and all the $\phi_{j,P}$'s; for the mass-conservation equation and the definition of a's, b's, c's and α's imply:

$$a_N - b_N + a_S - b_S + a_E - b_E + a_W - b_W + a_H - b_H + a_L - b_L + a_P - b_P = \sum_{j=1}^{j=n} (\alpha_j + c_j) \qquad .(3.4\text{-}11)$$

3.5 Equation for r_i

The equation for r_i has a similar form to that of the equation for ϕ_i, (3.4-1); and it may therefore also be written as:

$$r_{i,P} = \frac{\left(a_N r_{i,N} + a_S r_{i,S} + a_E r_{i,E} + a_W r_{i,W} + a_H r_{i,H} + a_L r_{i,L} + a_P r_{i,P-} + S_{i,P} \right)}{b_N + b_S \pm b_E + b_W + b_H + b_L + b_P - S'_{i,P}} \qquad ,(3.5\text{-}1)$$

but the a's and b's must be defined without the r_i's which appear in equations (3.4-2) through (3.4-5); and S_i and S'_i must also be appropriate.

As a consequence of the last-mentioned difference, there is no such relation between the a's and b's as was just described for the ϕ_i equation; one cannot use the same mass-conservation principle twice. This is by way of being a nuisance; for, during the approach to the converged solution, the a's and b's may not be in proper balance; therefore non-physical values of $r_{i,P}$ can be generated, necessitating special over-writing procedures. For the $\phi_{i,P}$ equation, by contrast, the b's can actually be eliminated from the denominator of the right-hand side of equation (3.4-1) by use of equation (3.4-11); and this makes the generation of non-physical ϕ_i values less probable.

3.6 Velocity Equations

The finite-difference equations for the velocity components take a similar form to that of equation (3.4-1). Especial interest attaches to the role of the pressures, which appear in the source term. The finite-difference equations will not be written out in full so as to avoid the necessity of introducing and defining further nomenclature; however, attention to important features of the equation will be drawn by concentration on special aspects of a typical velocity, say $u_{i,e}$.

Application of an integration-and-interpolation procedure of the same kind as before leads to an equation which can be expressed for present purposes as:

$$u_{i,e} = \frac{\mu_i + \zeta_i(p_P - p_E) + \sum_{j=1}^{j=n} \xi_{ij} u_{j,e}}{\varepsilon_i} \quad , \qquad (3.6\text{-}1)$$

wherein μ_i, ζ_i, ξ_{ij} and ε_i contain coefficients and values of neighbouring u_i's that require no attention at present.

The slow convergence referred to in section 2.4(a) and (c) above occurs especially when the ξ_j's are so large as to dominate the numerator of equation (3.6-1); for large interaction coefficients "tie together" the interacting variables, and an iterative correction procedure which treats one variable at a time proceeds as slowly as does a walk with tightly hobbled feet. For rapid progress, one needs to "hop", that is to shift all the tightly linked variables simultaneously.

The numerical equivalent of hopping is to manipulate the equation of type (3.6-1), for <u>all</u> the phases present at point e, so as to eliminate all but the u_e from each equation. There results a set of equations which can be written as:

$$u_{i,e} = \frac{\eta_i^* + \zeta_i^*(p_P - p_E)}{\varepsilon_i^*} \quad , \qquad (3.6\text{-}2)$$

wherein the expressions for η_i^*, ζ_i^* and ε_i^* are easily derived.

The Partial Elimination Algorithm, referred to above, amounts simply to solving equation (3.6-2) instead of (3.6-1).

Before terminating discussion of the velocity equations, it is useful to note that differentiation of equation (3.6-2) with respect to p_P and p_E yields:

$$\frac{\partial u_{i,e}}{\partial p_P} = -\frac{\partial u_{i,e}}{\partial p_E} = \frac{\zeta_i^*}{\varepsilon_i^*} \qquad .(3.6\text{-}3)$$

These differential coefficients, and similar ones for the five remaining velocity components, will be useful when the time comes to find pressure corrections which will change the velocities so as to balance the overall continuity equation. This equation will now be discussed.

3.7 <u>Overall continuity equation</u>

The finite-difference form of equation (2.2-2) can be expressed in words as:

> Sum for all phases of ((volume divided by time interval) × {(volume average of $\rho_i r_i$ for end of time interval) - (volume average of $\rho_i r_i$ for start of time interval)} + (sum for all six faces of {(face area) × (area-and-time average for the face of the outward-normal component of $\rho_i r_i \vec{V}_i$)}) = 0 .(3.7-1)

If this is compared with equation (3.2-2), it will be seen that the diffusion and mass source terms on the right-hand side are replaced by a zero. This is because the diffusion terms for all phases sum to zero <u>by definition</u>; and, although mass may be transferred from one phase to another, the <u>net</u> mass source must be zero, from the law of mass conservation.

The interpolation rules recommended for this equation are the same as for equation (3.2-2), except that no question of choosing ϕ's arises.

The most useful algebraic form of equation (3.7-1) now follows; it is:

$$\sum_{i=1}^{i=n}\Big(\alpha_i(\rho_{i,P} - \rho_{i,P-}) + \beta_{i,n}V_{i,n} + \beta_{i,s}V_{i,s} + \beta_{i,e}U_{i,e} + \beta_{i,w}U_{i,w} + \beta_{i,h}W_{i,h} + \beta_{i,e}W_{i,e}\Big) = 0 \qquad .(3.7\text{-}2)$$

Here the definitions of α_i, $\beta_{i,n}$, $\beta_{i,s}$ etc. can be deduced from the verbal equation (3.7-1) and the interpolation rules; the particular form shown has been chosen so as to exhibit those terms which are especially sensitive to the pressures at grid-point P and its neighbours N, S, E, W, H, L.

4. SOLUTION PROCEDURE

4.1 Outline of the procedure

The Task, for a typical problem, is to determine sets of values, for all the points of the grid, for all the time instants considered, and for all n phases, of: p, r_i, h_i, $m_{i,k}$ at central grid points, and of $\breve{u}_i$, v_i, w_i at velocity grid points.

The finite-difference equations, together with the auxiliary relations, are equal (provided no mistake of judgement or implementation has been made) to the number of unknowns; so the task is a possible one.

Because of the high degree of non-linearity and inter-connectedness of the equations, the task must be performed by iterative, i.e. "guess-and-adjust" means. It is therefore necessary to establish a reliably convergent successive-adjustment procedure. This is what IPSA has been found to be.

The main features of IPSA are as follows:

- In a single cycle of the iteration, the equations are solved in the order: (1) enthalpy, (2) volume-fraction, (3) velocity, (4) pressure-correction.
- The enthalpy equations are solved by use of the partial-elimination algorithm because of the possibility of tight linkage between the enthalpies of the various phases.
- The volume-fraction equations incorporate the inter-phase mass transfers which, sometimes, are closely connected with the enthalpies and the heat transfer.
- The solution of the velocity equations proceeds by way of a guessed pressure field (which leads to velocities which do not satisfy the overall continuity equations), followed by the addition of velocity corrections (which restore satisfaction of the overall continuity equation).
- The velocity corrections are deduced from the solution of a pressure-correction equation.
- This cycle of adjustments is performed many times, until the errors remaining in all the equations are acceptably small.

How the finite-difference equations for a given variable are to be solved, over the integration domain, at a given stage of the whole solving procedure, will not be specified; for it is immaterial. Thus, the Gauss-Seidel point-by-point adjustment procedure could be employed; alternatively, and more advantageously from the point of computer-time economy, some version of the ADI (i.e. "alternating-direction-implicit") procedure can be made use of. Readers of the present article are expected to be satisfied with the statement that the usual equation-solvers of single-phase numerical fluid mechanics retain their applicability when IPSA is applied to multi-phase flows.

In the following sections, the various stages will be described in more detail.

4.2 Solution of the enthalpy equation

The enthalpy equations are best thought of as leading to values of the temperatures, T_i, of the various phases at all the grid points. To bring temperature into the forefront has many advantages, not least that of ensuring that the thermal interactions are proportional to the difference between the values of the calculated variables.

The enthalpy equations have been mentioned first because they are solved first in the solution cycle, as mentioned in section 4.1. However, their treatment here will be merely outlined, because they are not central to IPSA, and because space is short. The following statements, if carefully elaborated, may be sufficient to allow specialist readers to understand what is to be done.

- The partial-elimination algorithm is used because the interaction coefficients between the phases are often large. The manner of its application is not significantly different from that described above, in connexion with the velocity equation.

- When mass transfer occurs between the phases, it is usually the case that the rate of mass transfer is not known. However, it is possible to express the mass-transfer rate itself as a linear function of temperature; thus if only two phases are present, a vapour (1) and a liquid (2), the relations may well be:

$$\dot{q}_{1\to s} = \alpha_1(T_1 - T_s) \qquad ,(4.2\text{-}1)$$

$$\dot{q}_{s\to 2} = \alpha_2(T_s - T_2) \qquad ,(4.2\text{-}2)$$

$$\dot{m}_{2\to 1} = (q_{1\to s} - q_{s\to 2})/\lambda \qquad ,(4.2\text{-}3)$$

$$\dot{m}_{2\to 1} = \beta(T_s - T_{sat}) \qquad ,(4.2\text{-}4)$$

wherein:- $\dot{q}_{1\to s}$ and $\dot{q}_{s\to 2}$ represent the heat-transfer rates from vapour to surface (interface), and surface to liquid, respectively; α_1 and α_2 are heat-transfer coefficients; $\dot{m}_{2\to 1}$ is the rate of mass transfer from liquid to vapour, i.e., of vaporization; λ is the latent heat of vaporization; β is a kinetic coefficient; and T_{sat} is the saturation temperature at the pressure in question.

• Explicitly, the corresponding mass-transfer-rate relation is:

$$\dot{m}_{2\to1} = \beta\{\alpha_1(T_1 - T_{sat}) + \alpha_2(T_2 - T_{sat})\}/(\alpha_1 + \alpha_2 + \beta\lambda) \quad ; \quad (4.2\text{-}5)$$

and the associated values of T_s, $\dot{q}_{1\to s}$ and $\dot{q}_{s\to2}$ are:

$$T_s = (\alpha_1 T_1 + \alpha_2 T_2 + \beta\lambda T_{sat})/(\alpha_1 + \alpha_2 + \beta\lambda) \quad , \quad (4.2\text{-}6)$$

$$\dot{q}_{1\to s} = \alpha_1\{\alpha_2\ (T_1 - T_2) + \beta\lambda(T_1 - T_{sat})\}/(\alpha_1 + \alpha_2 + \beta\lambda) \quad , \quad (4.2\text{-}7)$$

$$\dot{q}_{s\to2} = \alpha_2\{\alpha_1(T_1 - T_2) + \beta\lambda(T_{sat} - T_2)\}/(\alpha_1 + \alpha_2 + \beta\lambda) \quad . \quad (4.2\text{-}8)$$

• Because of the linearity of these relations, it is possible to retain the linear form for the finite-difference equation for temperature. The unknown mass-transfer rates are indeed eliminated from these equations.

• After solution of the enthalpy (i.e. temperature) equations, the mass-transfer rates can be computed; they are then available for use in the next stage of the computation, namely the solution for the volume fractions.

• There are special cases which bring certain simplifications; e.g. β is often so large as to permit the equations to be simplified to:

$$\dot{m}_{2\to1} = \{\alpha_1(T_1 - T_{sat}) + \alpha_2(T_2 - T_{sat})\}/\lambda \quad , (4.2\text{-}9)$$

$$T_s = T_{sat} \quad , (4.2\text{-}10)$$

$$\dot{q}_{1\to s} = \alpha_1(T_1 - T_{sat}) \quad , (4.2\text{-}11)$$

$$\dot{q}_{2\to s} = \alpha_2(T_{sat} - T_2) \quad , (4.2\text{-}12)$$

• Further special cases arise when α_1, or α_2, or both, are very large; then appropriate modifications to the solution procedure permit this procedure to be appreciably shortened.

4.3 Solution of the r_i equations

Equations of the form (3.5-1) can be formed in a straightforward manner. There is one such equation for every phase and for every grid point; but, as a rule, it is preferable to solve for all but one of the phases; the r_i's of the remaining phase are then deduced from the knowledge that all the r_i's must sum to unity at every point.

As mentioned in section 3.5, it is not possible, in general, to replace the sum of b's in the denominator of equation (3.5-1) by an equivalent sum of a's; the reason for this difference from what can be done with equation (3.4-10) for example is that, whereas there is a law of mass conservation to constrain the coefficients of the latter equation, the only limit on the coefficients of equation (3.5-1) is that they may not lead to values of r_i lying outside the range from zero to unity.

It is therefore useful to build corresponding "traps" into the solution procedure; and sometimes an under-relaxation of the r_i's is introduced, in order that the velocity fields have time to adjust themselves to the momentum equations and to the overall continuity equation, before the r_i's have moved far from the initial guesses for these quantities.

Apart from the point just mentioned, solution of the r_i equations exhibits no peculiarities. It may be mentioned however that the r_i equation can often reveal especially clearly, through its solution, some of the shortcomings of fixed-grid numerical procedures. For example, suppose that, in a field of uniform velocity, a "front" exists across which r_i experiences a discontinuity from one value to another. In reality, in the absence of diffusion, this front will simply be translated by the fluid, retaining its discontinuous-r_i character.

Computation of the grid-point r_i's at successive times, however, by way of conservation equations of the kind presented in section 3, leads to the propagation of a front of finite width: the sharp discontinuity is "smeared" over several grid intervals.

This numerical-diffusion behaviour is not in the least specific to multi-phase flows; it is well known; and the only satisfactory remedies involve the employment of other means (e.g. "floating grids", "particle tracking") for the computation of the volume fractions and similar quantities. The topic is mentioned here only for one reason: to lead into the statement that IPSA is by no means confined to the r_i-computation method that has been described above: it does not matter how the r_i's are determined; IPSA can adjust the velocity fields to accord with them in any case.

4.4 Solution of the velocity equation

In order to solve the velocity equation; it is necessary to ascribe values to the pressures. There are several ways of doing this, the most straightforward of which is to use those values which are already in store, and which result from the pressure-correction operation conducted at the end of the previous cycle.

Alternatively, a pressure field may be deduced from the in-store velocities by integrating a selected set of momentum equations. Which momentum equations to choose, i.e. which phase to attend to, and in which directions to integrate, are questions which must be arbitrarily answered; and the pressure field which is obtained will depend on the answers which are given. Only as convergence is approached does it become immaterial which phase and which directions are turned to.

This is a problem that is familiar from single-phase flows; but the presence of many phases widens the possible choices. At the present stage of development, probably the simplest choice is best, namely to use the pressures generated by the pressure-correction equation. Certainly, hundreds of successful calculations of multi-phase flows have been based upon this prescription.

Once the pressures have been settled, the solution for the velocities (all the components, for all the phases) proceeds in a perfectly straightforward way. In general, because of the different densities of the phases, the velocities of the phases also differ. The differences are greatest in regions of strongly varying pressure, whether these differences are associated with inertia effects (e.g. strong curvature of light-phase streamlines, which the heavy phase cannot follow) or a gravitational field (in which the denser fluid tends to fall).

4.5 Corrections to the velocity and pressure fields

The velocity fields which have been generated by the solution of the momentum equations, based on an incorrect pressure field cannot satisfy the overall continuity equation: when combined with the in-store values of the r_i's, and the densities which are appropriate to the pressure and temperatures, there is found to be a net accumulation, or loss, of mass from each cell.

The most characteristic feature of the IPSA process concerns the response which is made to this discovery, the nature of which can be expressed by replacing the zero on the right-hand side of equation (3.7-2) by E_P, the symbol for the mass-continuity error for cell P.

It is now supposed that all the pressures in the field will be modified so as to change the densities and the velocities, and thereby to bring about the elimination of the errors. Differentiation of the appropriate terms of equation (3.7-2) therefore leads to:

$$\sum_{i=1}^{i=n}\left(\alpha_i \frac{\partial\rho_{i,P}}{\partial p_P} \cdot p'_P + \beta_{in}\left(\frac{\partial v_{in}}{\partial p_N} p'_N + \frac{\partial v_{in}}{\partial p_P} p'_P\right) + \right.$$

$$+ \beta_{is}\left(\frac{\partial v_{is}}{\partial p_S} p'_S + \frac{\partial v_{is}}{\partial p_P} p'_P\right) + \beta_{ie}\left(\frac{\partial u_{ie}}{\partial p_E} p'_E + \frac{\partial u_{ie}}{\partial p_P} p'_P\right) +$$

$$+ \beta_{iw}\left(\frac{\partial u_{iw}}{\partial p_W} p'_W + \frac{\partial u_{iw}}{\partial p_P} p'_P\right) + \beta_{ih}\left(\frac{\partial w_{ih}}{\partial p_H} p'_H + \frac{\partial w_{ih}}{\partial p_P} p'_P\right) +$$

$$\left. + \beta_{il}\left(\frac{\partial u_{il}}{\partial p_L} p'_L + \frac{\partial u_{il}}{\partial p_P} p'_P\right)\right) = -E_p \qquad .(4.5\text{-}1)$$

This equation can be recognised as one which connects the required pressure corrections at grid-point P, p'_P, with the six pressure corrections for neighbouring points: p'_N, p'_S, p'_E, p'_W, p'_H and p'_L. The relation is a linear one.

The differential coefficients of velocity, of course, are obtained from the momentum equations, as was explained in section 3.6; and the differential coefficient of density is obtained by differentiating the equation of state.

It is not vital that the differential coefficients should be especially precise; indeed, the skilled numerical analyst can often enhance convergence by changing their magnitudes, selectively, by quite large amounts. What is however important is that the coefficients have the right sign, and that they properly reflect the relative sensitivities of the various phases to changes of pressure. Straightforward employment of the equations of momentum and state does achieve this.

Once the pressure-correction-equation coefficients have been calculated, it remains to solve for the pressure-correction field. When this has been done, the associated changes to the densities and velocities can be computed, and added to the in-store values of these quantities. The result is a set of fields of velocities and densities which satisfy overall continuity at every point.

4.6 Iteration and convergence

The changes to the velocities and the densities which have resulted from the last stage, though they have satisfied overall continuity, will have thrown most of the other equations out of balance. It is therefore necessary to repeat the cycle, so as to undo the damage.

What is to be hoped is that, at each repetition of the cycle, the errors which are to be corrected become smaller and smaller.

This hope is usually realised; and, when it is not, the failure is attributable to unusual geometrical proportions of the cell, to inadequate under-relaxation, or to some non-standard coupling between equations which requires a non-standard amendment to the solution procedure.

IPSA has proved itself to be satisfactorily convergent without amendment in the vast majority of cases; and, when divergence has appeared, the remedy for it has been found without difficulty.

5. THE LAYERED-FLOW APPROXIMATION

5.1 Occurrence

There is a long and wise tradition in fluid mechanics that, whenever practicable, the dimensionality of the problem should be reduced by the making of simplifying (but not over-simplifying) assumptions. Thus "shallow-water theory", for example, successfully predicts the tidal movement of the oceans; and flows in channels, over broad-crested weirs, and in "hydraulic jumps" can be successfully predicted with the aid of the theory.

This theory, in essence, takes note of the fact that many fluid-mechanical phenomena exhibit much larger velocities, and have larger extents, in some directions (e.g. horizontal) than in others (e.g. vertical). As a consequence, they are adequately represented for many purposes by the so-called "vertically-averaged equations".

Most of the layered flows to which the "shallow-water-theory" has been applied can be regarded, from the purist's viewpoint, as two-phase flows; for the first fluid (e.g. water) flows beneath a second fluid (e.g. air). Of course, the second fluid has, in most cases, a negligible influence on the first; yet one has only to put both fluids in a duct of fixed cross-sectional area, and to let the lower fluid nearly fill that cross-section, for the pressure differences in the upper one to become significant and to transmit their effects to the lower fluid.

There exists, therefore, a practically important class of ducted flows in which two (or more) phases are in motion, in which gravity (or some other body force) separates the two phases in the direction normal to that (or those) of the main flow, and for which it is meaningful to base an analysis on equations which, having been integrated in the direction of the body force, are of diminished dimensionality.

Examples of such flows include:- the flow of a gas and of a

liquid within a nearly horizontal pipe of large diameter, such as occurs when a moderate amount of condensate forms in a vapour-carrying pipe line; the flow of two gases of differing density in a duct which is rotating about an axis parallel to itself at such a velocity that the cooler gas is flung to the outside of the duct and the hotter gas is displaced to the inside; and the intrusion of a salt-water layer along the bottom of a fjord, the surface waters of which are fresh.

The IPSA algorithm has been successfully applied to the calculation of such flows; but a slight modification to the procedure is required, which will now be described.

5.2 Finite-difference equations

The finite-difference equations which are to be solved, for gravity-stratified flows, are the same as for dispersed flows, except in the following particulars:-

- Because of the reduced dimensionality, the H and L ("high" and "low") neighbours of a grid point are absent, and the corresponding finite-difference coefficients likewise. Indeed, if a pipe flow is in question, the E and W links can also be cut off, to render the phenomenon one-dimensional in space.

- If the pressure which appears in the momentum equations for both phases is taken as that at the top (say) surface of the upper phase, there exists an extra momentum source in the equation for the lower phase which is proportional to the gravitational acceleration, to the difference of density between the two phases, and to the gradient of the elevation of the interface between the two phases.

- An auxiliary relation, deducible from a prescription of the shape of the cross-section of the duct, connects the elevation of the interface with the volume fraction of the second phase r_2. This puts the gradient of r_2 into the momentum equation, and provides another form of coupling between the two equations.

Two points may be mentioned here by way of discussion. First, this is a situation in which it _is_ meaningful to ascribe differing pressures to the two fluids at the same horizontal location. It is not especially _useful_ to do so; and the author's practice is to have a single pressure appearing in _both_ equations, and an additional gravity-component term in _one_ of the equations.

However, since some authors have expressed a desire to distinguish between the pressures of the two fluids, the point may be of some interest. Strictly speaking, of course, there is a _continuous variation_ of pressure, in both fluids, along a vertical; so even the two-pressure formulation hides some aspects of the matter.

Secondly, some examiners of the IPSA procedure have experienced difficulty in recognising that the r's which are used for stratified flow have precisely the same significance as those for dispersed flow; yet it is so: the fact that the two fluids "keep their distance" as much as possible, does not alter the nature of the r's.

5.3 The solution procedure

The solution procedure outlined above for dispersed flows will work satisfactorily for most cases of layered flow also. However it may become rather inefficient, requiring small time steps or heavy under-relaxation, when the density difference is large and the volume fraction of the lower fluid is small.

The reason for this is that pressure differences in the upper fluid are then very small; and the practice of computing pressure corrections from the overall continuity equation has little influence on the motion of the liquid.

In these circumstances, a further step is included in the IPSA procedure: after the r_2 equation has been solved (with coefficients including velocities deduced from presumed r_2 gradients), the discrepancies between the computed and the presumed r_2's are noted. Then corrections are computed to the presumed r_2's, which will entail changes to the velocities, which will in turn modify the computed r_2's so as to agree with the presumed ones.

In other words, an r_2-correction stage is introduced which is akin in purpose and nature to the pressure-correction stage. Inclusion of this stage enhances the speed of convergence when large time steps are taken.

It is also possible, and interesting, to combine the solution-for-r_2 and the correction-of-r_2 steps into one; and thought-provoking similarities to, and differences from, material-diffusion equations are thereby revealed. However, a discussion of this topic would lead too far beyond the intended scope of this paper.

6 DISCUSSION

6.1 Some publications

Space does not permit the presentation here of the results which have been published so far of work based upon use of the IPSA calculation procedure. However, a few remarks about this work will now be made.

Moult, Pratap and Spalding (Ref. 3) demonstrated the working of the procedure for one-dimensional unsteady problems. The grid was coarse and the results scanty; but the publication deserves mention because it was the first one.

Baghdadi, Rosten, Singhal, Spalding and Tatchell (Ref. 4) presented further one-dimensional unsteady computations, and were especially concerned with the gravity-stratified flow of liquid and gas in a pipe. These authors investigated, among other things, the influences of the sizes of time step and of inter-node distance on the accuracy of the predicted speed of propagation of waves on the interface.

Pun, Rosten, Spalding and Svensson (Ref 5) presented computations for two-dimensional steady flows chosen to simulate environmental processes, such as the falling of rain-drops or snow-flakes in a cross-wind, and the influence of a fence on the deposition pattern. The results were plausible; but they were not compared with experimental data.

Markatos, Moult, Phelps and Spalding (Ref. 6) reported the results of three-dimensional steady-flow computations for the steam generator of pressurized-water nuclear-reactor plant. In such steam generators, the heat source is pressurised primary water, circulated through an array of hair-pin-shaped tubes within a shell, to the base of which secondary water is supplied. The latter is partly converted into steam by contact with the tubes; and, though both steam and water rise within the shell, the steam rises the fastest. The predicted patterns of void fraction and velocity exhibited a complex three-dimensional pattern; and, although no measurements are yet available, all the indications so far confirm the realism of the predictions.

In the latter case, of course, account had to be taken of the influences of the primary-fluid tubes not merely in supplying heat, but also in blocking space and resisting flow. To do this, use was made of the technique used by Patankar and Spalding first for heat exchangers (Ref. 8) and later for steam generators (Ref. 9). The latter references were concerned with (what amounted to) single-phase flows; but the technique has proved to be equally applicable to multi-phase ones.

Forthcoming publications include one by Kurosaki and Spalding) (Ref. 10) continuing the one-dimensional transient studies, one by Marchand, Singhal and Spalding (Ref. 11) concerning steam generators, and a review paper by Spalding (Ref. 12).

Unpublished work, which may be published before long, concerns flow in under-ocean pipe-lines of petroleum gases and the condensate formed from them.

An extension of the latter work has been projected in a forthcoming paper by the present author (Ref. 13).

6.2 Future developments

Because of the number and importance of the practical manifestations of multi-phase flow phenomena, some of which were indicated in section 1, it can be expected that application of the IPSA procedure, or of such better alternatives as manifest themselves, will multiply rapidly. It is therefore useful, at the close of this paper, to look into the future so as to be aware of the obstacles to complete success which remain to be removed.

First of all, of course, the incompletely-solved problems of single-phase numerical fluid mechanics spread their influence in the multi-phase area. The problems of a computational character are:

(i) How to conduct fine-grid computations cheaply;
(ii) How to combat the influences of numerical diffusion by appropriate choice of grid configuration and finite-difference formulation.
(iii) How to preserve order and flexibility in the associated computer codes, as the complexity of their physical content increases.

Those of a physical character primarily concern the search for realistic and universal turbulence models.

The computational problems are aggravated, for multi-phase flows, by the increased number of equations which have to be solved; however, this is a change only of degree, not of nature.

On the physical front, current ignorance however casts a longer and darker shadow. Firstly, the ways in which a dispersed second phase influences the turbulence properties of a supporting first phase are almost entirely unexplored. Influences there must be; but no-one is able to quantify them on the basis of knowledge.

Secondly, however, multi-phase flows exhibit physical processes of their own, namely the complete range of heat-, mass- and momentum-transfer processes between the phases. When single spherical solid particles are suspended, at wide separations, in a supporting medium, the transfer-process laws may be regarded as sufficiently known; but, when a liquid ∿ vapour mixture is in question, with its possibilities of exhibiting droplet, bubble, slug, annular, reverse-annular and other flow regimes, prescriptions of the inter-phase transfer laws can be done only speculatively.

Lastly, it is wise to contemplate the distinction which was made in section 1.1: just because the thermodynamicist recognises only two fluid phases in a steam generator namely liquid and vapour, there is no certainty whatever that a two-phase model will be an adequate representation. The simpler example of vaporisation of a liquid spray into its own hot vapour illustrates the point: there are larger droplets and smaller droplets at any location; and the various groups do not all travel at the same velocity; nor do all the droplets in one size group travel at the same velocity.

While it remains wise to start by studying the two-phase model, neither surprise nor excessive disappointment must be occasioned by failure of measurements always, or even often, to provide the looked-for validation. Multi-phase models are likely to prove the rule rather than the exception; and it will remain a matter of trained judgement into how many computational phases a single thermodynamic phase should be divided.

REFERENCES

1. Patankar, S.V. and Spalding, D.B. (1972)
A calculation procedure for heat, mass and momentum transfer in three-dimensional parabolic flows"
Int. J. Heat Mass Transfer, Vol 15, pp 1787-1806.

2. Spalding, D.B. (1977)
The calculation of free-convection phenomenon in gas-liquid mixtures.
ICHMT Seminar, Dubrovnik, 1976. Published in "Turbulent Buoyant Convection" by Hemisphere, Washington, 1977.

3. Moult, A., Pratap, V.S. and Spalding D.B.(1978)
Calculation of unsteady one-dimensional two-phase flows.
First International Conference on PhysicoChemical Hydrodynamics, Advance Publications, Channel Islands.

4. Baghdadi, A.H.A., Rosten, H.I., Singhal, A.K., Spalding, D.B and Tatchell, D.G. (1978)
Finite-difference predictions of waves in stratified gas-liquid flows.
ICHMT Seminar, Dubrovnik, 1978. To be published by Hemisphere, Washington.

5. Pun, W.M., Rosten, H.I., Spalding, D.B. and Svensson, U. (1978)
Calculation of two-dimensional steady two-phase flows.
ICHMT Seminar, Dubrovnik. To be published by Hemisphere, Washington.

6. Markatos, N.C., Moult, A., Phelps, P.J. and Spalding, D.B. (1978)
The calculation of steady, three-dimensional, two-phase flow and heat transfer in steam generators.
ICHMT Seminar, Dubrovnik. To be published by Hemisphere.

7. Spalding, D.B. (1976)
Basic equations of fluid mechanics and heat and mass transfer.
Imperial College, Heat Transfer Section Report, HTS/76/6.

8. Patankar, S.V. and Spalding, D.B. (1976)
A calculation procedure for the transient and steady-state behaviour of shell-and-tube heat exchangers.
In: Heat Exchangers Design and Theory Sourcebook, Scripta, Washington.

9. Patankar, S.V. and Spalding. (1978)
Computer analysis of the 3D flow and heat transfer in a steam generator.
Forschung Ing.-Wes. Vol 44, pp 47-52.

10 Kurosaki, Y. and Spalding, D.B. (1979)
One-dimensional unsteady two-phase flows with interphase slip; a numerical study.
Second Multi-Phase Flow and Heat Transfer Symposium, Miami.

11 Marchand, E.O., Singhal, A.K. and Spalding, D.B. (1979)
Numerical modelling of two-phase flow in steam generators.
Second Multi-Phase Flow and Heat Transfer Symposium, Miami.

12 Spalding, D.B. (1979)
Multi-phase flow prediction in power-system equipment and components.
EPRI Workshop on Basic Two-Phase Flow Modelling in Reactor Safety and Performance. Tampa, Florida.

13 Spalding, D.B. (1978)
Numerical computation of two-phase flow.
Second International Conference on PhysicoChemical Hydrodynamics (the Levich Conference). Washington Nov 1978.

NOMENCLATURE

Subscripts

i (or j)	the i'th (or j'th) of n phases
k	the k'th chemical direction
l	the l'th velocity direction
P,N,S,E,W,H,L n,s,e,w,h,l	grid-point locations.

Variables (Latin letters)

c	specific heat at constant pressure
$\tilde{h}$	stagnation enthalpy
M	mass source per unit volume
m	mass fraction
p	pressure
r	volume fraction
T	temperature
t	time
u,v,w	momenta per unit mass, in x,y,z directions
$\vec{V}$	velocity vector
V	volume
x,y,z	coordinate dimensions

Variables (Greek letters)

Γ	exchange coefficient
ρ	density
λ	thermal conductivity
ϕ	a general extensive variable
Φ	source rate of ϕ per unit volume.

CHAPTER 6

THE INTERACTION OF AN OBLIQUE SHOCK WAVE WITH A FLAT PLATE BOUNDARY LAYER

M. Wolfshtein, V. Reitman and M. Hanin

Department of Aeronautical Engineering
Technion, Israel Institute of Technology, Haifa

ABSTRACT

The interaction of a shock wave with a laminar or turbulent boundary layer is studied. A mathematical model is formulated, and a numerical method for its solution is presented. The basic features are a turbulence energy model for the calculation of the eddy viscosity, and an ADI method for the solution of the numerical problem. Computations are performed for Reynolds numbers $3. \times 10^5$ and $\sim 1.2 \times 10^7$ at freestream Mach numbers 2÷4. The results compare with experimental data favourably.

MAIN SYMBOLS

c_v	Specific heat at constant volume
D	turbulent energy dissipation
e	kinetic energy of turbulent fluctuations (or turbulent energy)
E	specific energy
M	Mach number
M	number of finite difference grid points in x-direction
N	number of finite difference grid points in y-direction
p	static pressure
Q_i	heat flux
Re	$\rho_0 u_0 x_s / \mu_0$
Re_t	turbulent Reynolds number
t	time
T	temperature
u, v	velocity in x and y directions
x, y	coordinates parallel and normal to the plate
x_0	distance from the loading edge to the entry of the field of solution
x_s	distance from loading edge to the impingement point of the "inviscid" shock wave on the plate

δ	boundary layer thickness
Λ	scale of turbulence
μ	molecular viscosity coefficient
ρ	density
σ	Prandtl number
τ_{ij}	shearing stress
ψ	stream function

<u>Subscripts</u>

e	of the turbulent energy
ℓ	laminar
m, n	indexes in x and y direction
t	turbulent
w	on the plate
0	immediately upstream of the interaction region
1	conditions in front of the shock wave in the inviscid region
2	conditions behind the shock wave in the inviscid region
3	conditions behind the shock waves system

<u>Superscripts</u>

k	index of the time-step.

1. INTRODUCTION

When an oblique shock wave impinges onto a solid surface a complex flow field is generated. If viscous and turbulent boundary layer effects are neglected an exact solution may be obtained, by which the shock is reflected from the surface. A more realistic analysis shows that the boundary layer on the solid is subjected to a severe pressure rise when it passes below the shock. This pressure rise causes significant thickening of the boundary layer, and may even cause boundary layer separation when the shock is sufficiently strong. Fig. 1 shows the system of compression waves generated at the boundary layer separation and reattachment points. These compression waves merge into two shock waves which intersect with one another to form the reflected shock mentioned above. The picture is even more complicated, due to formation of expansion waves in the region of maximum boundary layer thickness.

The problem is of major importance in aerodynamics. In transonic flow the phenomenon increases the drag, and may cause sudden fluctuating forces on control surfaces with serious effects on the control of the airplanes. In supersonic flight, the shock wave-boundary layer interaction may decrease the efficiency of engine inlets, and increase the total drag.

The problem is mathematically complex. The governing equations are non-linear, and coupled with one another. A suitable turbulence model is usually necessary and the computer resources required are significant. Therefore, the solution of this problem has become practical only when large and fast digital computers have become widely spread.

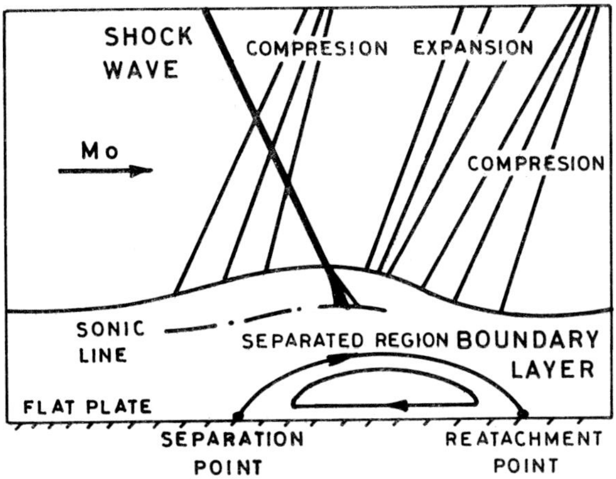

Figure 1. The Flow Field.

Some numerical methods have been developed for the solution of viscous supersonic flow fields. The two most common ones have been the explicit time split methods of MacCormack (1971) and Brailovskaya (1965). More recently Beam and Warming (1977) and MacCormack (1976), presented implicit methods. All these methods are very cumbersome, and require extensive computer resources. Hirsh (1976) showed that standard ADI techniques may be applied to supersonic flows, although he did not apply the method to flows with shock waves.

Solutions of the shock wave-boundary layer interaction problem have been obtained by various workers, e.g., MacCormack (1971), Wilcox (1973), Baldwin and MacCormack (1974), Shang et al (1976), Viegas and Coakley (1977), Shea (1978) and Hanin et al (1978). Most of these workers attempted the solution of shock wave interaction with a turbulent boundary layer as well. It appears that a mixing length model can not reproduce the flow accurately, and a turbulence energy model is required. Such models have been discussed by Wolfshtein et al (1975) for incompressible flows. At high speeds a compressibility correction may be required in the turbulence energy equation. Wilcox and Alber (1972) suggested such a modification to Soffman's (1970) turbulence energy model. However, other researchers did not find such corrections necessary.

In the present paper the problem is investigated using the Samarskii and Andreev (1963) ADI method, and the turbulence energy model proposed by Wolfshtein (1969). This combination appears to enable a good agreement between theory and experiments at a reasonable price, as is shown in the following sections of this paper.

2. THE GOVERNING EQUATIONS

2.1 The Basic Equations

Two-dimensional laminar unsteady flow in Cartesian coordinates is governed by the following conservation laws:

Continuity:

$$\frac{\partial \rho}{\partial t} + \frac{\partial}{\partial x}(\rho u) + \frac{\partial}{\partial y}(\rho v) = 0 \tag{2.1}$$

x-Momentum:

$$\frac{\partial}{\partial t}(\rho u) + \frac{\partial}{\partial x}(\rho u^2) + \frac{\partial}{\partial y}(\rho u v) = \frac{\partial}{\partial x}(-p + \tau_{xx}) + \frac{\partial}{\partial y}(\tau_{xy}) \tag{2.2}$$

y-Momentum

$$\frac{\partial}{\partial t}(\rho v) + \frac{\partial}{\partial x}(\rho u v) + \frac{\partial}{\partial y}(\rho v^2) = \frac{\partial}{\partial x}(\tau_{xy}) + \frac{\partial}{\partial y}(-p + \tau_{yy}) \tag{2.3}$$

Energy

$$\frac{\partial}{\partial t}(\rho E) + \frac{\partial}{\partial x}(\rho u E) + \frac{\partial}{\partial y}(\rho v E) = \frac{\partial}{\partial x}\left[Q_x + u(\tau_{xx} - p) + v\tau_{xy}\right] + \frac{\partial}{\partial y}\left[Q_y + u\tau_{xy} + v(\tau_{yy} - p)\right] \tag{2.4}$$

where

$$E = c_v T + \frac{1}{2}(u^2 + v^2) \tag{2.5}$$

The shear stresses are related to the rate of strain by

$$\tau_{ij} = + \lambda \frac{\partial u_m}{\partial x_m} \delta_{ij} + \mu \left(\frac{\partial u_i}{\partial x_j} + \frac{\partial u_j}{\partial x_i}\right) \tag{2.6}$$

where λ and μ are viscosity coefficients. The bulk viscosity λ is a quantity which is very difficult to determine, and it is customary to relate it to the viscosity by the Stokes hypothesis:

$$\lambda = -\frac{2}{3}\mu \tag{2.7}$$

The viscosity is a thermodynamic property, which is dependent mainly on the temperature. Various formulae for the viscosity-temperature relations have been published in the literature. A fairly general one is the Sutherland law

$$\frac{\mu}{\mu_0} = \left(\frac{T}{T_0}\right)^{3/2} \frac{T_0 + S}{T + S} \tag{2.8}$$

where μ_0 is the viscosity at temperature T_0.

Application of (2.7) to (2.6) and expansion of the resulting equation yields the following expressions

$$\tau_{xx} = \frac{2}{3}\mu\left(2\frac{\partial u}{\partial x} - \frac{\partial v}{\partial y}\right) \tag{2.9a}$$

$$\tau_{yy} = \frac{2}{3}\mu\left(2\frac{\partial v}{\partial y} - \frac{\partial u}{\partial x}\right) \tag{2.9b}$$

$$\tau_{xy} = \mu\left(\frac{\partial u}{\partial y} + \frac{\partial v}{\partial x}\right) \tag{2.9c}$$

The heat flux Q is related to the temperature gradients by

$$Q_x = k\frac{\partial T}{\partial x} \tag{2.10a}$$

$$Q_y = k\frac{\partial T}{\partial y} \tag{2.10b}$$

$$k = \frac{\mu c_p}{\sigma} \tag{2.11}$$

Equations (2.1), (2.2), (2.3) and (2.4) are the basic governing equations for a two-dimensional compressible laminar flow. The system of equations contains all the auxiliary relations outlined above as well, and it must be supplemented by the equation of state, which is, for an ideal gas

$$p = \rho RT \tag{2.12}$$

2.2 The Governing Equations in a Turbulent Flow

The present approach to the calculation of a turbulent flow field is based on the application of an eddy viscosity hypothesis. Thus, the governing equations remain similar to those applicable to laminar flows, and the numerical method is substantially identical.

A turbulence energy model has been used for the calculation of the eddy viscosity. Details of this model are discussed in the next section. The application of the eddy viscosity hypothesis to the turbulent form of the momentum and energy equations yields the following equations:

$$\frac{\partial}{\partial t}(\rho u) + \frac{\partial}{\partial x}(\rho u^2) + \frac{\partial}{\partial y}(\rho uv) = \frac{\partial}{\partial x}(-p + \tau_{xx} + \tau_{t,xx}) + \frac{\partial}{\partial y}(\tau_{xy} + \tau_{t,xy}) \tag{2.13}$$

$$\frac{\partial}{\partial t}(\rho v) + \frac{\partial}{\partial x}(\rho uv) + \frac{\partial}{\partial y}(\rho v^2) = \frac{\partial}{\partial x}(\tau_{xy} + \tau_{t,xy}) + \frac{\partial}{\partial y}(-p + \tau_{yy} + \tau_{t,yy}) \tag{2.14}$$

$$\frac{\partial}{\partial t}(\rho E) + \frac{\partial}{\partial x}(\rho uE) + \frac{\partial}{\partial y}(\rho vE) = \frac{\partial}{\partial x}\left[Q_x + Q_{t,x} + u(\tau_{xx} - p) + v\tau_{xy} + \Gamma_e\frac{\partial e}{\partial x}\right] + \frac{\partial}{\partial y}\left[Q_y + Q_{t,y} + u\tau_{xy} + v(\tau_{yy} - p) + \Gamma_e\frac{\partial e}{\partial y}\right] \tag{2.15}$$

where

$$\tau_{t,xx} = 2\mu_t \frac{\partial u}{\partial x} + \rho e \tag{2.16a}$$

$$\tau_{t,yy} = 2\mu_t \frac{\partial v}{\partial y} + \rho e \tag{2.16b}$$

$$\tau_{t,xy} = \mu_t \left(\frac{\partial u}{\partial y} + \frac{\partial v}{\partial x}\right) \tag{2.16c}$$

$$Q_{t,x} = \frac{\mu_t c_p}{\sigma_{E,t}} \frac{\partial T}{\partial x} \tag{2.17a}$$

$$Q_{t,y} = \frac{\mu_t c_p}{\sigma_{E,t}} \frac{\partial T}{\partial y} \tag{2.17b}$$

$$e = (\overline{u'^2} + \overline{v'^2} + \overline{w'^2})/2 \tag{2.18a}$$

$$E = c_v T + (u^2 + v^2)/2 + e \tag{2.18b}$$

$$\Gamma_e = \mu + \mu_t/\sigma_{e,t} \tag{2.19}$$

The above equations and definitions are identical with the incompressible form of the equations. No compressibility corrections are included, and therefore such a formulation is suitable only for cases where the density and viscosity do not change significantly over the scales of the corresponding turbulent eddies. The velocity and temperature are the mean values. The turbulent Prandtl numbers $\sigma_{e,t}$ and $\sigma_{E,t}$ are constants which will be specified in the next section, dealing with the determination of the turbulent properties.

2.3 Calculation of the Turbulent Properties

The calculation of the flow in a turbulent recirculation region requires the application of a two parameter turbulence model. Such a model usually accounts for the amplitude of the turbulent fluctuations, represented by the turbulent energy e, and a turbulent length scale Λ. The eddy viscosity is then given by

$$\mu_t = c_\mu \rho \sqrt{e} \Lambda \tag{2.20}$$

where c_μ is a universal function of the Reynolds number of turbulence, Re_t. The turbulent energy and scale are governed by suitable transport equations. These equations are derived from the Navier-Stokes equations (Wolfshtein et al, 1975). Yet, the final form of these equations requires substantial modelling and careful comparison with experimental data. This problem is particularly difficult for the length scale equation. Yet the length scale is somewhat related to the Prandtl mixing length constant, which is specified geometrically for thin shear layers. We have decided to extend the geometrical definition of the length scale to the present case as well, in view of the above discussion. Thus, it is required to add only a single transport equation for the turbulent energy e, as follows:

$$\frac{\partial}{\partial t}(\rho e) + \frac{\partial}{\partial x}(\rho u e) + \frac{\partial}{\partial y}(\rho v e) = \frac{\partial}{\partial x}\left(\Gamma_e \frac{\partial e}{\partial x}\right) + \frac{\partial}{\partial y}\left(\Gamma_e \frac{\partial e}{\partial y}\right) + G_e - D \qquad (2.21)$$

$$G_e = \mu_t \left[\left(\frac{\partial u}{\partial y} + \frac{\partial v}{\partial x}\right)^2 + 2\left(\frac{\partial u}{\partial x}\right)^2 + 2\left(\frac{\partial v}{\partial y}\right)^2\right] \qquad (2.21a)$$

$$D = c_1 \rho \frac{e^{3/2}}{\Lambda} \qquad (2.21b)$$

This equation is very similar to the governing equation for the velocities and energy and will be solved by a similar method. The universal constants and functions used were defined by Wolfshtein (1969) as follows:

$$c_\mu = a_\mu [1 - \exp(-A_\mu \mathrm{Re}_t)] \qquad (2.22a)$$

$$c_1 = a_1 [1 - \exp(-A_1 \mathrm{Re}_t)]^{-1} \qquad (2.22b)$$

$$\mathrm{Re}_t = \rho \sqrt{e}\, \Lambda / \mu \qquad (2.22c)$$

$a_\mu = 0.220$; $A_\mu = 0.016$; $\sigma_{E,t} = 0.9$

$a_1 = 0.416$; $A_1 = 0.263$; $\sigma_{e,t} = 1.53$

The turbulent length scale Λ is specified by the following formulae:

$$\left.\begin{array}{ll} \Lambda = \alpha y & 0 < y \leq \frac{\beta}{\alpha}\delta \\ \Lambda = \beta\delta & \text{elsewhere} \end{array}\right\} \qquad (2.23)$$

The boundary layer thickness δ is estimated by application of the definition that inside the boundary layer the velocity gradient is large, or in a mathematical form,

$$\frac{\partial u}{\partial y} > \frac{1}{\theta}\frac{u}{y} \quad \text{inside the boundary layer.} \qquad (2.24)$$

The necessary constants are as follows:

$\alpha = 1$

$\beta = 0.1$

$\theta = 10.$

3. BOUNDARY CONDITIONS

The boundary conditions are specified on four boundaries. On the solid surface the no-slip condition for the velocity and a temperature boundary condition are applied. Upstream the velocity and temperature profiles are obtained from solution of a laminar or turbulent boundary layer under a compressible flow. Downstream, the flow should reduce again to a boundary layer.

On the main-stream boundary, away from the wall, the flow is non-viscous, and therefore the Euler equations may be used as boundary conditions on this boundary.

The boundary conditions may be specified only when various assumptions and restrictions are specified. In the present paper the following assumptions are made:

(i) Flat plate, with zero pressure gradient everywhere apart from the region adjacent to impingement point of the shock wave onto the boundary layer.

(ii) Adiabatic wall.

(iii) In the boundary layer upstream of the interaction region the temperature profile is related to the velocity profile by the Crocco integral.

The boundary conditions may be specified now on all boundaries. This will be done in the next sections.

3.1 Solid Wall Boundary Conditions

On a solid wall parallel to the x axis, the parallel velocity vanishes

$$u_w = 0$$

The normal velocity must be specified. When the wall is impervious the normal velocity vanishes as well. The wall temperature or heat flux must be specified. For an adiabatic wall the normal temperature gradient vanishes:

$$\left.\frac{\partial T}{\partial y}\right|_w = 0 \tag{3.1}$$

These boundary conditions are sufficient to prescribe the solution. In particular, the density, which is governed by the first order continuity equation does not require any specification on the wall, as the solution of this equation extends to the wall. Thus, on a solid impermeable wall this equation reduces to

$$\left.\frac{\partial \rho}{\partial t}\right|_w = -\rho\left.\frac{\partial v}{\partial y}\right|_w \tag{3.2}$$

However, the numerical solution of this equation is not convenient, as it is unstable, and in particular it cannot be used for steady state solutions. An alternative method for the calculation of the wall density is by the equation of state (2.12). In this case the wall pressure must be approximated. Usually this is done by the following condition

$$\left.\frac{\partial p}{\partial y}\right|_w = 0 \tag{3.3}$$

This condition is exact if a boundary layer exists on the wall. It should be emphasized again that this condition is simply a prescription of the wall pressure and density, but it is not a boundary condition for a partial differential equation.

All the above conditions apply to laminar and turbulent flows. If the flow is turbulent the wall turbulent energy is given by

$$e_w = 0 \tag{3.4}$$

3.2 Upstream Boundary Conditions

Upstream of the shock wave a regular boundary layer under a compressible main stream prevails. Therefore the boundary conditions upstream of the shock wave must be the outcome of a calculation of this compressible boundary layer. An exact solution of the boundary layer is expensive. As the solution of the problem is numerical, it is sufficient to prescribe an approximate upstream boundary condition. Such a solution may be easily obtained by the integral boundary layer method, if the Howarth-Dorodnytzyn mass-weighted cross-stream distance, η, is used instead of the physical distance normal to the wall, y. Application of the method requires specification of a velocity profile in terms of η.

For a compressible boundary layer the density (and temperature) must be prescribed as well. The Crocco integral allows the following algebraic relation between the temperature and the velocity:

$$\frac{T}{T_0} = \frac{\rho_0}{\rho} = 1 + r\,\frac{\gamma-1}{2}\,M_0^2\left(1 - \frac{u^2}{u_0^2}\right) \tag{3.5}$$

where the recovery factor, r, is given by

$$r = \sigma^n \tag{3.6}$$

and $n = 1/2$ for laminar flow, $n = 1/3$ for turbulent flow. The Crocco integral is exact for an adiabatic, laminar flow of fluid with Prandtl number of unity, at a zero pressure gradient. It is a good approximation for the present case as well, as has been discussed by van-Driest (1951).

The velocity and turbulent energy profiles must be specified in different forms for laminar or turbulent flows. The velocity profile is easily prescribed in terms of the normalized coordinate, η. However, the physical coordinate y may be calculated only when the velocity has been calculated. Finally, the turbulent energy must be calculated as well. Unless transition calculations are attempted, this quantity is zero for the laminar case.

The various profiles were chosen from the literature and they are similar to those usually used for incompressible flows, but in terms of η rather than y. They are presented in the next two sections for the laminar and turbulent case respectively. Details are given in Appendix A, and only the main results are given here.

3.2.1 Laminar Flow In this case the velocity is given by the Karman-Pohlhausen profile

$$\frac{u}{u_0} = 2\eta - 2\eta^3 + \eta^4 \tag{3.7}$$

where η is a normalized cross-stream coordinate given by

$$\frac{y}{\delta_1} = \eta + r\,\frac{\gamma-1}{2}\,M_0^2\,\eta\left(1 - \frac{4}{3}\eta^2 + \frac{8}{5}\eta^4 - \frac{2}{3}\eta^5 - \frac{4}{7}\eta^6 + \frac{1}{2}\eta^7 - \frac{1}{9}\eta^8\right) \tag{3.8}$$

The normalizing boundary layer thickness δ_1 is related to the actual boundary layer thickness δ and to the distance from the leading edge by

$$\frac{\delta}{\delta_1} = 1 + \frac{263}{630}\,r\,\frac{\gamma-1}{2}\,M_0^2 \tag{3.9}$$

$$\frac{\delta_1}{x} = 5.836\left[\frac{\left(1 + r\,\frac{\gamma-1}{2}\,M_0^2\right)^{1/2}\left(1 + (\gamma-1)M_0^2\,\frac{c_p S}{u_0^2}\right)}{1 + (\gamma-1)M_0^2\left(\frac{r}{2} + \frac{c_p S}{u_0^2}\right)}\left(\frac{\mu_0}{\rho_0 u_0 x}\right)\right]^{1/2} \tag{3.10}$$

In this case, the turbulent energy is zero and its governing equation is redundant. The skin friction is given by

$$c_f = 4\mu_w\rho_w/(\rho_0^2 u_0 \delta_1) \tag{3.11}$$

3.2.2 Turbulent Flow In this case equations (3.7)-(3.11) are not suitable, and should be replaced by a corresponding turbulent set of equations. Such a set is proposed in Appendix A. The main results are given below. The velocity is given by

$$\frac{u}{u_0} = \eta^{1/7} \tag{3.12}$$

The cross-stream distance is given by:

$$\frac{y}{\delta_1} = \eta + r\,\frac{\gamma-1}{2}\,M_0^2\left(\eta - \frac{7}{9}\,\eta^{9/7}\right) \tag{3.13}$$

and the boundary layer thickness is given by

$$\frac{\delta}{\delta_1} = 1 + \frac{2}{9}\,r\,\frac{\gamma-1}{2}\,M_0^2 \tag{3.14}$$

$$\frac{\delta_1}{x} = 0.367\left(1 + r\,\frac{\gamma-1}{2}\,M_0^2\right)^{-0.3}\left[\frac{1 + (\gamma-1)M_0^2\,\dfrac{c_pS}{u_0^2}}{1 + (\gamma-1)M_0^2\left(\dfrac{r}{2} + \dfrac{c_pS}{u_0^2}\right)}\;\frac{\mu_0}{\rho_0 u_0 x}\right]^{0.2} \tag{3.15}$$

The skin friction is given by

$$c_f = 0.0444\,\frac{\rho_w}{\rho_0}\left(\frac{\mu_w}{\rho_w u_0 \delta_1}\right)^{0.25} \tag{3.16}$$

The turbulent energy profile is discussed in Appendix B. It is summarized by the following equations:

$$\frac{e}{e_{max}} = \left(\frac{\eta}{\eta_{max}}\right)^2 \qquad 0 \le \eta \le \eta_{max}$$

$$\frac{e}{e_{max}} = \frac{1-\eta}{1-\eta_{max}} + \frac{e_0}{e_{max}}\,\frac{\eta-\eta_{max}}{1-\eta_{max}} \qquad \eta_{max} \le \eta \le 1 \tag{3.17}$$

$$e_{max} = 3\,\frac{c_f}{2}\,\frac{\rho_0}{\rho_w}\,u_0^2 \tag{3.18}$$

$$\eta_{max} = \left(60\,\frac{2}{c_f}\,\frac{\rho_0}{\rho_w}\right)^{1/2}\frac{\mu_w}{\delta_1 u_0 \rho_0} \tag{3.19}$$

3.3 The Main Stream Boundary

Further away from the wall the viscous effects become negligible, and the flow is governed by the Euler equations. If the outer boundary of the computational region is placed sufficiently far from the wall, then a simple external, non-viscous, solution is usually available. In the present case this solution is simply that of three uniform property fields separated from one another by the impinging and reflected shocks. Unfortunately such a simple definition of the boundary conditions requires a large computational domain, and many finite difference mesh points. Such a mesh covers an extensive non-viscous supersonic region in which various compression and expansion waves prevail, and the flow is non-homogeneous. Yet, the detailed understanding of this region is not important usually, and therefore we need an approximation by which this non-homogeneous non-viscous region may be left out of the computational domain.

The Euler equations governing the external non-viscous flow are hyperbolic and in their steady state form they may be solved by the method of characteristics. Two assumptions facilitate a simple such solution:

(i) The flow properties are uniform along the downward facing characteristic.

(ii) The characteristic is approximately straight.

The angle between the characteristic and the local stream lines is given by

$$\operatorname{tg}\mu = (M^2-1)^{-1/2} \tag{3.20}$$

The angle between the local stream lines and the x direction is given by

$$\operatorname{tg}\theta = v/u \tag{3.21}$$

Therefore the boundary condition is

$$\frac{\partial\phi}{\partial x} + \frac{\partial\phi}{\partial y}\operatorname{tg}(\mu+\theta) = 0 \tag{3.22}$$

where ϕ stands for ρ, u, v, E, and e.

This equation is applicable in all the external region, apart from the shock wave.

3.4 The Shock Relations

An oblique shock wave is defined by two parameters. In the present case the convenient parameters are M_1, the Mach number upstream of the shock, and the pressure ratio $s = p_2/p_1$. All the other parameters are related to these two ones. For instance the shock angle relative to the local stream lines, σ, is given by

$$M_1^2 \operatorname{Sin}^2\sigma = \frac{\gamma+1}{2\gamma}s + \frac{\gamma-1}{2\gamma} \tag{3.23}$$

Other properties are related to σ and to α, the local stream line angle upstream of the shock given by

$$\operatorname{tg}\alpha = u_1/v_1 \tag{3.24}$$

$$u_2 = t^{-1}[u_1(t + 2\operatorname{Sin}^2\sigma) + v_1 \operatorname{Sin} 2\sigma] \tag{3.25}$$

$$v_2 = t^{-1}[u_1(t+1)\operatorname{Sin} 2\sigma + v_1(t \operatorname{Cos} 2\sigma + 2\operatorname{Cos}^2\sigma)] \tag{3.26}$$

$$T_2 = T_1 \frac{2(\gamma-1)^3 M_1^2}{(\gamma+1)^2 t^2}\left[1 + \frac{t^2}{2(\gamma-1)M_1^2}\right]\left[\frac{2\gamma t}{(\gamma-1)^2} - 1\right] \tag{3.27}$$

where $t = (\gamma-1)M_1^2 \operatorname{Sin}^2(\sigma+\alpha)$ (3.28)

$$p_2 = s p_1 \tag{3.29}$$

$$\rho_2 = \frac{p_2}{RT_2} \tag{3.30}$$

It should be noted that the properties at "1" upstream of the shock are dependent on the flow field, and thus the properties of the flow on both sides of the shock form a part of the solution. Moreover, the specified pressure ratio $s = p_2/p_1$ is the pressure ratio across the shock only. The total pressure ratio across the entire shock wave system is different from s and is not known in advance.

3.5 The Downstream Boundary

Downstream of the shock impingement boundary layer flow is re-established. Physically it means that no downstream influence is spread from this region in the upstream direction. This fact may be approximated by the condition that x-gradients of all quantities vanish in this region.

3.6 Initial Conditions

Theoretically the initial condition should not influence the steady state which is obtained after marching of the solution in the time direction for very long times. Practically it is advantageous to choose an initial condition which resembles the steady state as much as possible. Such a solution may be obtained by spreading the upstream condition along the entire field. Even better approximation is obtained if the recirculation region is approximated by a small triangular region of low velocity below the shock.

4. NON-DIMENSIONAL PRESENTATION

The normalizing quantities are those of the undisturbed free stream upstream of the impinging shock system, denoted by subscript "0", namely u_0, ρ_0, μ_0. The normalizing distance, x_S, was chosen as the distance from the leading edge to the intersection of the continuation of the straight undisturbed shock with the surface. With these normalizing quantities the following dimensionless groups may be defined:

$$\left.\begin{aligned} &\hat{x}_i = x_i/x_S \;;\quad \hat{t} = u_0 t/x_S \\ &\hat{u}_i = u_i/u_0 \quad \hat{T} = c_p T/u_0^2 \quad \hat{E} = E/u_0^2 \quad \hat{S} = c_p S/u_0^2 \\ &\hat{\rho} = \rho/\rho_0 \quad \hat{p} = p/\rho_0 u_0^2 \\ &\hat{\mu} = \mu/\mu_0 \\ &\hat{\tau}_{ij} = \tau_{ij}/\rho_0 u_0^2 \quad \hat{Q}_i = Q_i/\rho_0 u_0^3 \end{aligned}\right\} \qquad (4.1)$$

The definitions of the usual dimensionless quantities are:

$$\left.\begin{aligned} &\mathrm{Re} = \rho_0 u_0 x_S/\mu_0 \\ &\sigma = \mu c_p/k \\ &M = \hat{u}^2/(\gamma-1)\hat{T} \end{aligned}\right\} \qquad (4.2)$$

and the equation of state is

$$\hat{p} = \frac{\gamma}{\gamma-1}\,\hat{\rho}\,\hat{T} \tag{4.3}$$

From this point and on all non-denoted quantities are non-dimensional.

The non-dimensional governing equations will be presented in a matrix form, as follows

$$\frac{\partial W}{\partial t} = -\frac{\partial}{\partial x}(uW) - \frac{\partial}{\partial y}(vW) + \frac{\partial}{\partial x}\left[F\,\frac{\partial}{\partial x}\left(\frac{W}{\rho}\right)\right] + $$

$$+ \frac{\partial}{\partial y}\left[G\,\frac{\partial}{\partial y}\left(\frac{W}{\rho}\right)\right] + \frac{S_1}{\mathrm{Re}} + S_2 \tag{4.4}$$

where

$$W = \begin{bmatrix} \rho \\ \rho u \\ \rho v \\ \rho E \\ \rho e \end{bmatrix} \tag{4.5}$$

$$\mathrm{Re}\,F = \begin{bmatrix} 0 & 0 & 0 & 0 & 0 \\ 0 & \frac{4}{3}\mu + 2\mu_t & 0 & 0 & 0 \\ 0 & 0 & \mu+\mu_t & 0 & 0 \\ 0 & 0 & 0 & \gamma\left(\frac{\mu}{\sigma_E} + \frac{\mu_t}{\sigma_{E,t}}\right) & 0 \\ 0 & 0 & 0 & 0 & \mu + \frac{\mu_t}{\sigma_{e,t}} \end{bmatrix}$$

$$\mathrm{Re}\; G = \begin{bmatrix} 0 & 0 & 0 & 0 & 0 \\ 0 & \mu+\mu_t & 0 & 0 & 0 \\ 0 & 0 & \frac{4}{3}\mu+2\mu_t & 0 & 0 \\ 0 & 0 & 0 & \gamma\left(\frac{\mu}{\sigma_E}+\frac{\mu_t}{\sigma_{E,t}}\right) & 0 \\ 0 & 0 & 0 & 0 & \mu+\frac{\mu_t}{\sigma_{e,t}} \end{bmatrix} \tag{4.7}$$

$$S_1 = \begin{bmatrix} 0 \\ -\frac{2}{3}\frac{\partial}{\partial x}\left(\mu\frac{\partial v}{\partial y}\right) - \frac{2}{3}\frac{\partial}{\partial x}(\rho e) + \frac{\partial}{\partial y}\left[(\mu+\mu_t)\frac{\partial v}{\partial x}\right] \\ \frac{\partial}{\partial x}\left[(\mu+\mu_t)\frac{\partial u}{\partial y}\right] - \frac{2}{3}\frac{\partial}{\partial y}\left(\mu\frac{\partial u}{\partial x}\right) - \frac{2}{3}\frac{\partial}{\partial y}(\rho e) \\ \begin{gathered} \frac{\partial}{\partial x}\left[\left(\mu+\frac{\mu_t}{\sigma_{e,t}}\right)\frac{\partial e}{\partial x}\right] + \frac{\partial}{\partial x}\left[u\tau_{xx}+v\tau_{xy}\right] + \\ + \frac{\partial}{\partial y}\left[\left(\mu+\frac{\mu_t}{\sigma_{e,t}}\right)\frac{\partial e}{\partial y}\right] + \frac{\partial}{\partial y}\left[u\tau_{xy}+v\tau_{yy}\right] - \\ - \gamma\frac{\partial}{\partial x}\left\{\left(\frac{\mu}{\sigma_E}+\frac{\mu_t}{\sigma_{E,t}}\right)\frac{\partial}{\partial x}\left(\frac{1}{2}u^2+\frac{1}{2}v^2+e\right)\right\} - \\ - \gamma\frac{\partial}{\partial y}\left\{\left(\frac{\mu}{\sigma_E}+\frac{\mu_t}{\sigma_{E,t}}\right)\frac{\partial}{\partial y}\left(\frac{1}{2}u^2+\frac{1}{2}v^2+e\right)\right\} \end{gathered} \\ 0 \end{bmatrix} \tag{4.8}$$

$$S = \begin{bmatrix} 0 \\ \hline -\dfrac{\partial p}{\partial x} \\ \hline -\dfrac{\partial p}{\partial y} \\ \hline -\dfrac{\partial}{\partial x}(up) - \dfrac{\partial}{\partial y}(vp) \\ \\ \hline G_e - D \end{bmatrix} \tag{4.9}$$

$$\left.\begin{aligned} \mathrm{Re}\,\tau_{ij} &= \partial u_i/\partial x_j + \partial u_j/\partial x_i - \tfrac{2}{3}\delta_{ij}\,\partial u_m/\partial x_m \\ \mathrm{Re}\,G_e &= \mu_t\,[\,2(\partial u/\partial x)^2 + 2(\partial v/\partial y)^2 + (\partial u/\partial y + \partial v/\partial x)^2] \\ D &= c_1\,\rho\,e^{3/2}/\Lambda \\ \mu_t &= \mathrm{Re}\,\rho\,e^{1/2}\,\Lambda \end{aligned}\right\} \tag{4.10}$$

The non-dimensional boundary conditions are easily derived and need not be specified here, with the exception of the boundary layer thickness at x_0 which may be expressed as

$$\delta_0 = k_1\,x_0^{\,1-q} \tag{4.11}$$

For laminar flow

$$\left.\begin{aligned} k_1 &= 5.836\,\frac{(1+a^2)^{1/4}\;b(1+263\,a^2/630)}{(a^2+b^2)^{1/2}\,\mathrm{Re}^{1/2}} \\ q &= 0.5 \end{aligned}\right\} \tag{4.12}$$

For turbulent flow

$$\left.\begin{aligned} k_1 &= 0.367\,\frac{(1+a^2)^{1/2}\;b^{0.4}\,(1+2a^2/9)}{(a^2+b^2)^{0.2}\;\mathrm{Re}^{0.2}} \\ q &= 0.2 \end{aligned}\right\} \tag{4.13}$$

where

$$\left.\begin{aligned} a^2 &= r\,\frac{\gamma-1}{2}\,M_0^2 \\ b^2 &= 1 + S(\gamma-1)M_0^2 \end{aligned}\right\} \qquad (4.14)$$

5. THE NUMERICAL METHOD

5.1 The Computational Region

The computational region should include the region where significant changes occur. On the other hand, too large a computational region results in waste of computer resources. Therefore it is preferable to limit the computational domain to the region where the boundary layer thickens, and the boundary layer equations do not apply.

Figure 2 shows a typical computational field. A field geometry which satisfies the above requirements is given by:

$$l_x = 150\ \delta_0 \qquad (5.1a)$$

$$l_y = 3\ \delta_0 \qquad (5.1b)$$

$$x_0 = x_s - l_x/2 \qquad (5.1c)$$

These three equations must be solved for the four variables, l_x, l_y, x_0, δ_0. The fourth equation is the relation

$$\delta_0 = k_1\ x_0^{1-q} \qquad (5.1d)$$

where k_1 and q are given in (4.12) or (4.13). By substitution of (5.1d) and (5.1a) in (5.1c) the governing equation for x_0 is obtained.

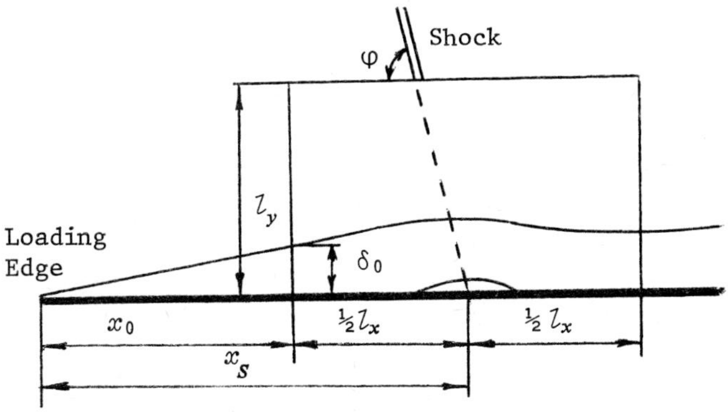

Figure 2. The Computational Region.

$$75\ k_1\ x_0^{1-q} + x_0 - x_S = 0 \tag{5.2}$$

where in non-dimensional form $x_S = 1$.

This equation is solved numerically for x_0. The mesh size must be sufficiently fine as to allow a reasonable resolution inside the boundary layer, and to give a reasonable aspect ratio to the mesh cells. A suitable definition is

$$\left.\begin{array}{l} \delta_0/h_y \geq 8 \\ h_x/h_y \leq 16 \end{array}\right\} \tag{5.3}$$

Combining these requirements with (5.1) the smallest possible mesh is about 75 points in the x-direction and 25 points in the y-direction.

5.2 Finite Difference Formulation

The finite differences grid is shown in Figure 3, where the mesh size is h_x and h_y in directions x and y respectively.

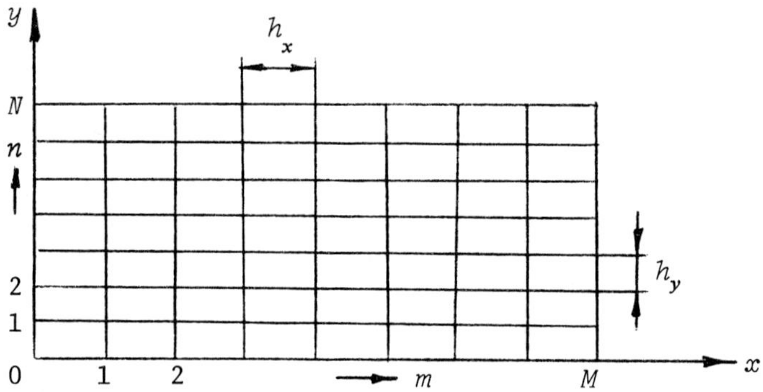

Figure 3. Finite Differences Grid.

The values of the independent variables in the solution field are:

$$x = m\,h_x$$

$$y = n\,h_y$$

$$t = k\,h_t$$

and the dependent variables are

$$W(x,y,t) = W^k_{m,n}$$

Inside the solution field central differences are used to calculate the derivatives (with second order accuracy), as follows:

$$\frac{\partial}{\partial x}(FW) = \frac{1}{2h_x}\left(F_{m+1,n}W_{m+1,n} - F_{m-1,n}W_{m-1,n}\right) + O(h_x^2) \qquad (5.3)$$

$$\frac{\partial}{\partial x}\left[F\frac{\partial}{\partial x}(TW)\right] = \frac{1}{2h_x^2}\left[\left(F_{m+1,n} + F_{m,n}\right)\circ\right.$$
$$\circ\left(T_{m+1,n}W_{m+1,n} - T_{m,n}W_{m,n}\right) + \left(F_{m-1,n} + F_{m,n}\right)\cdot$$
$$\left.\cdot\left(T_{m-1,n}W_{m-1,n} - T_{m,n}W_{m,n}\right)\right] + O(h_x^2) \qquad (5.4)$$

$$\frac{\partial}{\partial x}\left(F\frac{\partial}{\partial y}W\right) = \frac{1}{4h_xh_y}\left[F_{m+1,n}\left(W_{m+1,n+1} - W_{m+1,n-1}\right) -\right.$$
$$\left. - F_{m-1,n}\left(W_{m-1,n+1} - W_{m-1,n-1}\right)\right] + O(h_xh_y) \qquad (5.5)$$

5.3 Application of the ADI Method to the Problem

The finite difference form of equation (4.4) is presented in the following form

$$\frac{\partial W}{\partial t} = (A_x + A_y)\,W + S_D + O(h^2) \qquad (5.6)$$

where A_x and A_y are matrix operators representing derivatives, and S_D is a finite difference form of S_1 and S_2. Equation (5.6) is solved by the Samarskii and Andreev (1963) ADI method as described by Isenberg and de Vahl Davis (1975):

$$\left[I - \frac{h_t}{2}A_x\right]Q = [A_x + A_y]W^k + S_D^k \qquad (5.7a)$$

$$\left[I - \frac{h_t}{2}A_y\right]Q^* = Q \qquad (5.7b)$$

$$W^{k+1} = W^k + h_t\,Q^* \qquad (5.7c)$$

where I denotes the unit matrix, and A_x and A_y are given by

$$A_xW = \frac{1}{2}\left\{\left[\frac{1}{h_x}u_{m-1,n} + \frac{1}{\rho_{m-1,n}h_x^2}\left(F_{m-1,n} + F_{m,n}\right)\right]\cdot W_{m-1,n} -\right.$$
$$- \frac{1}{\rho_{m,n}h_x^2}\left[F_{m-1,n} + 2\cdot F_{m,n} + F_{m+1,n}\right]\cdot W_{m,n} +$$
$$\left. + \left[-\frac{1}{h_x}u_{m+1,n} + \frac{1}{\rho_{m+1,n}h_x^2}\left(F_{m+1,n} + F_{m,n}\right)\right]\cdot W_{m+1,n}\right\}$$

(5.8a)

$$A_y W = \frac{1}{2}\left\{\left[\frac{1}{h_y} v_{m,n-1} + \frac{1}{\rho_{m,n-1} h_y^2}\left(G_{m,n-1} + G_{m,n}\right)\right] \cdot W_{m,n-1} - \right.$$
$$- \frac{1}{\rho_{m,n} h_y^2}\left[G_{m,n-1} + 2\ G_{m,n} + G_{m+1,n}\right] \cdot W_{m,n} +$$
$$\left. + \left[- \frac{1}{h_y^2} v_{m,n+1} + \frac{1}{\rho_{m,n+1} h_y^2}\left(G_{m,n+1} + G_{m,n}\right)\right] \cdot W_{m,n+1}\right\} \quad (5.8b)$$

The solution method consists of the following steps:

1. Solution of Equation (5.7a) in the x-direction for each row successively and for each variable, by a standard tri-diagonal matrix algorithm.
2. Solution of Equation (5.7b) in the y-direction for each column successively and for each variable, by a standard tri-diagonal matrix algorithm.
3. Calculation of the new value of W, by Equation (5.7c).

The steady state solution is the asymptotic solution for long times.

5.4 Assessment of the Problem

The matrix equation (5.6) represents five partial differential equations. In this form the equations are apparently uncoupled: the matrices F and G forming A_x and A_y are diagonal matrices and therefore each equation is solved for one unknown. Yet, coupling between the equations is caused by non-linear mixed coefficients, and by the source term, S_D. Therefore the matrices F and G, the vectors S_1 and S_2, and the density, ρ, are calculated from W^k before W^{k+1} is calculated.

ADI solution was possible because the mixed second derivatives in the viscous terms were removed from the A_x and A_y matrices, to form the vector S_1. This essential step is possible only when the mixed derivatives are not too large.

6. RESULTS

The calculations were performed on the Technion IBM 370/168 computer for the following cases:

No.	M_0	Re × 10^7	P_3/P_0	x_s (cm)
1	2.96	1.2	2	0.174
2	2.96	1.2	3	0.174
		1.29		
3	2.96	1.2	4	0.174
		1.38		
4	2	0.03	1.4	0.645
5	3	0.03	1.4	0.430
6	4	0.03	1.4	0.323

Usually the calculations were performed for turbulent flow, although in case 4 a laminar solution was obtained as well. The computer time required for a turbulent solution with 76 × 25 mesh points was 4.2 seconds per time-step. The computer time may probably be shortened by program optimization. In order to examine the approach towards a steady state, the following convergence coefficients for u, e, and T were defined:

$$\text{SRSU} = \left(\sum_{m=1}^{M} \sum_{n=1}^{N} \frac{\left(u_{m,n}^{k+1} - u_{m,n}^{k}\right)^2}{MN\left(\max u_{i,j}^{k+1}\right)^2} \right)^{1/2} \tag{6.1}$$

$$\text{SRSET} = \left(\sum_{m=1}^{M} \sum_{n=1}^{N} \frac{\left(e_{m,n}^{k+1} - e_{m,n}^{k}\right)^2}{MN\left(\max e_{i,j}^{k+1}\right)^2} \right)^{1/2} \tag{6.2}$$

$$\text{SRST} = \left(\sum_{m=1}^{M} \sum_{n=1}^{N} \frac{\left(T_{m,n}^{k+1} - T_{m,n}^{k}\right)^2}{MN\left(\max T_{i,j}^{k+1}\right)^2} \right)^{1/2} \tag{6.3}$$

These coefficients are shown in Figure 4, for the standard case of $M_0 = 2.96$, $\text{Re} = 1.2 \times 10^7$, $p_3/p_0 = 3$. It may be seen that all the coefficients approach some low value asymptotically. After 100 time-steps the r.m.s. variation between consecutive time-steps relative to the maximum value of the corresponding variable is about 5% for u, 7% for T, and 9% for e. Hanin et al (1978) suggested that these results can not be significantly improved without mesh refinement. Therefore, it was decided to use only 100 time-steps per run in all the parametric studies to be described below.

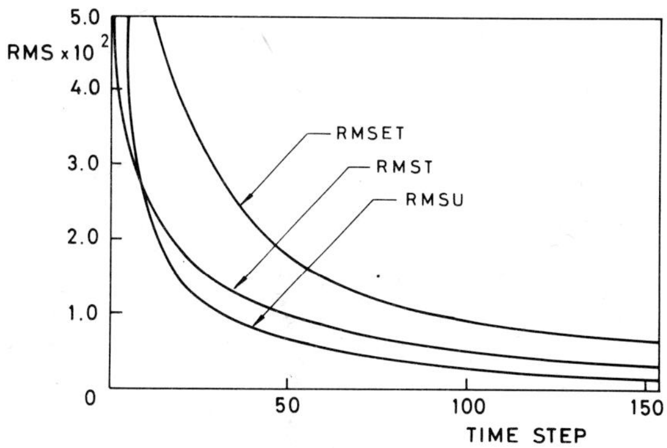

Figure 4. Convergence Coefficients for $M_0 = 2.96$, $\text{Re} = 1.2 \times 10^7$, $p_3/p_0 = 3$.

The accuracy and reliability of the results were tested by comparison with the experimental data of Hakkinen et al (1959) and Law (1976), and with the theoretical solutions of Shang et al (1976). These comparisons are shown in Figures 5, 6 and 7 for a weak, medium and strong shocks. In Figure 5 the computed results compare favourably with the experimental data of Hakkinen et al (1959). In the same figure the difference between the computed laminar and turbulent flow is shown as well. Apparently, in this case the difference between these two cases is not very large.

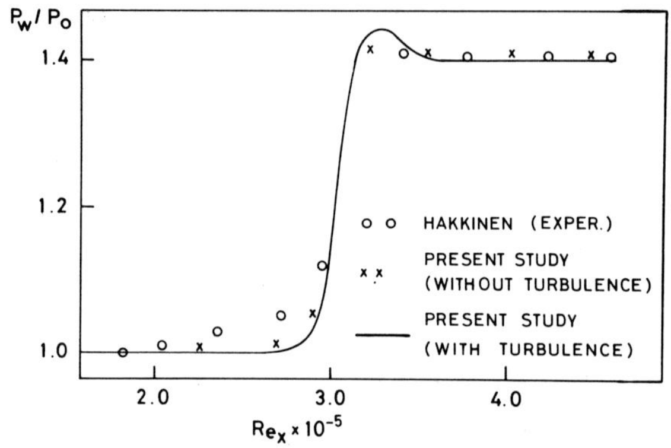

Figure 5. The Wall Pressure Variation, $M_0 = 2$, Re $= 3 \times 10^5$, $p_3/p_0 = 1.4$.

In Figures 6 and 7 the present computed results for medium and strong shock are compared with Law's experimental data and with Shang et al's computations. The agreement with the experiments is as good as Shang et al's, and sometimes even better in most areas.

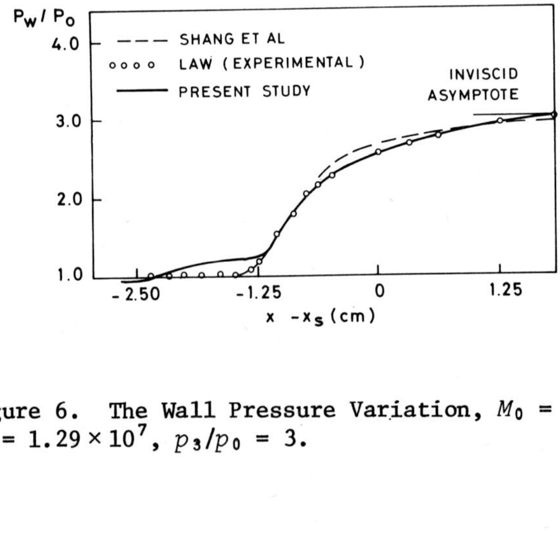

Figure 6. The Wall Pressure Variation, $M_0 = 2.96$, Re $= 1.29 \times 10^7$, $p_3/p_0 = 3$.

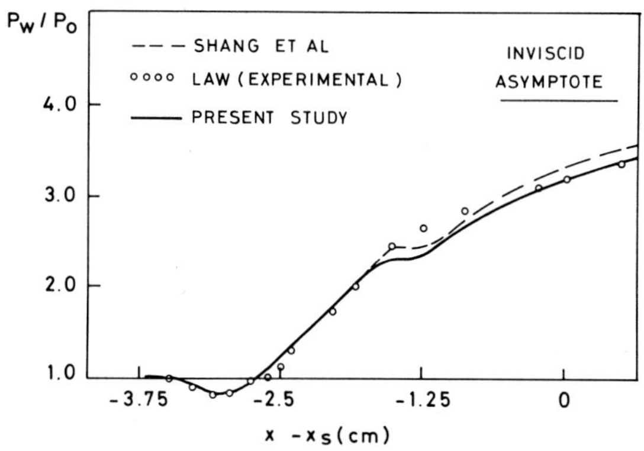

Figure 7. The Wall Pressure Variation, M_0 = 2.96, Re = 1.38×10^7, p_3/p_0 = 4.

The wall pressure distribution for various overall pressure ratios is shown in Figure 8. The wall pressure growth is clearly dependent on the pressure ratio, or, in other words, on the shock strength. On the other hand, variation of the Mach number upstream of the shock waves system, which is shown in Figure 9 does not appear to have much influence on the shock structure.

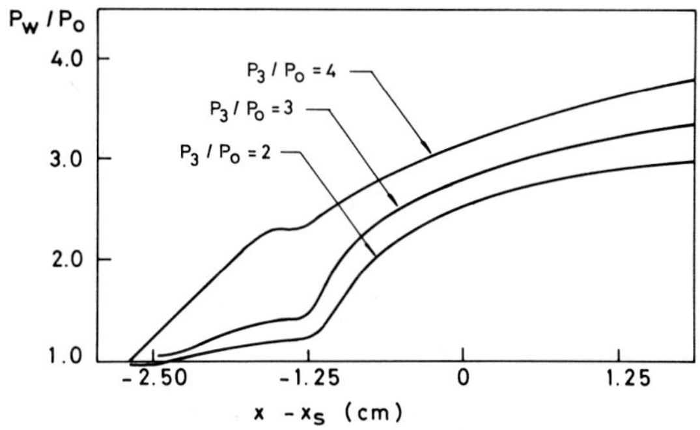

Figure 8. The Wall Pressure Variation, M_0 = 2.96, Re = 1.2×10^7, Various p_3/p_0.

For a better understanding of the flow field, attention must be directed towards the full distribution of the important quantities. The stream lines are shown in Figure 10. The recirculating bubble is clearly seen, as well as the widening of the shear layer below the shock wave.

The isobars are shown in Figure 11. The impinging and reflected shocks are clearly seen. It should be noted that the figure is distorted as the scale in the y-direction is five times smaller than the scale in the x-direction.

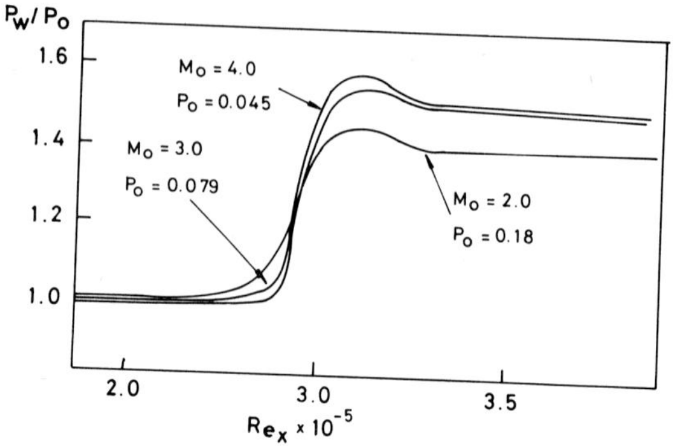

Figure 9. The Wall Pressure, various M_0, Re = 3×10^5, $p_3/p_0 = 1.4$.

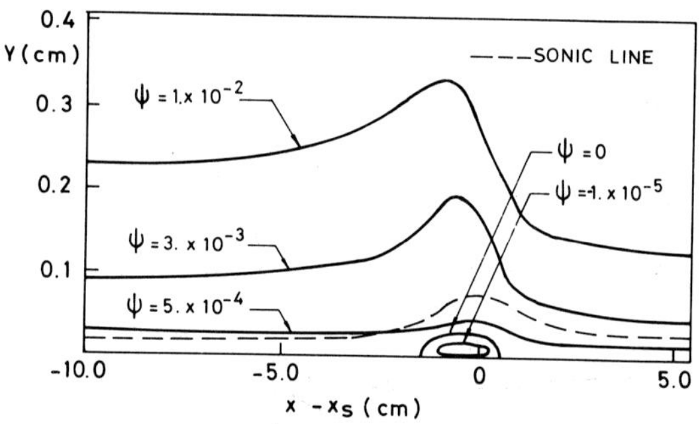

Figure 10. The Stream Lines, $M_0 = 2.96$, Re = 1.2×10^7, $p_3/p_0 = 3$.

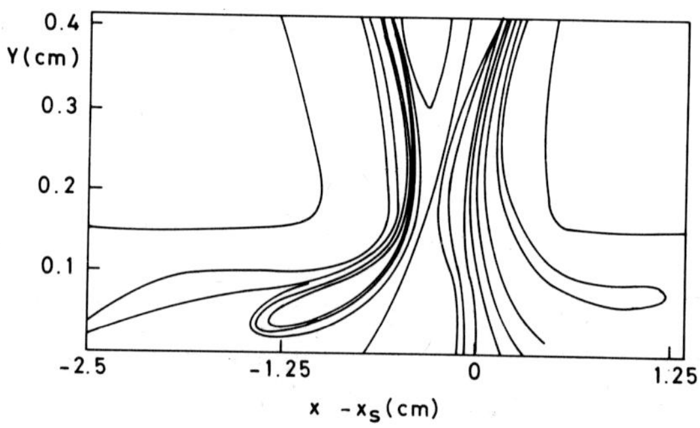

Figure 11. Isobars, $M_0 = 2.96$, Re = 1.2×10^7, $p_3/p_0 = 3$.

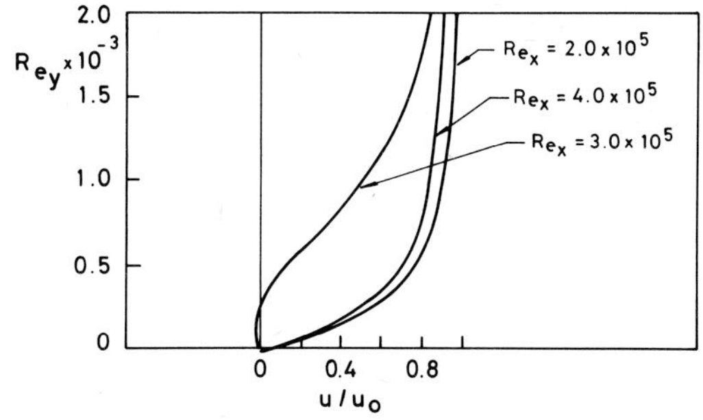

Figure 12. Velocity Profiles, M_0 = 2, Re = 3×10^5, p_3/p_0 = 1.4.

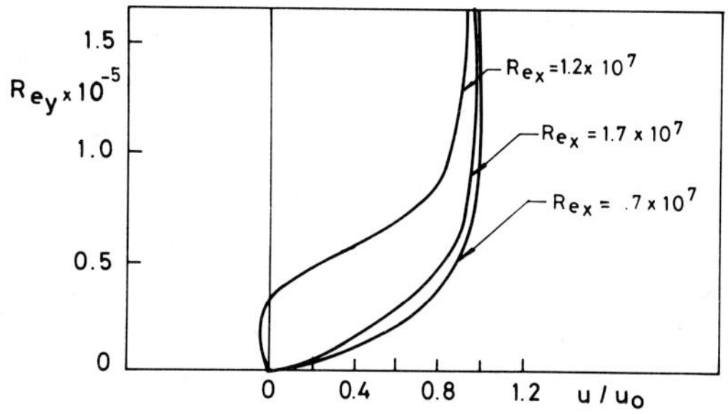

Figure 13. Velocity Profiles, M_0 = 2.96, Re = 1.2×10^7, p_3/p_0 = 3.

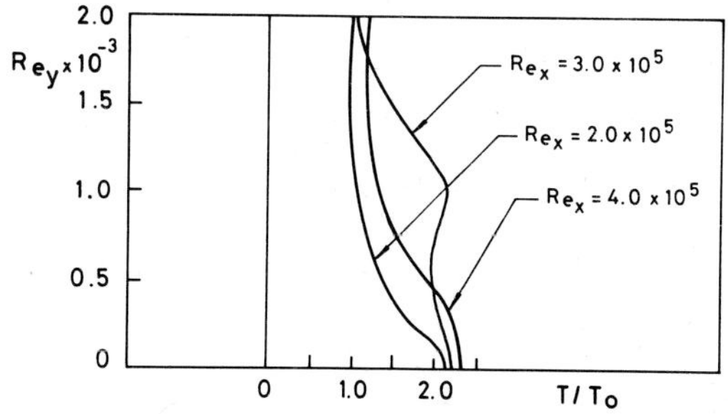

Figure 14. Temperature Profile, M_0 = 2, Re = 3×10^5, p_3/p_0 = 1.4.

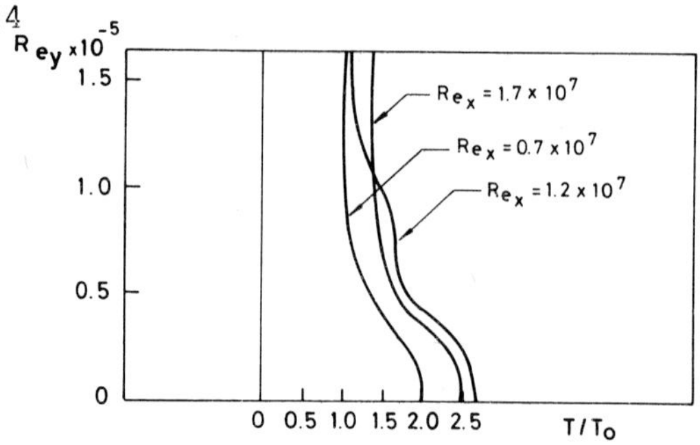

Figure 15. Temperature Profiles, $M_0 = 2.96$, Re $= 1.2 \times 10^7$, $p_3/p_0 = 3$.

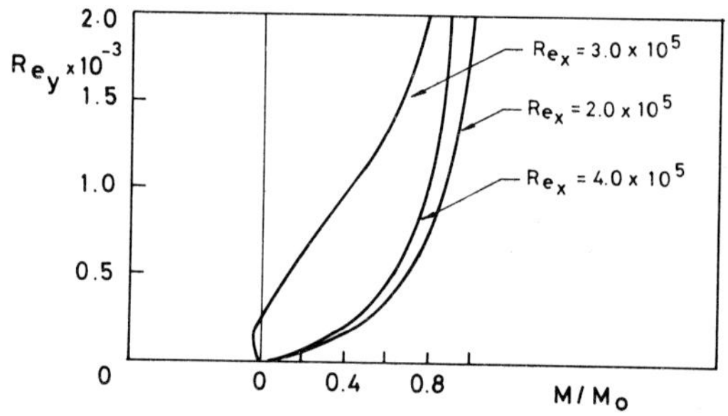

Figure 16. Mach Number Profiles, $M_0 = 2$, Re $= 3 \times 10^5$, $p_3/p_0 = 1.4$.

Finally, velocity, temperature and Mach number profiles are shown in Figures 12, 13, 14, 15, 16. The velocity reversal near the wall below the shock is seen to increase for a stronger shock, while the velocity profiles in other regions are fairly similar to one another. The temperature profiles are more complex. The recirculation bubble is warmer than the rest of the flow, as the convection is weak inside it. Otherwise the temperature profiles are typical to those formed by aerodynamic heating on an adiabatic wall. The last figure shows Mach number profiles. These are very similar to the velocity profiles.

7. DISCUSSION AND CONCLUSIONS

7.1 Formulation of the Problem

The problem of shock wave interaction with a laminar or turbulent boundary layer is formulated. The main features of this formulation are:

1. A turbulence energy model is used to calculate the eddy viscosity. The turbulence energy equation is not corrected for compressibility effects.

2. The turbulent length scale is algebraically prescribed.
3. Upstream boundary conditions are calculated by a simple integral method. Although the present paper describes flat plate flow only, the integral method may be easily extended to flows with pressure gradients.
4. The application of an approximate solution by characteristics for the Euler equations in the outer part of the flow enables a significant reduction of the computational region.

7.2 The Numerical Method

An ADI method is used to solve the system of differential equations, governing compressible two-dimensional turbulent flow. The main features of this method are:

1. For linear differential equations this method is unconditionally stable for any time-step. For the present non-linear equations it is still very stable.

2. This method is programmed more easily than the other methods (MacCormack, Brailovskaya). Moreover, calculation of intermediate values of the variables on the boundaries is not required.

3. As the stability of the present method is high, it is possible to use larger time-steps and to reach the steady state more quickly. Typically, MacCormack's method requires about 600 time-steps, while the present ADI method calculation requires only 120 time-steps. However, one ADI time-step requires more calculation time and so the saving in computer time is in the order of 50%.

4. The size of computer memory required is relatively small. Moreover, the ADI method does not require additional matrices. While MacCormack's method requires 500 kbits memory, the present ADI method of calculation requires 256 kbits memory.

7.3 Some Possible Extensions

The method described in this paper is a general method for the calculation of laminar and turbulent elliptic flows at high Mach numbers. It is not suitable for incompressible flows. Extension to three dimensions is straight forward, as the Samarskii and Andreev ADI method is readily applicable to three-dimensional problems, and is very efficient for such problems. Calculation of flows with non-adiabatic walls is simple as well, although the upstream boundary condition must be slightly varied as equation (3.5) is not suitable anymore and must be replaced by its non-adiabatic version. Mass transfer and chemical reaction may be handled as well, if suitable mass concentration equations are added. These equations are similar to the present governing transport equation, and may be easily added to the matrix equation (4.4).

Another line of applications is that of improved turbulence modelling. The present model may be applied to transitional problems, and as such may be used to calculate cases where the boundary layer undergoes a transition from laminar to turbulent flow under the shock wave. The general framework of the computational method allows such extensions as compressibility corrections to the turbulent energy equation, the addition of a length-scale (or dissipation) equation, and even the addition of Reynolds stress equations.

APPENDIX A: CALCULATION OF A FLAT PLATE BOUNDARY LAYER

A.1 The Governing Equations

The simplest method to calculate a boundary layer is by the integral method, for a uniform flow over a flat plate the momentum deficit integral equation is

$$\frac{d\theta}{dx} = \frac{c_f}{2} = \frac{\tau_w}{\rho_0 u_0^2} \tag{A.1}$$

where

$$\theta = \int_0^\delta \frac{\rho u}{\rho_0 u_0}\left(1 - \frac{u}{u_0}\right)dy \tag{A.2}$$

and the subscripts "0" and "w" denote the main stream and wall conditions. The density and viscosity distributions are dependent on the temperature distribution which is given by the Crocco integral:

$$\frac{\rho_0}{\rho} = \frac{T}{T_0} = 1 + r\,\frac{\gamma-1}{2}\,M_0^2\left[1 - \left(\frac{u}{u_0}\right)^2\right] \tag{A.3}$$

where the recovery factor r is given by

$$\left.\begin{aligned} r &= \sigma^{1/2} \quad \text{for laminar flow} \\ r &= \sigma^{1/3} \quad \text{for turbulent flow} \end{aligned}\right\} \tag{A.4}$$

The wall density is easily derived from (A.3) by substitution of $u = 0$. The wall viscosity is obtained by substitution of (A.3) in (2.9), to yield:

$$\frac{\mu_w}{\mu_0} = \left(1 + r\,\frac{\gamma-1}{2}\,M_0^2\right)^{3/2} \frac{1 + (\gamma-1)M_0^2\,\dfrac{c_pS}{u_0^2}}{1 + (\gamma-1)M_0^2\left(\dfrac{r}{2} + \dfrac{c_pS}{u_0^2}\right)} \tag{A.5}$$

Equation (A.2) may be simplified by the application of the Howarth-Dorodnytzyn transformation from the cross-stream coordinate y to a new variable η:

$$\rho_0\delta_1 d\eta = \rho dy \tag{A.6}$$

where

$$\delta_1 = \int_0^{\delta} \frac{\rho}{\rho_0}\, dy \tag{A.7}$$

The new cross-stream coordinate η is a non-dimensional quantity, varying between zero and unity. For zero Mach number $\eta = y/\delta$. For high Mach numbers η acts as a stretching transformation, where the outer part of the boundary layer is stretched, while the region adjacent to the wall shrinks. This inconvenient stretching results from the fact that the temperature increases and the density decreases as the wall is approached. However, this transformation allows the application of incompressible boundary layer methods to the compressible case, and may therefore serve as a useful approximation. The following relations may be easily derived as well:

$$y = \delta_1 \int_0^{\eta} \frac{\rho_0}{\rho}\, d\eta \tag{A.8}$$

$$\delta = \delta_1 \int_0^{1} \frac{\rho_0}{\rho}\, d\eta \tag{A.9}$$

$$\eta(y{=}0) = 0 \tag{A.10}$$

$$\eta(y{=}\delta) = 1 \tag{A.11}$$

$$\theta = \delta_1 \int_0^{1} \frac{u}{u_0}\left(1 - \frac{u}{u_0}\right) d\eta \tag{A.12}$$

The new coordinate, η, varies between zero and unity across the boundary layer. It varies linearly near the wall and the outer boundary. Clearly, specification of the velocity profile in terms of η rather than y makes the treatment of (A.2) much easier. The velocity profile is different for laminar and turbulent flows, and these two cases will be treated in the next two sections.

A.2 The Laminar Case

The Karaman-Pohlhausen velocity profile for a flat plate boundary layer in terms of η is

$$\frac{u}{u_0} = 2\eta - 2\eta^3 + \eta^4 \tag{A.13}$$

and the skin friction and momentum thickness may be calculated:

$$\left(\frac{\partial u}{\partial y}\right)_w = 2\,\frac{u_0}{\delta_1}\,\frac{\rho_w}{\rho_0} \tag{A.14}$$

$$\theta = \frac{37}{315}\,\delta_1 \tag{A.15}$$

After substitution of (A.3), (A.14) and (A.15) in (A.1), the following expressions are obtained:

$$\frac{\delta_1}{x} = 5.836\left(1 + r\,\frac{\gamma - 1}{2}\,M_0^2\right)^{1/4}\left[\frac{1+(\gamma-1)M_0^2\,\dfrac{c_p S}{u_0^2}}{1+(\gamma-1)M_0^2\left(\dfrac{r}{2}+\dfrac{c_p S}{u_0^2}\right)}\,\frac{\mu_0}{\rho_0 u_0 x}\right]^{1/2} \tag{A.16}$$

$$\frac{y}{\delta_1} = \eta + r\,\frac{\gamma-1}{2}\,M_0^2\left(\eta - \frac{4}{3}\eta^3 + \frac{8}{5}\eta^5 - \frac{2}{3}\eta^6 - \frac{4}{7}\eta^7 + \frac{1}{2}\eta^8 - \frac{1}{9}\eta^9\right) \tag{A.17}$$

$$\frac{\delta}{\delta_1} = 1 + \frac{263}{630}\,r\,\frac{\gamma-1}{2}\,M_0^2 \tag{A.18}$$

A.3 The Turbulent Case

In this case it is reasonable to express u as a power law in η:

$$\frac{u}{u_0} = \eta^{1/7} \tag{A.19}$$

This profile does not account for the viscous sublayer and therefore the skin friction can not be calculated as the gradient of the velocity near the wall. A simple way to overcome this difficulty is by the application of an empirical skin friction law. The following law may be obtained if equation (A.19) is written in the form

$$\frac{u}{V^*} = A\left(\frac{\delta_1 \eta V_* \rho_w}{\mu_w}\right)^{1/7} \tag{A.20}$$

where $V^{*2} = \tau_w/\rho_w$. When η is unity, u is equal to u_0. Moreover, by definition

$$c_f = \frac{\tau_w}{\frac{1}{2}\rho_0 u_0^2} = 2\,\frac{\rho_w}{\rho_0}\left(\frac{V^*}{u_0}\right)^2 \tag{A.21}$$

and by substitution of V^* from (A.20) the following skin friction law is obtained

$$c_f = c_1\,\frac{\rho_w}{\rho_0}\left(\frac{\mu_w}{\rho_w \delta_1 u_0}\right)^{1/4} \tag{A.22}$$

It may be noted that the normalizing density and viscosity are calculated on the wall and not in the main stream. This is done so because Equation (A.22) is basically an incompressible flow equation, which has been confirmed for the region adjacent to the wall of pipe and duct flows. The constant c_1 has the value of 0.0444.

Substitution of (A.19) in (A.12) yields

$$\theta = \frac{7}{72}\,\delta_1 \tag{A.23}$$

Equation (A.1) may now be solved, to give the following solution:

$$\frac{\delta_1}{x} = c_2 \left(1 + r\frac{\gamma-1}{2} M_0^2\right)^{-0.3} \left[\frac{1+(\gamma-1)M_0^2 \frac{c_p S}{u_0^2}}{1+(\gamma-1)M_0^2\left(\frac{r}{2}+\frac{c_p S}{u_0^2}\right)} \frac{\mu_0}{\rho_0 u_0 x}\right]^{0.2} \tag{A.24}$$

$$\frac{y}{\delta_1} = \eta + r\frac{\gamma-1}{2} M_0^2\left(\eta - \frac{7}{9}\eta^{9/7}\right) \tag{A.25}$$

$$\frac{\delta}{\delta_1} = 1 + \frac{2}{9} r \frac{\gamma-1}{2} M_0^2 \tag{A.26}$$

$$c_2 = 0.367$$

APPENDIX B: THE TURBULENT ENERGY PROFILE

In an incompressible boundary layer the turbulent energy profile is divided usually into two main points. Very near to the wall, in the viscous sublayer, it grows as the square of the distance

$$e = a\left(\frac{\tau_w}{\mu_w}\right)^2 y^2 \tag{B.1}$$

up to a maximum value

$$e_{max} = b\frac{\tau_w}{\rho_w} \tag{B.2}$$

which is obtained at

$$y_{max} = \left(\frac{b}{a}\right)^{1/2} \frac{\mu_w}{(\rho_w \tau_w)^{1/2}} \tag{B.3}$$

In the outer part of the boundary layer the turbulent energy decays monotonically to its main stream value. This decay is approximately linear, and may be represented by

$$e = e_{max} - (e_{max} - e_0)\frac{y - y_{max}}{\delta - y_{max}} \tag{B.4}$$

In the compressible case these equations will be assumed to hold, when y is replaced by $\eta\delta_1$ and the wall shear stress is given by

$$\tau_w = c_f \rho_0 u_0^2/2 \tag{B.5}$$

Some rearrangement yields the following relations:

$$\left.\begin{aligned} e/e_{max} &= (\eta/\eta_{max})^2 && 0 \le \eta \le \eta_{max} \\ \frac{e}{e_{max}} &= 1 - \left(1 - \frac{e_0}{e_{max}}\right)\frac{\eta - \eta_{max}}{1 - \eta_{max}} && \eta_{max} < \eta \le 1 \end{aligned}\right\} \tag{B.6}$$

$$e_{max} = b \frac{c_f}{2} \frac{\rho_0}{\rho_w} u_0^2 \tag{B.7}$$

$$\eta_{max} = \left(\frac{b}{a} \frac{2}{c_f} \frac{\rho_0}{\rho_w}\right)^{1/2} \frac{\mu_w}{\rho_0 u_0 \delta_1} \tag{B.8}$$

The constants a and b must be assigned their values from experiments. Typical values are

$a = 0.05$

$b = 3$.

APPENDIX C: OBLIQUE SHOCK WAVE RELATION

An oblique shock wave is defined by two parameters. Usually the Mach number upstream of the shock and the shock angle are prescribed, and all the other shock wave relations are given in terms of these two parameters. However, in the numerical procedure described in this paper it is much more convenient to prescribe the Mach number, M_1, and the shock pressure ratio, $s = p_1/p_2$. Denoting the conditions upstream and downstream of the shock by 1 and 2 respectively, the other shock relations are given by the following equations:

The shock angle:

$$M_1^2 \operatorname{Sin}^2\sigma = s \frac{\gamma+1}{2\gamma} + \frac{\gamma-1}{2\gamma} \tag{C.1}$$

The velocity components in the x and y direction, downstream of the shock:

$$u_2 = \frac{(\gamma-1)M_1^2 \operatorname{Sin}^2(\sigma+\alpha) + 2}{(\gamma-1)M_1^2 \operatorname{Sin}^2(\sigma+\alpha)} (u_1 \operatorname{Sin} \sigma + v_1 \operatorname{Cos} \sigma)\operatorname{Sin} \sigma + $$

$$+ (u_1 \operatorname{Cos} \sigma - v_1 \operatorname{Sin} \sigma)\operatorname{Cos} \sigma \tag{C.2}$$

$$v_2 = \frac{(\gamma-1)M_1^2 \operatorname{Sin}^2(\sigma+\alpha) + 2}{(\gamma-1)M_1^2 \operatorname{Sin}^2(\sigma+\alpha)} (u_1 \operatorname{Sin} \sigma + v_1 \operatorname{Cos} \sigma)\operatorname{Cos} \sigma +$$

$$+ (u_1 \operatorname{Cos} \sigma - v_1 \operatorname{Sin} \sigma)\operatorname{Sin} \sigma \tag{6.3}$$

where

$$\operatorname{tg} \alpha = u_1/v_1 \tag{6.4}$$

The temperature ratio:

$$\frac{T_2}{T_1} = \frac{[2+(\gamma-1)M_1^2 \operatorname{Sin}^2(\sigma+\alpha)][2\gamma M_1^2 \operatorname{Sin}^2(\sigma+\alpha)-(\gamma-1)]}{(\gamma+1)^2 M_1^2 \operatorname{Sin}^2(\sigma+\alpha)} \tag{6.5}$$

The pressure ratio:

$$\frac{p_2}{p_1} = s \tag{C.6}$$

The density ratio:

$$\frac{\rho_2}{\rho_1} = \frac{p_2}{p_1}\frac{T_1}{T_2} \tag{C.7}$$

It may be noted that in the present case the shock wave is placed in a yawed flow field relative to the flat surface over which the boundary layer flows.

Equations (C.2) and (C.3) may be transformed to a more compact form:

$$\frac{u_2}{u_1} = 1 + \frac{2\ \mathrm{Sin}^2\ \sigma}{(\gamma-1)M_1^2\ \mathrm{Sin}^2(\sigma+\alpha)} + \frac{v_1}{u_1}\frac{2\ \mathrm{Sin}\ \sigma\ \mathrm{Cos}\ \sigma}{(\gamma-1)M_1^2\ \mathrm{Sin}^2(\sigma+\alpha)} \tag{C.8}$$

$$\frac{v_2}{u_1} = \left[1 + \frac{1}{(\gamma-1)M_1^2\ \mathrm{Sin}^2(\sigma+\alpha)}\right] 2\ \mathrm{Sin}\ \sigma\ \mathrm{Cos}\ \sigma + \\ + \frac{v_1}{u_1}\left[(\mathrm{Cos}^2\sigma - \mathrm{Sin}^2\sigma) + 2\,\frac{\mathrm{Cos}^2\ \sigma}{(\gamma-1)M_1^2\ \mathrm{Sin}^2(\sigma+\alpha)}\right] \tag{C.9}$$

REFERENCES

1. Baldwin, B.S., and MacCormack, R.W. (1974) Numerical Solution of the Interaction of a Strong Shock Wave with a Hypersonic Turbulent Boundary Layer, *AIAA Paper 74-558*, 7th Fluid and Plasma Dynamics Conference.
2. Beam, R.M. and Warming, R.F. (1977) An Implicit Factored Scheme for the Compressible Navier-Stokes Equations, *AIAA Paper 77-645*, 3rd Computational Fluid Dynamics Conference.
3. Brailovskaya, I.Y. (1965) A Difference Scheme for Numerical Solution of the Two-Dimensional Non-Steady Navier-Stokes Equations for a Compressible Fluid, *Sov. Phys. Doklady, 10* 2 : 107-110.
4. Hakkinen, R.J., Greler, I., Trilling, L., Abarbanel, S.S. (1959) The Interaction of an Oblique Shock Wave with a Laminar Boundary Layer, *NASA Memo. 2-18-59W.*
5. Hanin, M., Wolfshtein, M., and Landau U.E. (1978) Numerical Navier-Stokes Solution for the Effects of Suction on Shock Wave-Boundary Layer Interaction, *Proc. 1st Int. Conf., Numerical Methods in Laminar and Turbulent Flows*, Pentech Press, Plymouth, 399-409.
6. Hirsh, R.S. (1976) Numerical Solution of Supersonic Three-Dimensional Free-Mixing Flows Using the Parabolic-Eliptic Navier-Stokes Equations, *NASA TN D-8195.*
7. Isenberg, J. and de Vahl Davis, G. (1975), Finite Difference Methods in Heat and Mass Transfer, in *Topics in Transport Phenomena*, ed. C. Gutfinger, John Wiley & Sons, N.Y., 457-553.

8. Law, C.H. (1976) Supersonic Shock Wave Turbulent Boundary-Layer Interactions, *AIAA Journal, 14*, 6 : 730-734.
9. MacCormack, R.W. (1971) Numerical Solutions of the Interaction of a Shock Wave with a Laminar Boundary Layer *Lecture Notes in Physics, 8,* Springer-Verlag, New York, 151.
10. MacCormack, R.W. (1976) An Efficient Numerical Method for Solving the Time-Dependent Compressible Navier-Stokes Equations at High Reynolds Numbers, *NASA TM X-73,* 129.
11. Saffman, P.G. (1970) A Model for Inhomogeneous Turbulent Flow, *Proceedings of the Royal Society,* London *A317*, 417-433.
12. Samarskii, A.A. and Andreev, V.B. (1963) On a Hagh-Accuracy Difference Scheme for an Elliptic Equation with Several Space Variables, *USSR Comp. Math.Phys., 3,* 1373.
13. Shang, J.S., Hankey, W.L., and Law, C.H. (1976) Numerical Simulation of Shock Wave-Turbulent Boundary-Layer Interaction, *AIAA Journal, 14,* 10 : 1451-1457.
14. Shea, J.R. (1978) A Numerical Study of Transonic Normal Shock-Turbulent Boundary Layer Interactions, *AIAA Paper 78-1170,* AIAA 11th Fluid and Plasma Dynamics Conference.
15. van Driest, E.R. (1951) Turbulent Boundary Layer in Compressible Fluids, *J. Aero. Sci., 18,* 3 : 145-160.
16. Viegas, J.R., and Coakley, T.J. (1977) Numerical Investigation of Turbulence Models for Shock Separated Boundary-Layer Flows, *AIAA Paper 77-44,* AIAA 15th Aerospace Sciences Meeting.
17. Wilcox, D.C. and Alber, I.E. (1972) A Turbulence Model for High Speed Flows, *Proceedings of the 1972 Heat Transfer and Fluid Mechanics Institute*, 231-252.
18. Wilcox, D.C. (1973) Calculation of Turbulent Boundary-Layer Shock-Wave Interaction, *AIAA Journal, 11,* 11 : 1592-1594.
19. Wolfshtein, M. (1969) The Velocity and Temperature Distribution in One-Dimensional Flow with Turbulence Augmentation and Pressure Gradient, *Int. J. Heat Mass Transfer, 12,* 301-318.
20. Wolfshtein, M., Naot, D., and Lin, A. (1975) Models of Turbulence, in *Topics in Transport Phenomena,* ed. C. Gutfinger, John Wiley & Sons, N.Y. 3-45.

CHAPTER 7

NUMERICAL METHODS FOR INCOMPRESSIBLE FLOW STUDIES IN TWO DIMENSIONS

D.J. Evans

Department of Computer Studies, University of Technology, Loughborough, Leicestershire.

ABSTRACT

In this paper, new computational methods are presented for the solution of the large sparse linear systems which are derived from the discretisation of the parabolic and elliptic partial differential equations of incompressible flow studies in many dimensions by finite difference and finite element methods.

Both direct and iterative methods of solution are studied and strategies exploiting the sparseness of the system thus leading to semi-direct implicit methods with reduced storage requirements, fast algorithms for specialised systems, pre-conditioned iterative methods with improved convergence rates as well as new matrix factorisation procedures for the parallel computers of the future are presented.

1. INTRODUCTION

The basic Eulerian equations for the two dimensional incompressible flow of a Newtonian fluid in rectangular geometry (Roache, 1972) are the Momentum (Navier-Stokes) equations,

$$\frac{\partial u}{\partial t} + u\frac{\partial u}{\partial x} + v\frac{\partial u}{\partial y} = -\frac{1}{\rho}\frac{\partial p}{\partial x} + \nu\nabla^2 u + F_x \tag{1}$$

$$\frac{\partial v}{\partial t} + u\frac{\partial v}{\partial x} + v\frac{\partial v}{\partial y} = -\frac{1}{\rho}\frac{\partial p}{\partial y} + \nu\nabla^2 v + F_y \tag{2}$$

and continuity equation,

$$\frac{\partial u}{\partial x} + \frac{\partial v}{\partial y} = 0 \tag{3}$$

where u and v are the velocity components in the x and y directions respectively, p is the pressure, ν is the kinematic viscosity, ρ is the fluid density and F_x and F_y are the external forces which are assumed zero (Nakamura, 1977) .

By differentiating Equation 1 with respect to y, differentiating Equation 2 with respect to x and subracting from each other the pressure in Equations 1 and 2 can be eliminated.

Using the vorticity defined by

$$\zeta = \frac{\partial u}{\partial y} - \frac{\partial v}{\partial x}, \tag{4}$$

we obtain the vorticity equation in non-conservative form as

$$\frac{\partial \zeta}{\partial t} = -\frac{\partial (u\zeta)}{\partial x} - \frac{\partial (v\zeta)}{\partial y} + \nu \nabla^2 \zeta, \tag{5}$$

or in compact form as

$$\frac{\partial \zeta}{\partial t} = -\nabla (V\zeta) + \nu \nabla^2 \zeta \tag{6}$$

where V represents the velocity vector (u,v).

If now the st⁻ᵉam function ψ is defined as

$$\frac{\partial \psi}{\partial y} = u, \quad \frac{\partial \psi}{\partial x} = -v, \tag{7}$$

then Equation 4 becomes the Poisson equation:

$$\nabla^2 \psi = \zeta. \tag{8}$$

Thus, we see that when the vorticity function and the stream function are used as field functions rather than velocities and pressure, the flow equations are transformed to the set of parabolic and elliptic partial differential Equations 6 and 8 respectively.

Two basic problems now arise in the numerical solution of Equations 6 and 8.

a) The formulation of a procedure called discretisation for approximating the solution to the continuous problem such that the approximate solution can be obtained by solving a system of linear algebraic equations. Most discretisation procedures use either the

   1. finite difference method in which the discretisation is essentially based on the approximation of derivatives by divided differences. Mesh elements of a regular character (i.e., rectangular mesh elements) are usually associated with the use of the finite difference method.

   2. finite element method in which the discretisation is based on finding a stationary point in an approximating subspace to a variational (minimisation) formulation of the boundary value problem. The approximating sub-space is usually constructed by assuming the solution to be of a particular form (i.e., a polynomial of

degree n) within each element. The finite element approach allows more flexibility for the systematic development of higher order approximation methods and in the choice of mesh elements e.g. triangular or general quadrilateral mesh elements with curved sides can be used to describe the problem geometry more accurately.

b) The development of an efficient algorithmic technique for solving the large, sparse and generally banded systems of linear equations which are derived from the discretisation process.

In this paper, we shall generally adhere to the finite difference approach but ensure that the algorithmic techniques derived could be extended in principle for use in finite element methods as well.

In deriving a numerical solution process for the vorticity Equation 6, it is usual to replace the time derivative by its finite difference equivalent for which we have the following alternatives:

1. the forward difference (or explicit) scheme,

$$(\zeta^{(k+1)}-\zeta^{(k)})/\Delta t = -\nabla(V^{(k)}\zeta^{(k)}) + \nu\nabla^2\zeta^{(k)} \qquad (9)$$

or

2. the backward difference (or fully implicit) scheme,

$$(\zeta^{(k+1)}-\zeta^{(k)})/\Delta t = -\nabla(V^{(k+1)}\zeta^{(k+1)}) + \nu\nabla^2\zeta^{(k+1)} \qquad (10)$$

and

$$\nabla^2\psi^{(k+1)} = \zeta^{(k+1)}. \qquad (11)$$

Basically, then the solution process involves the calculation of the vorticity $\zeta^{(k+1)}$ by either Equations 9 or 10 followed by the solution of the Poisson Equation 11 to obtain $\psi^{(k+1)}$. Updated values of u and v are now calculated from Equation 7 and the new boundary values of ζ determined. The whole process is repeated for the next time step (k+1) with the updated values of u and v and the new boundary values of ζ.

Thus, we see that the numerical solution of Equations 6 and 8 involves the solution of a parabolic and elliptic partial differential equation at each time step. In two and three space dimensions this represents a formidable computational problem.

2. FORMULATION OF THE FINITE DIFFERENCE EQUATIONS

In deriving the finite difference schemes for the vorticity Equation 6, the region $\bar{R}\equiv R\times[0<t\leq T]$ where $R=\{(x,y);\ 0<x<X, 0<y<Y\}$ is covered by a rectilinear net with mesh spacings h_x, h_y

Δt in the x,y and t directions respectively and mesh points (x_i, y_j, t_k) where:

$$\begin{aligned} x_i &= ih_x, & i &\in [0,p] \\ y_j &= jh_y, & j &\in [0,m] \\ t_k &= k\Delta t, & k &\in [0,T/\Delta t] \end{aligned} \tag{12}$$

it is usual to use central differences to replace the space derivatives which immediately gives the alternative schemes,

1. the explicit scheme which yields the equations

$$\begin{aligned}(\zeta_{i,j}^{(k+1)}-\zeta_{i,j}^{(k)})/\Delta t = {} & (u\zeta_{i-1,j}^{(k)}-u\zeta_{i+1,j}^{(k)})/2h_x+(v\zeta_{i,j-1}^{(k)}- \\ & -v\zeta_{i,j+1}^{(k)})/2h_y+\nu[(\zeta_{i-1,j}^{(k)}-2\zeta_{i,j}^{(k)}+\zeta_{i+1,j}^{(k)})/h_x^2 + \\ & (\zeta_{i,j-1}^{(k)}-2\zeta_{i,j}^{(k)}+\zeta_{i,j+1}^{(k)})/h_y^2]\end{aligned} \tag{13}$$

   corresponding to Equation 9;

2. the fully implicit scheme which yields

$$\begin{aligned}(\zeta_{i,j}^{(k+1)}-\zeta_{i,j}^{(k)})/\Delta t = {} & (u\zeta_{i-1,j}^{(k+1)}-u\zeta_{i+1,j}^{(k+1)})/2h_x+(v\zeta_{i,j-1}^{(k+1)} \\ & -v\zeta_{i,j+1}^{(k+1)})/2h_y+\nu[(\zeta_{i-1,j}^{(k+1)}-2\zeta_{i,j}^{(k+1)}+\zeta_{i+1,j}^{(k+1)})/h_x^2+ \\ & (\zeta_{i,j-1}^{(k+1)}-2\zeta_{i,j}^{(k+1)}+\zeta_{i,j+1}^{(k+1)})/h_y^2],\end{aligned} \tag{14}$$

   corresponding to Equation 10;

whilst the Poisson Equation (11) is replaced by

$$\begin{aligned}[(\psi_{i-1,j}^{(k+1)}-2\psi_{i,j}^{(k+1)}+\psi_{i+1,j}^{(k+1)})/h_x^2+(\psi_{i,j-1}^{(k+1)}-2\psi_{i,j}^{(k+1)} \\ +\psi_{i,j+1}^{(k+1)})/h_y^2] = \zeta_{i,j}^{(k+1)}\end{aligned} \tag{15}$$

where $\zeta_{i,j}^{(k)}, \psi_{i,j}^{(k)}$ denotes the values of $\zeta(x_i,y_j,t_k), \psi(x_i,y_j,t_k)$ on the chosen grid.

By inspection of the explicit scheme 13 we see that the solution process for this case is a simple time stepping procedure. However, from the Fourier stability analysis for this scheme, (Richtmyer & Morton, 1967), we have that the time step Δt is governed by the conditions,

$$\nu[\Delta t/h_x^2 + \Delta t/h_y^2] \leqslant 1/2,$$

and

$$u\,\Delta t/h_x + v\,\Delta t/h_y \leqslant 1, \tag{16}$$

in order to avoid numerical instability in the computation. Although the use of a small time step Δt is a severe disadvantage, often little change occurs in the Stream function and hence the solution of Equation 15 is only necessary after several time steps which can reduce overall computation time.

However, for problems involving large integration times, the explicit method is unacceptable and the implicit scheme which is unconditionally stable thus allowing a larger time step Δt has to be considered. The disadvantage however is that the values of $\zeta_{i,j}^{(k+1)}$ and $\psi_{i,j}^{(k+1)}$ have to be determined simultaneously which involves either a direct inversion of a large sparse generally banded matrix or an iterative method of solution for Equations 14 and 15 at each time step. Thus, the time saved by using a larger time step can be offset by a longer or more complicated computational procedure to solve the implicit equations. In the following sections we shall consider some new computational strategies for the solution of Equations 14 and 15.

An alternative to the explicit and implicit methods is the alternating direction implicit (ADI) method [Peaceman and Rachford, 1955] for not only is it unconditionally stable but no iteration or direct inversion of the large order matrix is involved. Instead only the solution of tridiagonal systems alternatively along the rows or columns of the rectilinear grid is required. In the case of the vorticity equation, the ADI method is given by the following two half-steps:

$$2(\zeta_{i,j}^{(k+\frac{1}{2})}-\zeta_{i,j}^{(k)})/\Delta t = (u\zeta_{i-1,j}^{(k+\frac{1}{2})}-u\zeta_{i+1,j}^{(k+\frac{1}{2})})/2h_x + (v\zeta_{i,j-1}^{(k)}- v\zeta_{i,j+1}^{(k)})/2h_y+\nu[(\zeta_{i-1,j}^{(k+\frac{1}{2})}-2\zeta_{i,j}^{(k+\frac{1}{2})}+\zeta_{i+1,j}^{(k+\frac{1}{2})})/h_x^2+ (\zeta_{i,j-1}^{(k)}-2\zeta_{i,j}^{(k)}+\zeta_{i,j+1}^{(k)})/h_y^2], \tag{17}$$

and

$$2(\zeta_{i,j}^{(k+1)}-\zeta_{i,j}^{(k+\frac{1}{2})})/\Delta t = (u\zeta_{i-1,j}^{(k+\frac{1}{2})}-u\zeta_{i+1,j}^{(k+\frac{1}{2})})/2h_x + (v\zeta_{i,j-1}^{(k+1)}- v\zeta_{i,j+1}^{(k+1)})/2h_y+\nu[(\zeta_{i-1,j}^{(k+\frac{1}{2})}-2\zeta_{i,j}^{(k+\frac{1}{2})}+\zeta_{i+1,j}^{(k+\frac{1}{2})})/h_x^2+ (\zeta_{i,j-1}^{(k+1)}-2\zeta_{i,j}^{(k+1)}+\zeta_{i,j+1}^{(k+1)})/h_y^2]. \tag{18}$$

Since Equation 17 includes 3 unknowns $\zeta_{i-1,j}^{(k+\frac{1}{2})}, \zeta_{i,j}^{(k+\frac{1}{2})}, \zeta_{i+1,j}^{(k+\frac{1}{2})}$ on a row of grid points, then only the inversion of a tridiagonal matrix is involved. Similarly Equation 18 only involves 3 unknowns of $\zeta^{(k+1)}$ along a column of the grid.

In these methods, the advection terms $\frac{\partial(u\zeta)}{\partial x}$ and $\frac{\partial(v\zeta)}{\partial y}$ are often replaced by 'upwind differences' which involves the approximations,

$$\frac{\partial(u\zeta)}{\partial x} = \begin{cases} (u\zeta^{(k)}_{i+1,j} - u\zeta^{(k)}_{i,j})/h_x, & u<0 \\ (u\zeta^{(k)}_{i,j} - u\zeta^{(k)}_{i-1,j})/h_x, & u>0 \end{cases} \tag{19}$$

and a similar representation for the term $\partial(v\zeta)/\partial y$ which although less accurate are known to possess better stability properties by ensuring 'diagonal dominance' in the coefficient matrix.

Finally, an alternative implicit method is obtained by averaging the time derivative $\partial\zeta/\partial t$ between the $(k+1)^{st}$ and k^{th} time levels, i.e.,

$$\frac{\partial\zeta}{\partial t} = \frac{1}{2}\left(\frac{\partial\zeta^{(k+1)}}{\partial t} + \frac{\partial\zeta^{(k)}}{\partial t}\right) \tag{20}$$

on the left-hand side of Equation 6 accompanied by the usual central difference approximations for the first and second space derivatives on the right hand side again evaluated at the $(k+1)^{st}$ and k^{th} time levels. The resulting formula [Crank-Nicolson, 1947] is extremely accurate and unconditionally stable, although extra arithmetic is involved in the computation at each step.

Now given initial and boundary conditions such that we can determine

$$\zeta(x,y,0) = f(x,y), \tag{21}$$

and

$$\zeta(x,y,t) = \gamma(x,y,t), \ (x,y,t)\in C\times(0<t<T), \tag{22}$$

where C is the boundary of R, then the finite difference discretisation of Equation 20 on the chosen grid, can be written as a series of five-point linear finite difference equations of the form:

$$d_{i,j}\zeta^{(k+1)}_{i,j-1} + a_{i,j}\zeta^{(k+1)}_{i-1,j} + b_{i,j}\zeta^{(k+1)}_{i,j} + c_{i,j}\zeta^{(k+1)}_{i+1,j} + e_{i,j}\zeta^{(k+1)}_{i,j+1} = s^{(k)}_{i,j}$$

$$\text{for } i\in[1,p-1],\ j\in[1,m-1], \tag{23}$$

where

$$d_{i,j} = -r_y(2\nu+h_y v)/4; \quad e_{i,j} = -r_y(2\nu-h_y v)/4;$$

$$a_{i,j} = -r_x(2\nu+h_x v)/4; \quad c_{i,j} = -r_x(2\nu-h_x v)/4;$$

$$b_{i,j} = 1+\nu(r_x+r_y); \tag{24}$$

and

$$s_{i,j}^{(k)} = \zeta_{i,j}^{(k)} + r_x h_x [u\zeta_{i-1,j}^{(k)} - u\zeta_{i+1,j}^{(k)}]/4 + r_y h_y [v\zeta_{i,j-1}^{(k)} - v\zeta_{i,j+1}^{(k)}]/4$$

$$+ r_x \nu [\zeta_{i-1,j}^{(k)} - 2\zeta_{i,j}^{(k)} + \zeta_{i+1,j}^{(k)}]/2 + r_y \nu [\zeta_{i,j-1}^{(k)} - 2\zeta_{i,j}^{(k)} + \zeta_{i,j+1}^{(k)}]/2;$$

with $r_x = \Delta t/h_x^2$, $r_y = \Delta t/h_y^2$.

Grouping the above system of finite difference equations when ordered in a columnwise ordering into a matrix form, we obtain a sparse, quindiagonal matrix of order $n=(m-1)(p-1)$ of the following form:

$$A = \begin{bmatrix} \multicolumn{5}{l}{\xleftarrow{\qquad}\; m \;\xrightarrow{\qquad}} & & \\ b_1 & c_1 & & & e_1 & & 0 \\ a_2 & b_2 & c_2 & & & \ddots & \\ & \ddots & \ddots & \ddots & 0 & & e_{n-m+1} \\ d_m & \ddots & 0 & \ddots & \ddots & \ddots & \\ & \ddots & & & \ddots & \ddots & c_{n-1} \\ 0 & & d_n & & & a_n & b_n \end{bmatrix}, \qquad (25)$$

where $c_q=0$, if $q=t(m-1)$, $1 \leq t \leq p-2$, $e_q=0$, $n-m+2 \leq q \leq n$,
$a_q=0$, if $q=t(m-1)+1$, $1 \leq t \leq p-2$, $d_q=0$, $1 \leq q \leq m-1$,

and the derived linear system,

$$A\underline{\zeta}^{(k+1)} = \underline{s}^{(k)} \qquad (26)$$

has to be solved at each time step k, to give the required vorticity solution $\underline{\zeta}$.

Thus we see that the implicit methods lead to the requirement that large sparse generally unsymmetric banded matrices have to be solved for each time interval. For multi-dimensional problems (>2) we can naturally expect a similar structured matrix of even larger order.

It is essential for the stability of the solution processes which are to follow that the matrix A remains diagonally dominant, since then in the Gaussian elimination procedure no interchange of rows occurs and the matrix retains its band structure form throughout. Thus, appropriate changes to the coefficients in Equation 24 are made if the upwind differences given by Equation 19 are used for the advection terms.

However, when we concern ourselves with the solution of the discretised Poisson Equation 15, then a similar columnwise grouping of the equations yields the equations,

$$-\psi_{i,j-1}^{(k+1)}-r^2\psi_{i-1,j}^{(k+1)}+2(1+r^2)\psi_{i,j}^{(k+1)}-r^2\psi_{i+1,j}^{(k+1)}-\psi_{i,j+1}^{(k+1)}=-h_y^2\zeta_{i,j}^{(k+1)}$$

$$\text{for } i\in[1,p-1],\ j\in[1,m-1], \qquad (27)$$

where $r=h_y/h_x$,

which then gives the simpler linear system,

$$\tilde{A}\underline{\psi}^{(k+1)} = \underline{\zeta}^{(k+1)} , \qquad (28)$$

where $\underline{\psi}$ is the required vector of mesh values of ψ, and $\underline{\zeta}^{(k+1)}$ is a vector of known quantities involving the mesh values of ζ and the boundary conditions on ψ and $\tilde{A}$ is a large, sparse, symmetric, positive definite matrix of order $n=(m-1)(p-1)$ with the following block partitioned form,

$$\tilde{A}=\begin{bmatrix} B & -I & & & 0 \\ -I & B & -I & & \\ & \ddots & \ddots & \ddots & \\ & & \ddots & \ddots & -I \\ 0 & & & -I & B \end{bmatrix}_{n\times n} \text{ and } B = \begin{bmatrix} 2(1+r^2) & -r^2 & & & 0 \\ -r^2 & \ddots & \ddots & & \\ & \ddots & \ddots & \ddots & \\ & & \ddots & \ddots & -r^2 \\ 0 & & & -r^2 & 2(1+r^2) \end{bmatrix}_{(m-1)\times(m-1)} \qquad (29)$$

As previously, similar remarks apply concerning the matrix structure for multi-dimensional problems.

3. COMPUTATIONAL METHODS OF SOLUTION.

For the computer solution of large order linear systems such as those given by Equations 26 and 28, a direct solution of the finite difference or finite element equations is usually preferred, especially, where as here, the same matrix has to be solved at each time step or for a sequence of different right hand sides. The preferred method of solution i.e. the Gaussian elimination method is equivalent to the matrix decomposition,

$$A = L.U \qquad (30)$$

where L is a lower triangular and U is an upper triangular matrix. The solution of Equations 26 and 28 is obtained from the equivalent 'easily solvable' triangular systems,

$$L\underline{y} = \underline{s} \quad \text{and} \quad U\underline{\zeta} = \underline{y} \qquad (31)$$

by recursive sequences of forward and backward substitution processes. Here, the computation involved in Equation 30 is the major part of the computational effort and this needs only to be completed once.

However, it is well known that in any elimination process, the zero elements within the outer band envelope of the matrix are likely to be 'filled in' by non-zero values. This 'fill in' is often a considerable amount for wide banded matrix (~O(nm) where m and n are the bandwidth and order of the matrix respectively) and leads to considerable computer storage problems as well as associated commitments such as extra work and round-off errors. Thus, in order to derive more efficient computer methods of solution, it is necessary to look at ways in which the 'fill in' can be reduced or its effects reduced. Naturally, if the 'fill in' procedure is to be analysed, and exploited it is essential that the band structure of the matrix be maintained throughout the elimination process. Thus we must ensure that the coefficient matrix is diagonally dominant and no interchange of rows occurs. An initial strategy was outlined by Stone (1968) with the Strongly Implicit Method whilst an alternative approach has been given in Evans (1974) and some extensions to this work are given in Section 4.

In addition, recently there have been some important new developments in the direct methods for solving special systems of the sparse linear systems such as those given by Equation 28. These systems because they consist of constant elements and possess a special structure, often facilitate a rapid or 'fast' solution which involves considerably less work than the known elimination methods. Early algorithms in this class have been given by Golub (in Hockney, 1965) and Evans (1972). In Section 5, we present some extensions to this work which widens the range of applicability to more varied boundary conditions.

For the larger systems of linear equations that occur in multi-dimensional problems (>2) there is substantial evidence (Hageman, 1975) that suggests that the direct methods are uneconomic due to the storage involved and iterative methods of solution have to be considered.

Most methods in this class involve a 'splitting' of the coefficient matrix and in a typical linear system $A\underline{z}=\underline{b}$ the splitting usually takes the form:

$$A = M-N \tag{32}$$

and successive iterates are obtained from the equation,

$$M\underline{z}^{(\ell+1)} = N\underline{z}^{(\ell)} + \underline{b} , \tag{33}$$

where the index ℓ indicates the iteration number and $\underline{z}^{(\ell)}$ denotes the ℓth approximation to the solution vector $\underline{z}$. The iteration sequence (33) is continued until the condition $|\underline{z}^{(\ell+1)}-\underline{z}^{(\ell)}|/|\underline{z}^{(\ell)}|<\varepsilon$, where ε is a prescribed small quantity determined by the accuracy required by the solution (usually 10^{-6})

Since Equation 33 has to be solved at each step of the iteration, then the choice of M is restricted to easily solvable systems such as diagonal, triangular or triple diagonal forms and the methods they relate to are given as follows:

$$\begin{array}{llll} 1. & M=\hat{D} & ; \; N=\hat{L}+\hat{U} & ; \text{ Jacobi} \\ 2. & M=\hat{D}-\hat{L} & ; \; N=\hat{U} & ; \text{ Gauss-Seidel} \\ 3. & M=(\hat{D}-\omega\hat{L})/\omega ; & N=[\omega\hat{U}+(1-\omega)\hat{D}]/\omega ; & \text{ Successive Over-relaxation} \end{array} \tag{34}$$

where $\hat{D}$,$\hat{L}$ and $\hat{U}$ are the diagonal, lower triangular and upper triangular sparse components of the coefficient matrix A.

It can be easily seen that only the non-zero elements are used in these iterative processes and thus any sparseness in the matrix is retained throughout (no fill in). Convergence of these methods is dependent on the largest eigenvalue of the iteration matrix $M^{-1}N$ satisfying the criterion $|\lambda_{max}|<1$. For the problems under discussion in this paper, λ_{max} is very close to 1 and the methods 1. and 2. above can be discarded as they converge too slowly. However, the Successive Overrelaxation method 3. represents an extremely useful and simple method especially if the optimal value of the overrelaxation parameter ω is chosen (Varga, 1962). Further, by noticing the block partitioned form of $\hat{D}$ in Equation 29, the concept of block iterative methods which possess even greater convergence rates can be introduced. In these methods, groups of points along a line or peripheral circuit of the chosen grid can be solved simultaneously (Varga, 1963; Evans, 1972A) within the iterative process by special direct algorithms. Alternative methods involving a sequence of different acceleration parameters as well as second order methods involving 2 previous iterates $\underline{z}^{(\ell)}$ and $\underline{z}^{(\ell-1)}$ have also been proposed (Young, 1971).

Finally, an alternative strategy was considered by Peaceman & Rachford (1955) where the 'splitting'

$$A = H+V \tag{35}$$

was introduced where H and V represent tridiagonal matrices derived from second order differential operators discretised along the rows and columns of the two dimensional grid.

This splitting leads to the alternating direction implicit iterative (ADI) process,

$$\begin{aligned} (H+\rho I)\underline{z}^{(\ell+\frac{1}{2})} &= (\rho I-V)\underline{z}^{(\ell)} + \underline{b} \\ \text{and} \quad (V+\rho I)\underline{z}^{(\ell+1)} &= (\rho I-H)\underline{z}^{(\ell+\frac{1}{2})} + \underline{b} \end{aligned} \tag{36}$$

where ρ is an acceleration parameter chosen to maximise the rate of convergence.

In Section 6, we extend the concept of 'preconditioning' as introduced in Evans (1967) to these ADI techniques to yield more powerful iterative methods.

To summarise, it is well known that the iterative solution techniques require much less storage for sparse matrix problems than the direct methods, whilst the latter often requires less computational work. In an analysis comparing the efficiencies of the two basic solution approaches [Hageman, 1975] has given the following results:

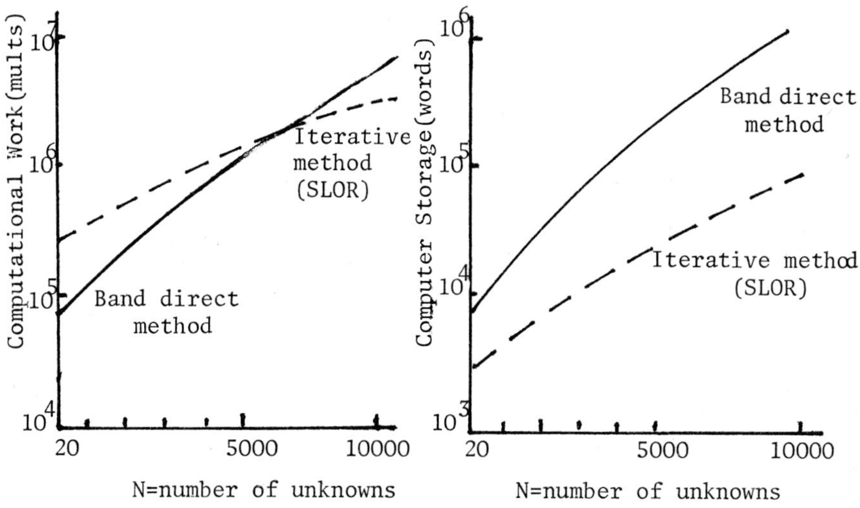

FIGURE 1

From Figure 1 we see that for the one dimensional problems, the band direct algorithms are always more efficient whilst the iterative methods are better for 3 dimensional problems. For the 2 dimensional problems, there is usually a 'cross over point' in the number of unknowns for which a good iterative method becomes more efficient than a good direct method. Thus, a knowledge of the 'cross over point' can serve as a guide in the choice of solution method for a particular class of problems.

Thus, the factors that are going to influence future methods of solution are:

1. The exploitation of the sparseness of the matrix in direct methods.

2. The development of fast algorithmic methods for special problems, rectangular geometries, constant coefficients, etc.

3. The development of new iterative methods of solution with faster convergence rates.

4. The development of new algorithms for use on the future computers with specialised hardware features such as parallel processing capabilities.

In the following sections new strategies in the above categories will be presented.

4. A SPARSE LU FACTORISATION PROCEDURE.

In Evans (1974), an algorithmic procedure equivalent to the Gaussian elimination process for solving sparse quindiagonal linear systems as in Equation 25 was presented.

We now consider a diagonally dominant, sparse, structured matrix A and partition it following Lipitakis (1978) such that A,L and U have the matrix forms:

$$A \equiv \left[\begin{array}{ccccc|cccc}
\multicolumn{5}{c|}{\longleftarrow m \longrightarrow} & & & & \\
b_1 & c_1 & & & & e_1 & & & \\
a_2 & b_2 & c_2 & & & & \ddots & & 0 \\
 & \ddots & \ddots & \ddots & & & & \ddots & \\
 & & \ddots & \ddots & c_{m-2} & & 0 & & e_{n-m+1} \\
 & & & a_{m-1} & b_{m-1} & c_{m-1} & & & \\
\hline
d_m & & & & a_m & b_m & c_m & & \\
 & \ddots & & & & a_{m+1} & \ddots & \ddots & \\
 & & 0 & & & & \ddots & \ddots & c_{n-1} \\
0 & & & \ddots & d_n & & & a_n & b_n
\end{array}\right] \equiv \left[\begin{array}{c|c} A_{11} & A_{12} \\ \hline A_{21} & A_{22} \end{array}\right] \qquad (37)$$

$L \equiv$

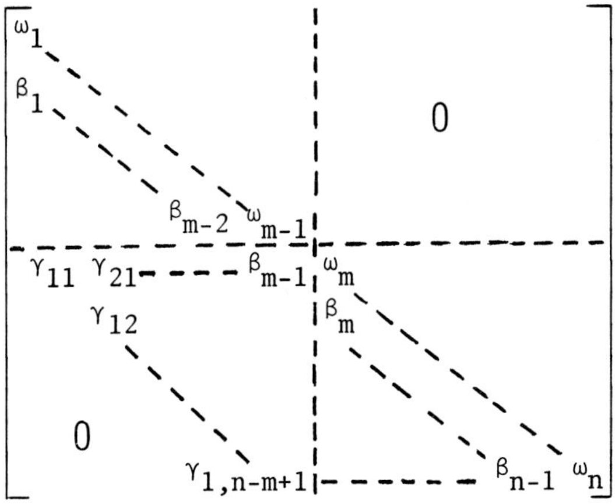

$$\equiv \begin{bmatrix} L_{11} & 0 \\ L_{21} & L_{22} \end{bmatrix} \tag{38}$$

and

$U \equiv$

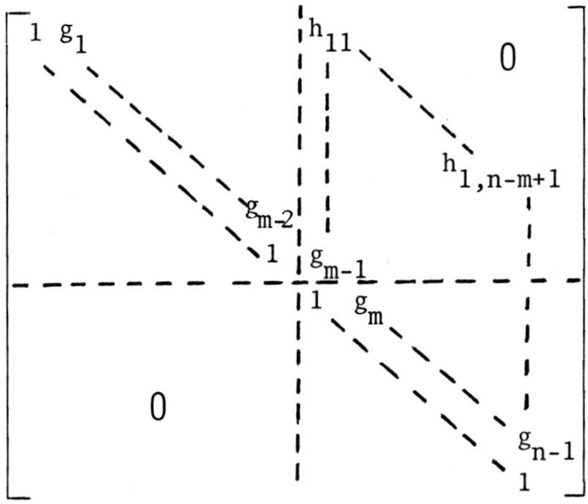

$$\equiv \begin{bmatrix} U_{11} & U_{12} \\ 0 & U_{22} \end{bmatrix} \tag{39}$$

By equating the terms in the partitioned forms of A,L and U we can obtain the elements of $L_{11}, U_{11}, L_{21}, L_{22}$ and U_{22} from the following identities,

$$\begin{aligned} L_{11}U_{11} &= A_{11} \quad ; \qquad U_{21} = L_{11}^{-1}U_{12} \quad ; \\ L_{21} &= A_{21}U_{11}^{-1}; \qquad L_{22}U_{22} = A_{22}-L_{21}U_{12}. \end{aligned} \tag{40}$$

Now the elements of L_{11}, U_{11} may be easily derived from the relationships,

$$\begin{aligned} &\omega_1 = b_1 \ ; \ \beta_1 = a_2 \ ; \ \ g_1 = c_1/\omega_1, \\ &\qquad\qquad\qquad\qquad \text{for } i=2,3,\ldots,m-2, \\ &\omega_i = b_i-\beta_{i-1}g_{i-1}; \ \beta_i = a_{i+1}; \ g_i = c_i/\omega_i \end{aligned} \tag{41}$$

and $\qquad \omega_{m-1} = b_{m-1}-\beta_{m-2}g_{m-2}.$

The elements of L_{21}, U_{21}, L_{22} and U_{22} can now be obtained in the following manner,

for $j=1,2,\ldots,n-m+1$

$$h_{1,j} = e_j/\omega_j \; ; \quad \gamma_{1,j} = d_{j+m-1}$$

and $$g_{m+j-2} = c_{m+j-2}/\omega_{m+j-2} \; ; \; \beta_{m+j-2} = a_{m+j-1}, \tag{42}$$

whilst for $j \leq m-2$

$$h_{i,j} = -\beta_{i+j-2}h_{i-1,j}/\omega_{i+j-1}, \text{ for } i=2,3,\ldots,m-j$$

and $$\gamma_{i,j} = -g_{i+j-2}\gamma_{i-1,j}, \qquad \text{for } i=2,3,\ldots,m-j.$$

Then, for all $j>1$

$$\gamma_{i,j} = -g_{i+j-2}\gamma_{i-1,j} - \sum_{k=1}^{i-1}\gamma_{k,j}h_{k-i+m,i+j-m}$$

and $$h_{i,j} = (-\beta_{i+j-2}h_{i-1,j} - \sum_{k=1}^{i-1}\gamma_{k-i+m,i+j-m}h_{k,j})/\omega_{i+j-1}, \tag{43}$$

for either $i=(m-j+1),(m-j+2),\ldots,(m-1)$ and all $j \leq m-1$
or $i=2,3,\ldots,m-1$ for $j>m-1$.

Then, for $i=m-1$

$$\omega_{m+j-1} = b_{m+j-1}-\beta_{i+j-1}h_{i,j}-g_{i+j-1}\gamma_{i,j}-\beta_{i+j-1}g_{i+j-1} - \sum_{k=1}^{i}\gamma_{i,j}h_{k,j}.$$

The original linear system,

$$A\underline{\zeta} = \underline{s} \; ,$$

can then be solved directly for $\underline{\zeta}$ in terms of an auxiliary vector $\underline{y}$, where

$$L\underline{y} = \underline{s} \quad \text{and} \quad U\underline{\zeta} = \underline{y}. \tag{44}$$

The initial processes yields the algorithmic form,

$$y_1 = s_1/\omega_1$$

$$y_i = (s_i-\beta_{i-1}y_{i-1})/\omega_i, \text{ for } i=2,3,\ldots,m-1$$

and $$y_i = (s_i-\beta_{i-1}y_{i-1} - \sum_{k=i-m+1}^{i-1}\gamma_{k-i+m,i-m+1}y_k)/\omega_i \tag{45}$$

for $i=m,m+1,\ldots,n$.

Then, a back substitution process yields $\underline{\zeta}$ in terms of $\underline{y}$, the components being given by the equations:

$$\zeta_n = y_n$$

and
$$\zeta_i = y_i - g_i y_{i+1} - \sum_{j=p}^{q} h_{i-j+m,j-m+1} y_j, \qquad i=n-1,n-2,\ldots,2,1. \tag{46}$$

where p,q are given by:

if $m>[\frac{n}{2}+1]$

$p=i+1,\ q=n$,for $i=n-1,n-2,\ldots,m$ (except if m=n)

$p=m,\ q=n$,for $i=m-1,m-2,\ldots,n-m+1$

$p=m,\ q=i+m-1$, for $i=n-m,n-m+1,\ldots,2,1$ (except if m=n)

if $m<[\frac{n}{2}+1]$

<u>if</u> n is even and $m=n/2$

<u>then</u>

$p=i+1,\ q=n$, for $i=n-1,n-2,\ldots,m+1$

$p=m+1,\ q=n-1$, for $i=m$

$p=m,\ q=i+m-1$, for $i=m-1,m-2,\ldots,2,1$

<u>else</u> (47)

$p=i+1,\ q=n$, for $i=n-1,n-2,\ldots,n-m+1$

$p=i+1,\ q=i+m-1$, for $i=n-m,n-m+1,\ldots,m$

$p=m,\ q=i+m-1$, for $i=m-1,m-2,\ldots,2,1$

and by [s] we denote the greatest integer not exceeding s.

This algorithm requires storage for the 4(n×1) vectors ω,β g and s are well as $2(m-1)\times(n-m+1)$ rectangular arrays γ and h giving a maximum total of approximately (2m+4)N computer words.

The amount of work involved for each major part of the algorithm is given as:

i) decompose $A_{11}=L_{11}U_{11}$ - O(m) operations,
ii) form $U_{12}=L_{11}^{-1}A_{12}$ - $O(m^2)$ operations, (48)
iii) solve for L_{21}: $L_{21}U_{11}=A_{21}$ - $O(m^2)$ operations,
iv) decompose $A_{22}=L_{21}U_{12}+L_{22}U_{22}$ - $O(m^2N)$ operations,

giving an approximate total of $O(m^2N)$ multiplication operations where we have assumed that the division and multiplication operations are comparable.

In the previous computational procedure although some exploitation of the sparseness in the original matrix A has taken place in A_{11}, it can be seen that for large order matrices and m in the range $n/4 \leq m \leq n/2$, the storage requirements and computational work is still prohibitively high for computers with limited core memory. A further stage of reduction is now considered where we attempt to minimise the effects of the 'fill in' process which occurred in the factorisation procedure, i.e., the $(m-1)\times(n-m+1)$ rectangular arrays γ and h.

The strategy adopted similar to Evans (1974) was to achieve an approximate factorisation procedure of the form:

$$A \approx L_s^r U_s^r \tag{49}$$

by retaining only r rows of the outermost elements of the and h 'fill in' arrays.

Thus, the sparse triangular matrices L_s^r, U_s^r now have the form:

$$L_s^r \equiv \left[\begin{array}{cccc|cccc} \omega_1 & & & & & & & \\ \beta_1 & \omega_2 & & & & 0 & & \\ & \ddots & \ddots & & & & & \\ & & \beta_{m-2} & \omega_{m-1} & & & & \\ \hline \gamma_{1,1}\cdots\gamma_{r,1} & & & \beta_{m-1} & \omega_m & & & \\ & \ddots & 0 & & \ddots & \ddots & & \\ 0 & & \ddots & & & \ddots & \ddots & \\ & & \gamma_{1,n-m+1}\cdots & & \gamma_{r,n-m+1} & & \beta_{n-1} & \omega_n \end{array}\right]$$

$$U_s^r = \left[\begin{array}{cccc|cccc} 1 & g_1 & & & h_{1,1} & & 0 & \\ & 1 & g_2 & & h_{r,1} & \ddots & & \\ & & \ddots & \ddots & & 0 & h_{1,n-m+1} & \\ & & & 1 & g_{m-1} & & \vdots & \\ \hline & & & & 1 & g_m & h_{r,n-m+1} & \\ & 0 & & & & \ddots & \ddots & \\ & & & & & & 1 & g_{n-1} \\ & & & & & & & 1 \end{array}\right] \tag{50}$$

Since the neglected terms are usually quite small, it seems a sensible approach to follow to achieve the two objectives of reducing the storage and minimising the computational work. Quite small values of r are envisaged, possibly in the range $2<r<5$, which has the consequential effect of producing a very competitive algorithm. The sparse form of the algorithm can be easily obtained from Equations 41-46 by reducing the inner loops of the recursive procedures from (m-1) to r respectively as given in Lipitakis (1978).

The total memory requirement for the sparse factorisation algorithm now reduces to $(2r+6)n$ words and the amount of work involved is $O(r^2)n$ operations which are both acceptable amounts on present day computers.

Naturally, with such an approximate factorisation of the matrix A we cannot expect to obtain a direct method of solution and we now consider a class of iterative procedures of the form,

$$L_s^r U_s^r \delta \underline{\zeta}^{(\ell+1)} = \alpha \underline{r}^{(\ell)} , \tag{51}$$

where L_s^r and U_s^r are the known sparse triangular matrices given by Equation 50, α is a predetermined acceleration parameter, $\underline{r}^{(\ell)}=\underline{s}-A\underline{\zeta}^{(i)}$ and $\delta\underline{\zeta}^{(i+1)}=\underline{\zeta}^{(i+1)}-\underline{\zeta}^{(\ell)}$. For the optimal choice of parameter α and an initial guess of the solution $\underline{\zeta}^{(0)}$, a sequence of approximate solutions $\underline{\zeta}^{(1)},\underline{\zeta}^{(2)},\ldots,\underline{\zeta}^{(\ell)}$ can be obtained from the Strongly Implicit Simultaneous Displacement method (51).

Similarly, the following strongly implicit second order iterative procedures can be formulated for the Richardson and Chebyshev methods respectively,

$$L_s^r U_s^r \delta \underline{\zeta}^{(\ell+1)} = \alpha \underline{r}^{(\ell)} + \beta \delta \underline{\zeta}^{(\ell)} \tag{52}$$

and

$$L_s^r U_s^r \delta \underline{\zeta}^{(\ell+1)} = \alpha_\ell \underline{r}^{(\ell)} + \beta_\ell \delta \underline{\zeta}^{(\ell)} , \tag{53}$$

where α,β are the Richardson acceleration parameters and α_ℓ,β_ℓ the Chebyshev sequences of parameters.

In Lipitakis (1978), both theoretical and experimental analysis was carried out and the best choice of r with regard to the minimum of computational effort (i.e. number of iterations $\times$ work per iteration) was determined as 4. In addition, the above sparse factorisation procedure has been extended to the matrices derived from three space dimensional problems and this is given in Lipitakis and Evans (1979).

Finally, the above iterative algorithmic procedures have been applied to a wide variety of linear and non-linear boundary value problems in Lipitakis & Evans (1979A) and the results confirm that the above methods are competitive with other solution methods based on the strongly implicit approach of Stone (1968) but uses standard well-known accelerated iterative procedures with the flexibility of choosing the order of factorisation r.

Further, it can be shown that for constants e_1, e_2 such that $e_1 < (x.LUx)/(x.L_s^r U_s^r x) < e_2$, then the iteration (51) converges in the range $0 < \alpha < 2/e_2$. The optimum value of α is then $2/(e_1+e_2)$ and the convergence factor is $(e_2-e_1)/(e_2+e_1)$. Since, by choice of r, the range $[e_1, e_2]$ can be made very small, indeed then the number of iterations required to reduce the error norm by a factor ε is $O(h^{-1/r}\log\varepsilon^{-1})$.

However for the boundary value problem Equation 29, where A is now a positive definite matrix, a more competitive algorithm can be obtained with reduced storage requirements. In Benson & Evans (1977), it has been shown that there exists a unique factorisation of the form

$$\tilde{A} = DTT'D, \tag{54}$$

where D is a positive diagonal matrix,
T is an upper triangular matrix with unit diagonal elements and T' denotes the matrix transpose of T.

Then, with $\tilde{A}$ given as:

$$\tilde{A} = \begin{array}{c} \xleftarrow{\quad\quad} m \xrightarrow{\quad\quad} \\ \left[\begin{array}{cccccc} a_1 & b_1 & & c_1 & & 0 \\ b_1 & a_2 & b_2 & & \ddots & \\ & \ddots & \ddots & \ddots & 0 & c_{n-m-1} \\ c_1 & & \ddots & \ddots & \ddots & \\ & \ddots & 0 & b_{n-2} & a_{n-1} & b_{n-1} \\ 0 & & c_{n-m+1} & & b_{n-1} & a_n \end{array}\right] \end{array}$$

and D and T are of the form:

$$D = \mathrm{diag}[d_1, d_2, \ldots, d_n],$$

$$T = \begin{array}{c} \xleftarrow{\quad\quad\quad} m \xrightarrow{\quad\quad\quad} \\ \left[\begin{array}{cccccc} 1 & e_1 & & t_{1,1} & & 0 \\ & 1 & e_2 & t_{2,1} & t_{1,2} & \\ & & \ddots\ \ddots & 0 \quad \vdots & \ddots & \\ & & e_{m-2} & \vdots & & t_{1,n-m+1} \\ & & 1 & e_{m-1}+t_{m-1,1} & & \vdots \\ & 0 & & 1 & \ddots & e_{n-1}+t_{m-1,n-m+1} \\ & & & & \ddots & 1 \end{array}\right] \end{array}, \tag{55}$$

we can carry out the required matrix multiplications and equate corresponding matrix elements in Equation 54 to obtain the required algorithm.

The following normalised algorithm [Benson and Evans, 1972] gives the elements of D and T,

$$d_1 = \sqrt{a_1};\ d_i = [a_i-(b_{i-1}/d_{i-1})^2]^{\frac{1}{2}};\ e_{i-1} = b_{i-1}/d_{i-1}d_i$$

$$\text{for } i=2,3,\ldots,m-1. \qquad (56)$$

Then, for $j=1,2,3,\ldots,n-m+1$

$$z_1 = c_j/d_j, \qquad v = b_{m+j-2}/d_{m+j-2} \qquad (57)$$

whilst for $j \leq m-2$

$$z_i = -e_{i+j-2}z_{i-1},\ \text{for } i=2,3,\ldots,m-j. \qquad (58)$$

Then, for all $j>1$

$$z_i = -e_{i+j-2}z_{i-1}-\sum_{k=1}^{i-1} z_k t_{k-i+m,i+j-m} \qquad (59)$$

for either $i=m-j+1,m-j+2,\ldots,m-1$ and all $j \leq m-1$.

Then,

$$d_{m+j-1} = [a_{m+j-1}-\sum_{k=1}^{m-2} z_k^2-(z_{m-1}+v)^2]^{\frac{1}{2}};\ e_{m+j-2}=v/d_{m+j-1} \qquad (60)$$

and $$t_{i,j} = z_i/d_{m+j-1}, \quad i=1,2,\ldots,m-1. \qquad (61)$$

The linear system $A\underline{\psi}=\underline{\zeta}$ can then be solved by rewriting it in the equivalent form

$$T'T(D\underline{\psi}) = D^{-1}\underline{\zeta} \qquad (62)$$

and introducing the auxiliary vectors $\underline{y}$ and $\underline{g}$ where

$$\underline{y} = D\underline{\psi}\ ;\quad \underline{g} = D^{-1}\underline{s} \qquad (63)$$

the problem is reduced to solving the normalised system

$$T'T\underline{y} = \underline{g}\ . \qquad (64)$$

Then $\underline{y}$ is obtained in terms of an auxiliary vector $\underline{h}$, where

$$T'\underline{h} = \underline{g} \text{ and } T\underline{y} = \underline{h}\ , \qquad (65)$$

i.e., $$h_1=g_1;\ h_i=g_i-e_{i-1}h_{i-1}, \qquad i=2,3,\ldots,m-1.$$

and $$h_i = g_i-e_{i-1}h_{i-1}-\sum_{k=i-m+1}^{i-1} t_{k-i+m,i-m+1}h_k,$$

$$i=m,m+1,\ldots,n. \qquad (66)$$

A back substitution process gives $\underline{y}$ in terms of $\underline{h}$, the components being given by the equations

$$y_n = h_n \,,$$

$$y_i = h_i - e_i y_{i+1} - \sum_{j=p}^{q} t_{i-1+m,\,j-m+1}\, y_j \tag{67}$$

where p and q are as specified previously in Equations 47.

Finally, the required solution is obtained from,

$$\underline{\psi} = D^{-1}\underline{y} \tag{68}$$

which is an operation involving only one division per component.

The total memory requirement now reduces to (m+3)n words with only the d,e and s vectors and the t 'fill in' array to consider whilst the amount of work involved in the factorisation is $O(m^2/2)n$ multiplications and n square roots.

In a similar manner as for the previous algorithm, we can consider the approximate factorisation

$$\tilde{A} \simeq D_s T_s^{r'} T_s^r D \,, \tag{69}$$

where D_s, T_s^r are the sparse forms of the D and T matrices given in Equation 54 and $T_s^{r'}$ denotes the transpose of T_s^r.

Thus D_s and T_s^r will have the following forms:

$$D_s \equiv [d_1, d_2, \ldots, d_n],$$

and

$$T_s^r \equiv \left[\begin{array}{ccccc|ccc}
1 & e_1 & & & & t_{1,1} & & 0 \\
 & 1 & e_2 & & & \vdots & \ddots & \\
 & & \ddots & \ddots & & t_{r,1} & & t_{1,n-m+1} \\
 & & & \ddots & e_{m-2} & & \ddots & \vdots \\
 & & & & 1 & e_{m-1} & 0 & \\
\hline
 & & & & & 1 & e_m & t_{r,n-m+1} \\
 & & 0 & & & & \ddots & \ddots \\
 & & & & & & \ddots & e_{n-1} \\
 & & & & & & & 1
\end{array}\right] \tag{70}$$

obtained, yet again, by retaining the r-outermost matrix elements in the T matrix.

The resulting sparse form of the algorithm can be obtained by reducing the inner loops in Equations 57-67 from (m-1) to r and its final form is given in Lipitakis (1978). Correspondingly, the storage requirements and computational work are also reduced in a similar manner.

Obviously, similar iterative procedures to Equations 51,52 and 53 can be derived using this sparse normalised factorisation procedure. However, in this section, we indicate its suitability for use with the variational methods and choose by way of illustration, the Conjugate Gradient method.

Thus, instead of considering the linear system given by Equation 28, we now consider the equivalent linear system

$$(D_s T_s^{r'} T_s^r D_s)^{-1} A\underline{\psi} = (D_s T_s^{r'} T_s^r D_s)^{-1} \underline{\zeta} \tag{71}$$

where D_s and T_s^r are the known sparse factors of A.

Then, let $\underline{\psi}_0$ be an arbitrary initial approximation of the solution $\underline{\psi}$ and form the residual and normalised residual vectors respectively, given by

$$\underline{r}_0 = \underline{s} - A\underline{\psi}_0 \; ; \qquad \underline{\tilde{r}}_0 = D_s^{-1} \underline{r}_0 \; . \tag{72}$$

We then solve the system,

$$(T_s^{r'} T_s) \underline{\bar{r}}_0 = \underline{\tilde{r}}_0 \; , \tag{73}$$

which by virtue of the sparse triangular forms of $T_s^{r'}$ and T_s are simple forward and backward recursive procedures.

Then, we set

$$\underline{\tilde{\sigma}}_0 = \underline{\bar{r}}_0 \tag{74}$$

and for $\ell=0,1,2,\ldots$ calculate the vectors $\underline{\psi}^{(\ell+1)}, \underline{\tilde{r}}^{(\ell+1)}, \underline{\tilde{\sigma}}^{(\ell+1)}$ and the scalar quantities $\tilde{\alpha}_i, \tilde{\beta}_{i+1}$ as follows:

$$\tilde{\alpha}_\ell = (\underline{\tilde{r}}_i^{(\ell)} \underline{\bar{r}}^{(\ell)}) / (\underline{\tilde{\sigma}}^{(\ell)} A \underline{\tilde{\sigma}}^{(\ell)}) \; , \tag{75}$$

$$\underline{\psi}^{(\ell+1)} = \underline{\psi}^{(\ell)} + \tilde{\alpha}_i \underline{\tilde{\sigma}}^{(\ell)} , \quad \underline{\tilde{r}}^{(\ell+1)} = \underline{\tilde{r}}^{(\ell)} - \tilde{\alpha}_i A \underline{\tilde{\sigma}}^{(\ell)} . \tag{76}$$

Then, we solve

$$(T_s^{r'} T_s^r) \underline{\bar{r}}^{(\ell+1)} = \underline{\tilde{r}}^{(\ell+1)} \; , \tag{77}$$

and evaluate

$$\tilde{\beta}_{i+1} = (\underline{\tilde{r}}^{(\ell+1)} \underline{\bar{r}}^{(\ell+1)}) / (\underline{\tilde{r}}^{(\ell)} \underline{\bar{r}}^{(\ell)}) \tag{78}$$

and finally

$$\underline{\tilde{\sigma}}^{(\ell+1)} = \underline{\tilde{r}}^{(\ell+1)} + \beta_{i+1} \underline{\tilde{\sigma}}^{(\ell)} \; , \tag{79}$$

which with exact arithmetic converges in n steps. However, recently its use has been confined to iterative methods and Young (1975) has shown that it can be related to the class of second order iterative methods given by Equation 53. In this context the algorithm converges quickly and when used in conjunction with the normalised implicit approach a powerful method emerges.

Other advantages of the Conjugate Gradient algorithm is that it makes no basic assumptions about the structure of the coefficient matrix A other than positive definiteness and requires no estimation of iterative parameters or the spectral radius of the iteration matrix. Furthermore, the new method, i.e., Normalised Implicit Conjugate Gradient, because of the normalisation and choice of optimum fill in parameter r compares favourably with other similar methods recently developed, Meijernik et al (1977), Chandra et al (1975), Gustafsson (1977), Kershaw (1977).

Numerical studies concerned with the determination of the optimal choice of r, the extension of the algorithm to multi-dimensional problems as well as application studies to a wide variety of linear and non-linear boundary value problems are given in Lipitakis (1978) and Evans and Lipitakis (1979B).

5. FAST SOLUTION METHODS FOR SEPARABLE EQUATIONS.

As an introduction, we consider the tridiagonal linear systems of special form that can occur in the numerical solution of one dimensional boundary value problems by finite difference methods (Evans, 1977), i.e.,

$$A\underline{x} = \underline{d} \ , \tag{80}$$

where

$$A \equiv \begin{bmatrix} b & -c & & & 0 \\ -a & b & -c & & \\ & \ddots & \ddots & \ddots & \\ & & \ddots & \ddots & -c \\ 0 & & & -a & b \end{bmatrix} \tag{81}$$

$\underline{x}^T = [x_1, x_2, \ldots, x_n]$ is the unknown vector,

and $\underline{d}^T = [d_1, d_2, \ldots, d_n]$ is the known right hand vector usually emanating from the boundary conditions and source terms in such problems.

The system given by Equation 80 can be rewritten in the form

$$\tilde{A}\underline{x} = \underline{g} \tag{82}$$

where

$$\tilde{A} \equiv \begin{bmatrix} 1+\beta\gamma & -\gamma & & & 0 \\ -\beta & 1+\beta\gamma & -\gamma & & \\ & \ddots & \ddots & \ddots & \\ & & \ddots & \ddots & -\gamma \\ 0 & & & -\beta & 1+\beta\gamma \end{bmatrix} \tag{83}$$

and we have completed the simple transformations

$$\delta = [b+(b^2-4ac)^{\frac{1}{2}}], \quad \gamma=2c/\delta, \quad \beta=2a/\delta \tag{84}$$

with $\qquad g_i = 2d_i/\delta \quad$ for $i=1,2,\ldots,n$.

Consider now the unique factorisation

$$\tilde{A} \equiv PQ \tag{85}$$

where P and Q are the $n\times(n+1)$ and $(n+1)\times n$ rectangular matrices of the form:

$$P = \begin{bmatrix} 1 & -\gamma & & & 0 \\ & 1 & -\gamma & & \\ & & \ddots & \ddots & \\ 0 & & & 1 & -\gamma \end{bmatrix}_{n\times(n+1)}, \quad Q = \begin{bmatrix} 1 & & & 0 \\ -\beta & 1 & & \\ & \ddots & \ddots & \\ & & \ddots & 1 \\ 0 & & & -\beta \end{bmatrix}_{(n+1)\times n} \tag{86}$$

Thus, the system given by Equation 82 is more easily solved by using the two alternative systems,

$$P\underline{y} = \underline{g} \quad \text{and} \quad Q\underline{x} = \underline{y} \tag{87}$$

where we have introduced the auxiliary $(n+1)$-dimensional vector $\underline{y}$ $(y_i{=}1,2,\ldots,n,n+1)$.

Now the matrices P and Q although rectangular have upper triangular and lower triangular forms respectively and can be easily inverted and the linear systems Equation 87 can be shown [Evans, 1972] to give the following efficient algorithmic solution

$$x_n = [(g_n+\beta g_{n-1}+\ldots+\beta^{n-1}g_1)-\beta^n(\gamma g_1+\ldots+\gamma^{n-1}g_{n-1}+\gamma^n g_n)]/[1-(\beta\gamma)^{n+1}] \tag{88}$$

and $\qquad y_{n+1} = -\beta x_n.$

Then, a simple back substitution process yields the components of the auxiliary solution,

$$y_i = g_i + \gamma y_{i+1}, \qquad i=n,n-1,\ldots,2,1. \tag{89}$$

whilst the solution vector $\underline{x}$ are given by a simple forward process,

$$x_1 = y_1;\; x_i = y_i + \beta x_{i-1}, \; i=2,3,\ldots,n-1. \tag{90}$$

Since, by definition from Equation 84 $|\beta|,|\gamma|<1$, then it follows immediately that the recursive processes given by Equations 89 and 90 are stable against the accumulation of rounding errors.

The algorithm (Equations 88-90) requires 4n additions and 5n multiplications assuming that Equation 88 is calculated in the most efficient manner.

For the special case $b=2a=2c$, then $\beta\gamma=-1$ and the calculation of x_n in Equation 88 becomes indeterminate. However, an alternative expression can easily be obtained in the form

$$x_n = [ng_n - (n-1)g_{n-1} + (n-2)g_{n-2} + \ldots + (-1)^{n-1}g_1]/(n+1) \tag{91}$$

together with

$$y_{n+1} = x_n;\quad y_i = g_i - y_{i+1};\quad i=n,(n-1),\ldots,2,1, \tag{92}$$

and

$$x_1 = y_1;\qquad x_i = y_i - x_{i-1};\quad i=2,3,\ldots,(n-1). \tag{93}$$

Similarly, for $b=-2a=-2c$, then $\beta,\gamma=1$ and the algorithm becomes

$$x_n = [ng_n + (n-1)g_{n-1} + (n-2)g_{n-2} + \ldots + g_1]/(n+1) \tag{94}$$

together with

$$y_{n+1} = -x_n;\quad y_i = g_i + y_{i+1};\; i=n,(n-1),\ldots,2,1 \tag{95}$$

and

$$x_1 = y_1;\qquad x_i = y_i + x_{i-1};\; i=2,3,\ldots,(n-1). \tag{96}$$

Similar algorithms (Evans, 1971) can be obtained for the periodic case, i.e. when

$$A = \begin{bmatrix} b & -c & & & & -a \\ -a & b & -c & & 0 & \\ & \ddots & \ddots & \ddots & & \\ & & \ddots & \ddots & \ddots & \\ & 0 & & \ddots & \ddots & -c \\ -c & & & & -a & b \end{bmatrix} \tag{97}$$

whilst the more generalised case, which occurs under Neumann and mixed boundary conditions, i.e., when

$$A = \begin{bmatrix} b_1 & a & & & & \\ a & b & a & & 0 & \\ & \ddots & \ddots & \ddots & & \\ & & \ddots & \ddots & \ddots & \\ & 0 & & \ddots & b & a \\ & & & & a & b_2 \end{bmatrix} \tag{98}$$

can be similarly treated, by considering the factorisation

$$A = PP^T$$

where

$$P = \begin{bmatrix} x & \alpha & & & & \\ & \beta & \ddots & & 0 & \\ & & \ddots & \ddots & & \\ & & & \ddots & \ddots & \\ & 0 & & & \ddots & \alpha \\ & & & & \beta & y \end{bmatrix}_{n\times(n+1)} \tag{99}$$

from which the quantities α, β, x and y can be determined by the relations,

$$\alpha = \tfrac{1}{2}[(b+2a)^{\frac{1}{2}}-(b-2a)^{\frac{1}{2}}]; \quad \beta = \tfrac{1}{2}[(b+2a)^{\frac{1}{2}}-(b-2a)^{\frac{1}{2}}] \tag{100}$$

and $\quad x = (b_1-\alpha^2)^{\frac{1}{2}}; \quad y = (b_2-\beta^2)^{\frac{1}{2}}.$

Further applications to linear elliptic and parabolic partial differential equations are given in Evans and Avdelas (1978) whilst similar analyses for special quindiagonal systems that occur in Biharmonic problems have recently been given by Evans and Hadjidimos (1979).

Finally, the usefulness of the algorithm can be readily extended to block tridiagonal systems, where A is a real (n×n) positive definite symmetric matrix of the form,

$$A = \begin{bmatrix} B & -I & & & & \\ -I & B & -I & & 0 & \\ & \ddots & \ddots & \ddots & & \\ & & \ddots & \ddots & \ddots & \\ & 0 & & \ddots & B & -I \\ & & & & -I & B \end{bmatrix} \tag{101}$$

which is obtained from the approximation of the five point finite difference approximation to a second order elliptic partial differential equation on a rectangular grid of (m×p) grid points as in Equation 28.

Consider the partitioned form (101) and suppose that the vectors

$$\underline{z}^T = [\underline{z}_1^T, \underline{z}_2^T, \ldots, \underline{z}_{p-1}^T], \quad \underline{b}^T = [\underline{b}_1^T, \underline{b}_2^T, \ldots, \underline{b}_{p-1}^T]$$

are partitioned similarly. B and I are square (m-1)×(m-1) symmetric matrices and A is square of size (m-1)×(p-1).

Now there is an orthogonal matrix Q which "diagonalises" B, that is,

$$Q^T B Q = \Lambda, \quad Q^T Q = I, \tag{102}$$

where Λ is a diagonal matrix whose elements are the eigenvalues of B, the columns of Q being the eigenvectors of B. Expansion of (101) in the form

$$\left.\begin{array}{ll} B\underline{z}_1 - \underline{z}_2 & = \underline{b}_1 \\ -\underline{z}_1 + B\underline{z}_2 - \underline{z}_3 & = \underline{b}_2 \\ \cdots\cdots\cdots\cdots\cdots & \\ -\underline{z}_{p-2} + B\underline{z}_{p-1} & = \underline{b}_{p-1} \end{array}\right\} \tag{103}$$

and use of (102) gives

$$\left.\begin{array}{ll} \Lambda\underline{\bar{z}}_1 - \underline{\bar{z}}_2 & = \underline{\bar{b}}_1 \\ -\underline{\bar{z}}_1 + \underline{\bar{z}}_2 - \underline{\bar{z}}_3 & = \underline{\bar{b}}_2 \\ \cdots\cdots\cdots\cdots\cdots & \\ -\underline{\bar{z}}_{p-2} + \Lambda\underline{\bar{z}}_{p-1} & = \underline{\bar{b}}_{p-1} \end{array}\right\} \tag{104}$$

where $\underline{\bar{z}}_i = Q^T\underline{z}_i$ and similarly for $\underline{\bar{b}}_i$. Then, if $\underline{\bar{z}}_i$ has components $\bar{z}_{1,i}, \bar{z}_{2,i}, \ldots, \bar{z}_{m-1,i}$ and similarly for $\underline{\bar{b}}_i$, we can write, for i=1,2,...,m-1, the scalar equations

$$\left.\begin{array}{ll} \lambda_i \bar{z}_{i,1} - \bar{z}_{i,2} & = \bar{b}_{i,1} \\ -\bar{z}_{i,1} + \lambda_i \bar{z}_{i,2} - \bar{z}_{i,3} & = \bar{b}_{i,2} \\ -\bar{z}_{i,2} + \lambda_i \bar{z}_{i,3} - \bar{z}_{i,4} & = \bar{b}_{i,3} \\ \cdots\cdots\cdots\cdots\cdots & \\ -\bar{z}_{i,p-2} + \lambda_i \bar{z}_{i,p-1} & = \bar{b}_{i,p-1} \end{array}\right\} \tag{105}$$

and, if we use the further notation

$$\Gamma_i = \begin{bmatrix} \lambda_i & -1 & & & 0 \\ -1 & \lambda_i & -1 & & \\ & \ddots & \ddots & \ddots & \\ & & \ddots & \ddots & -1 \\ 0 & & & -1 & \lambda_i \end{bmatrix}, \hat{\underline{z}}_i = \begin{bmatrix} \bar{z}_{i,1} \\ \bar{z}_{i,2} \\ \vdots \\ \bar{z}_{i,p-1} \end{bmatrix}, \hat{\underline{b}}_i = \begin{bmatrix} \bar{b}_{i,1} \\ \bar{b}_{i,2} \\ \vdots \\ \bar{b}_{i,p-1} \end{bmatrix} \tag{106}$$

We can write (105) in the condensed form

$$\Gamma_i \hat{\underline{z}}_i = \hat{\underline{b}}_i, \qquad i=1,2,\ldots,m-1. \tag{107}$$

The algorithm therefore consists of: (i) finding the eigenvalues and eigenvectors of B;

(ii) computing $\bar{\underline{b}}_i = Q^T \underline{b}_i$, $i=1,2,\ldots,p-1$;

(iii) solving (107) for $i=1,2,\ldots,m-1$ and

(iv) computing $\underline{z}_i = Q\bar{\underline{z}}_i$, $i=1,2,\ldots,p-1$.

Steps (ii) and (iv) can be performed rapidly by using the fast Fourier transform technique of Cooley and Tukey(1965) in certain cases, and Q is often known from analytical considerations. The solution of the special triple-diagonal systems (107) is also quick and easy by the method given by Equations 81-90.

6. PRECONDITIONED ITERATIVE METHODS.

In Evans (1967), a class of preconditioned iterative methods in which the convergence rate was maximised by the introduction of a new preconditioning factor was introduced. In these methods, the given linear system of finite difference equations

$$A\underline{x} = \underline{b}, \tag{108}$$

i.e.,

$$\sum_i a_{i,j} u_j = b_i, \quad 1 \leq i,j \leq N$$

is made equivalent to the following systems,

$$RAQ\underline{y} = R\underline{b}, \tag{109}$$

$$\underline{u} = Q\underline{y}, \tag{110}$$

by the introduction of the non-singular (N×N) matrices R and Q. Then, the alternative system,

$$B\underline{y} = \underline{d}, \tag{111}$$

where

$$B = RAQ, \quad \underline{d} = Q\underline{y}, \tag{112}$$

is solved by iterative methods which will be described later.

The problem now is to choose the matrices R and Q such that system (111) converges faster than if the original system (108) had been taken. Since the convergence rates of many iterative methods depends inversely upon the condition number of the matrix then we see that this objective will be achieved if the chosen matrices R and Q 'precondition' the given system (108).

Since Q must be a matrix of easily invertible form so that the final solution $\underline{u}$ can be recovered from (110), we have 2 alternative forms to consider for the R and Q matrices, i.e., triangular and tridiagonal. The former choice for R,Q was investigated fully in Evans (1967), and in this section we propose to consider the latter interesting case, i.e., when R and Q are tridiagonal matrices.

Consider the solution of the self-adjoint elliptic equation of 2nd order, i.e.,

$$\frac{\partial}{\partial x}\left(P(x,y)\frac{\partial U}{\partial x}\right) + \frac{\partial}{\partial y}\left(Q(x,y)\frac{\partial U}{\partial y}\right) + \sigma(x,y)U(x,y) + f(x,y) = 0, \quad (x,y)\in R, \tag{113}$$

where R is the rectangular region

$$0 \leq x \leq \ell_1, \quad 0 \leq y \leq \ell_2 .$$

For boundary conditions, we assume the problem to be of Dirichlet type so that

$$U(x,y) = g(x,y), \quad (x,y)\in B , \tag{114}$$

where $g(x,y)$ is a prescribed function on the boundary B of $\underline{R}$. We also have the conditions that P,Q,σ,f are continuous in $\bar{R}$ and satisfy

$$P(x,y),\ Q(x,y)>0;\ \sigma(x,y)\leq 0, \text{ for all } (x,y)\in R. \tag{115}$$

We superimpose a rectangular grid (of length Δx in the x-direction and Δy in the y-direction) on R, such that

$$(n+1)\Delta x = \ell_1, \qquad (m+1)\Delta y = \ell_2 , \tag{116}$$

and consider the numerical solution of the problem (113) subject to the boundary conditions (114) associated with the set of mesh points on the rectangular grid. Defining the operators,

$$\left.\begin{aligned} Hu(x,y) &= -a(x,y)u(x+h,y)+2b(x,y)u(x,y)-c(x,y)u(x-h,y) \\ Vu(x,y) &= -\alpha(x,y)u(x,y+h)+2\beta(x,y)u(x,y)-\gamma(x,y)u(x,y-h) \end{aligned}\right. \tag{117}$$

where

$$\left.\begin{aligned} a &= (\Delta y/\Delta x)P(x+h/2,y),\ c = (\Delta y/\Delta x)P(x-h/2,y),\ 2b=a+c \\ \alpha &= (\Delta x/\Delta y)Q(x,y+h/2),\ \gamma = (\Delta x/\Delta y)Q(x,y-h/2),\ 2\beta=\alpha+\beta \end{aligned}\right\} \tag{118}$$

we see that they are respectively the central difference approximations to the terms $\{-\Delta x\Delta y\ \frac{\partial}{\partial x}(P(x,y)\frac{\partial U}{\partial x})$ and $\{-\Delta x\Delta y\frac{\partial}{\partial y}(Q(x,y)\frac{\partial U}{\partial y})\}$ in (113).

Ordering the unknown mesh points along successive rows (x-lines) of the grid, the finite difference equations approximating (113) at each of these mesh points can be written as

$$A\underline{u} = (H+V+\Sigma)\underline{u} = (H_1+V_1)\underline{u} = \underline{b} , \tag{119}$$

where H_1 and V_1 are defined to be $H_1=H+\Sigma/2$, $V_1=V+\Sigma/2$ and are real, symmetric, diagonally dominant matrices with positive diagonal entries and non-positive off-diagonal entries. In order that H_1 and V_1 commute the differential equation (113) must be of self-adjoint form.

Let us first rewrite the coefficient matrix A as

$$\begin{aligned} A &= I+(H_1-I/2) + (V_1-I/2) \\ &= I+\hat{H}+\hat{V} , \end{aligned} \tag{120}$$

where

$$\hat{H} = H_1-I/2,\ \hat{V} = V_1-I/2 . \tag{121}$$

If (119) is premultiplied by $(I+\omega\hat{H})^{-1}$ then it is equivalent to the preconditioned system,

$$(I+\omega\hat{H})^{-1}A(I+\omega\hat{V})^{-1}(I+\omega\hat{V})\underline{u} = (I+\omega\hat{H})^{-1}\underline{b} \tag{122}$$

or $\qquad B_\omega \cdot \underline{y} = \underline{d}$,

where

$$\left.\begin{aligned} B_\omega = (I+\omega\hat{H})^{-1}A(I+\omega\hat{V})^{-1};\ \underline{y} &= (I+\omega\hat{V})\underline{u}; \\ \underline{d} &= (I+\omega\hat{H})\underline{b}. \end{aligned}\right\} \tag{123}$$

The objective now is identical to that described in the preconditioning techniques previously applied by Evans (1967). It is obvious that because H_1,V_1 are symmetric and commute, then the same applies to $\hat{H},\hat{V}$. Hence, B_ω is symmetric also. Because $A,\hat{H}$ and $\hat{V}$ are pairwise commutative, and symmetric, they have a common set of orthonormal eigenvectors, so that the eigenvalues λ, of B_ω can be expressed as

$$\lambda(\mu,\nu,\omega) = \frac{\mu+\nu}{(1-\omega/2+\omega\mu)(1-\omega/2+\omega\nu)} , \tag{124}$$

where μ,ν are the real positive eigenvalues of H_1,V_1 respectively. We see immediately from (124) that if $0\leq\omega\leq 2$, then $\lambda>0$ for all λ. Hence with ω in the range (0,2) the matrix B_ω is symmetric and positive definite.

The main motivation of preconditioning the original system is to obtain a new system, (122) whose coefficient matrix varies with a parameter ω which we can then minimise the P-condition number [denoted by $P(B_\omega)$] with respect to ω, for some "optimum" set of values $0\leq\omega\leq 2$.

In order to investigate the P-condition number of B_ω, we must examine

$$P(B_\omega) = \frac{\max\limits_{\substack{a\lesssim\mu\lesssim b\\ \alpha\lesssim\nu\lesssim\beta}} \{\lambda(\mu,\nu,\omega)\}}{\min\limits_{\substack{a\lesssim\mu\lesssim b\\ \alpha\lesssim\nu\lesssim\beta}} \{\lambda(\mu,\nu,\omega)\}} \tag{125}$$

where we assume that the eigenvalues μ of H and ν of V lie in the ranges $0<a\lesssim\mu\lesssim b$, $0<\alpha\lesssim\nu\lesssim\beta$.

To do this we consider the following continuous function

$$\lambda(x,y,\omega) = \frac{x+y}{(1-\omega/2+\omega x)(1-\omega/2+\omega y)}, \quad \left.\begin{matrix} 0<a\lesssim x\lesssim b \\ 0<\alpha\lesssim y\lesssim \beta \end{matrix}\right\} \tag{126}$$

initially for the case where μ and ν (and hence x and y) lie in the same range, i.e., $a=\alpha$, $b=\beta$.

By the substitution $\omega=2/(1+2r)$, (126) can be written in the form,

$$\lambda(x,y,r) = (x+y)/(r+x)(r+y)\omega^2 \tag{127}$$

and the problem of minimising (125) then becomes identical to the situation previously discussed by Hadjidimos (1977). However we follow the latter analysis who in the extrapolated A.D.I. method minimised the amplification factor

$$\kappa = \min_r \left\{ \frac{\max\limits_{\alpha\lesssim\mu,\nu\lesssim\beta} \lambda(\mu,\nu,r)}{\min\limits_{\alpha\lesssim\mu,\nu\lesssim\beta} \lambda(\mu,\nu,r)} \right\} \tag{128}$$

to obtain for the optimum value of $r=\sqrt{\alpha\beta}$.

Thus using this result, we have after some analysis, the results

$$\omega_{opt} = \bar{\omega} = 2/(1+2\sqrt{\alpha\beta}), \quad P(B_{\omega_{opt}}) = (\alpha+\beta)/2\sqrt{\alpha\beta} \tag{129}$$

from which we can determine the eigenvalue range for $B_{\omega_{opt}}$

$$\bar{a}=\lambda_{min}(B_{\bar{\omega}})=(1+2\sqrt{\alpha\beta})^2/2(\sqrt{\beta}+\sqrt{\alpha})^2, \quad b=\lambda_{max}(B_{\bar{\omega}})=(\alpha+\beta)(1+2\sqrt{\alpha\beta})^2/4\sqrt{\alpha\beta}(\sqrt{\beta}+\sqrt{\alpha})^2. \tag{130}$$

To date, we have assumed that $0<a\lesssim\mu\lesssim b$ and $0<\alpha\lesssim\nu\lesssim\beta$ and that $a=\alpha$ and $b=\beta$. This will not always occur in practice, e.g. when solving Laplace's equation on a rectangle with adjacent sides of different length, using a square mesh.

As far as the minimum eigenvalue of B_ω is concerned an approximate analysis can be achieved by noticing that $0<\alpha\lesssim\mu,\nu\lesssim\beta'$, where $\alpha'=\min(a,\alpha)$ and $\beta'=\max(b,\beta)$. Then

$$\min_{\substack{a\leq\mu\leq b\\ \alpha\leq\nu\leq\beta}} \lambda(\mu,\nu,\omega) \geq \min_{\alpha'\leq\mu,\nu\leq\beta'} \lambda(\mu,\nu,\omega),$$

and

$$\max_{\substack{a\leq\mu\leq b\\ \alpha\leq\nu\leq\beta}} \lambda(\mu,\nu,\omega) \leq \max_{\alpha'\leq\mu,\nu\leq\beta'} \lambda(\mu,\nu,\omega),$$

for $0<\omega<2$.

Hence, we can derive the result,

$$\min_{0<\omega<2} P(B_\omega) \leq \min_{0<\omega<2} \left\{ \frac{\max\limits_{\alpha'\leq\mu,\nu\leq\beta'} \lambda(\mu,\nu,\omega)}{\min\limits_{\alpha'\leq\mu,\nu\leq\beta'} \lambda(\mu,\nu,\omega)} \right\} = P(B_{\bar{\omega}'}),$$

where $\omega'=2/(1+2\sqrt{\alpha'\beta'})$.

The stationary case of 2 optimum parameters $\bar{\omega}_1$ and $\bar{\omega}_2$ is discussed in Young (1971) but interest has been centred for many years on the non-stationary case when a sequence of parameters ω_i, $i=1,2,\ldots,m$ is applied in a cycle resulting in an order of magnitude improvement in the rate of convergence, (Peaceman & Rachford (1955), Wachspress (1966)).

However here we are mainly concerned with the single parameter case and its application to 1st and 2nd order extrapolation methods and their comparison with existing methods whilst the non-stationary case is given in Evans & Missirlis (1979).

We now apply the results of the previous section to the Simultaneous Displacement and Richardson's and Chebyshev second degree methods, respectively as given previously in Evans (1967).

The equations for these basis iterative processes are then,

$$\underline{y}^{(k+1)} = \underline{y}^{(k)} + \bar{\alpha}(\underline{d} - B_{\bar{\omega}}\underline{y}^{(k)}), \qquad (131)$$

$$\underline{y}^{(k+1)} = \underline{y}^{(k)} + \bar{\alpha}(\underline{d} - B_{\bar{\omega}}\underline{y}^{(k)}) + \bar{\beta}(\underline{y}^{(k)} - \underline{y}^{(k-1)}), \qquad (132)$$

and

$$\underline{y}^{(k+1)} = \underline{y}^{(k)} + \alpha_k(\underline{d} - B_{\bar{\omega}}\underline{y}^{(k)}) + \beta_k(\underline{y}^{(k)} - \underline{y}^{(k-1)}), \qquad (133)$$

respectively, where the acceleration parameters $\bar{\alpha},\bar{\beta},\alpha_k$ and β_k are defined as in Evans (1967). The iterations proceed in the y variable until a specified degree of accuracy is achieved. The final solution is then obtained by one application of the formula,

$$\underline{u} = (I+\omega\hat{V})^{-1}\underline{y}. \qquad (134)$$

The optimum value of $\alpha=\bar{\alpha}$ for method (131) is given by,

$$\bar{\alpha} = \frac{2}{\bar{a}+\bar{b}} = \frac{8\sqrt{\alpha\beta}}{(1+2\sqrt{\alpha\beta})^2} , \tag{135}$$

where $0<\bar{a}\leq\lambda_i(\bar{\omega})\leq\bar{b}$, and the $\lambda_i(\bar{\omega})$ are the positive real eigenvalues of $B_{\bar{\omega}}$. With this choice of α, the spectral radius of the iteration matrix $(I-\bar{\alpha}B_{\bar{\omega}})$ is

$$\max_i |1-\bar{\alpha}\lambda_i(\bar{\omega})| \leq \frac{\bar{b}-\bar{a}}{\bar{b}+\bar{a}} = \frac{P(B_{\bar{\omega}})-1}{P(B_{\bar{\omega}})+1} . \tag{136}$$

The values $\bar{a}=\min\limits_{\alpha\leq\mu,\nu\leq\beta} \lambda(\mu,\nu,\bar{\omega})$ and $\bar{b}=\max\limits_{\alpha\leq\mu,\nu\leq\beta} \lambda(\mu,\nu,\bar{\omega})$ are given by equations (130) and from equation (136) the rate of convergence is

$$R(I-\alpha B_{\bar{\omega}}) = -\log\left\{\frac{P(B_{\bar{\omega}})-1}{P(B_{\bar{\omega}})+1}\right\} . \tag{137}$$

Similarly, for the preconditioned Richardson's second degree method equation (132) we have,

$$\bar{\alpha} = \left[\frac{2}{\sqrt{\bar{a}}+\sqrt{\bar{b}}}\right]^2 \quad \text{and} \quad \bar{\beta} = \left[\frac{\sqrt{\bar{a}}-\sqrt{\bar{b}}}{\sqrt{\bar{a}}+\sqrt{\bar{b}}}\right]^2 \tag{138}$$

and the spectral radius of the iteration matrix is $\sqrt{\bar{\beta}}$, so that the rate of convergence satisfies

$$-\log(\sqrt{\bar{\beta}}) = -\log\left\{\frac{\sqrt{P(B_{\bar{\omega}})}-1}{\sqrt{P(B_{\bar{\omega}})}+1}\right\} \approx 2/\sqrt{P(B_{\bar{\omega}})} . \tag{139}$$

Similarly, the Chebyshev acceleration defined by equation (133) has an identical asymptotic rate of convergence.

We now consider the computational procedure for the method developed in (131). Assume that the grid points are ordered horizontally along successive x-lines of the grid. Assume also that the optimum parameters $\bar{\omega},\bar{\alpha}$ have been obtained. The computation of the R.H.S. vector $\underline{d}$, i.e.,

$$(I+\omega\hat{H})\underline{d} = \underline{b} . \tag{140}$$

is obtained by applying a Gaussian elimination procedure to the tridiagonal system (140). Next, the vector $\underline{q}^{(k)}$, i.e.,

$$\underline{q}^{(k)} = (I+\omega\hat{H})^{-1}A(I+\omega\hat{V})^{-1}\underline{y}^{(k)}, \quad k\geq 0 \tag{141}$$

is generated as a 3 stage process using auxiliary vectors z_1,z_2 in the manner, i.e.,

$$(I+\omega\hat{V})\underline{z_1} = \underline{y}^{(k)} \tag{142}$$

$$\underline{z_2} = A\underline{z_1} \tag{143}$$

$$(I+\omega\hat{H})\underline{q}^{(k)} = \underline{z_2} \tag{144}$$

Equation (142) is solved by re-ordering the elements of the vector $\underline{y}^{(k)}$ and thus $\underline{z_1}$ also, to correspond with a vertical ordering of the grid points along successive y-lines which results in a similar tridiagonal systems of equations to solve for $\underline{z_1}$.

The vector $\underline{z_2}$ in equation (143) is generated by a simple matrix vector multiplication with A as a sparse matrix formed from the 5 point difference approximation to (113).

Finally, we solve (144) in exactly the same way as system (139) to obtain the vector $\underline{q}^{(k)}$.

Method (131) can now be written in the form,

$$\underline{y}^{(k+1)} = \underline{y}^{(k)} + \bar{\alpha}(\underline{d} - \underline{q}^{(k)}), \quad k \geq 0, \tag{145}$$

which is easily solved for $\underline{y}^{(k+1)}$, $k \geq 0$ since each vector on the right hand side is known. This process continues until convergence (to the desired accuracy) is achieved. The original solution vector $\underline{u}$ is then found from (134) in the same way as system (142) was obtained.

In a similar manner the second methods (132) and (133) can be evaluated with, of course, the storage of an additional vector $\underline{y}^{(k-1)}$.

We shall consider the model problem:

$$\frac{\partial^2 U}{\partial x^2} + \frac{\partial^2 U}{\partial y^2} = 0, \qquad 0<x,y<1 \tag{146}$$

in the unit square with boundary B, subject to the boundary conditions,

$$U(x,y) = \gamma(x,y), \qquad (x,y)\ B, \tag{147}$$

Choosing a square mesh size of side h (h=1/(n+1)), then the matrix of finite difference equations can be written as

$$A = H + V \quad (=H_1+V_1), \tag{148}$$

where

$$(H_1)H = \begin{bmatrix} B & & & \\ & B & & 0 \\ & & \ddots & \\ 0 & & & B \end{bmatrix}_{(n^2 \times n^2)} ; \; B = \begin{bmatrix} 1/2 & -1/4 & & & 0 \\ -1/4 & \ddots & \ddots & & \\ & \ddots & \ddots & \ddots & \\ & & \ddots & \ddots & -1/4 \\ 0 & & & -1/4 & 1/2 \end{bmatrix}_{(n \times n)} \tag{149}$$

and

$$(V_1)V=\begin{bmatrix} I/2 & -I/4 & & \\ -I/4 & I/2 & \ddots & \\ & \ddots & \ddots & -I/4 \\ & & -I/4 & I/2 \end{bmatrix}_{(n^2\times n^2)} \quad ; \text{ I is the unit matrix of order n.} \tag{150}$$

If again μ,ν denote the eigenvalues of $H(=H_1), V(=V_1)$ respectively, then

$$\mu_i=\sin^2(i\pi/2(n+1)))\}; \; \nu_j=\sin^2\{j\pi/(2(n+1))\}, \; 1\leq i,j\leq n. \tag{151}$$

Hence, if $a\leq\mu_i\leq b, \alpha\leq\nu_k\leq\beta$, $1\leq i,j\leq n$, then

$$a = \alpha = \sin^2\left(\frac{\pi}{2(n+1)}\right) \; ; \; b = \beta = \cos^2\left(\frac{\pi}{2(n+1)}\right). \tag{152}$$

Thus, for the model problem Equation (129) yields

$$\bar{\omega} = \frac{2}{1+\sin(\pi/(n+1))} = \frac{2}{1+\sin(\pi h)} \quad , \tag{153}$$

and

$$\min_{0<\omega<2} P(B_\omega) = P(B_{\bar{\omega}}) = \frac{1}{\sin(\pi/(n+1))} = \frac{1}{\sin(\pi h)} \; , \tag{154}$$

using the optimum acceleration parameter $\bar{\alpha}$ defined by (135). Thus, the preconditioned alternating direction implicit (PADI) simultaneous displacement method has the spectral radius

$$\max_{1\leq i,j\leq n} |1-\bar{\alpha}\lambda_{i,j}(\bar{\omega})| = \frac{1-\sin(\pi h)}{1+\sin(\pi h)} \; . \tag{155}$$

with the asymptotic rate of convergence given by

$$R_\infty = 2 \sin \pi h \simeq 2\pi h \qquad \text{as } h\to 0. \tag{156}$$

From this we can ascertain that in the class of optimised one parameter iterative methods, the Peaceman-Rachford method, the point S.O.R. process and the P.A.D.I. simultaneous displacement method have identical asymptotic rates of convergence for all $h>0$ for the model problem.

Also we note that the optimum preconditioning parameter $\bar{\omega}$ given by (153) is identical to the optimum over-relaxation parameter for the point S.O.R. method (Young, 1971).

Finally for the preconditioned alternating direction 2nd degree Richardson method defined by (132) with optimum preconditioning factor $\bar{\omega}$ and parameters and $\bar{\beta}$ defined in (138) we have for the asymptotic rate of convergence,

$$R_\infty = 2\sqrt{\sin(\pi h)} \approx 2\sqrt{\pi h} \qquad \text{as } h\to 0. \tag{157}$$

This result can be shown to be equivalent to the A.D.I. iterative method with $M=2^k$, $k=1$ parameters, (Wachpress, (1966).

7. THE PARALLEL SOLUTION OF LINEAR EQUATIONS.

The emergence of parallel computers has meant that many of the most important and common serial numerical algorithms in use, i.e., linear equation solvers must be re-evaluated as to their usefulness and viability on such computers. Inevitably, the restructuring of such well known algorithms may result in the appearance of many new algorithms (Evans and Hatzopoulos,1976).

This section investigates the adaptability of the direct methods for solving the linear system of equations

$$A\underline{x} = \underline{b} \tag{158}$$

where A is a compact dense (n×n) matrix on a parallel computer. The emphasis will be based on the type of parallel computer often referred to as a single-instruction stream computer in which there are p independent processors capable of performing p tasks on different sets of data.

A new factorisation method for the solution of a linear system is proposed. The method is similar to an LU type factorisation of the coefficient matrix A where the factors are interlocking matrix quadrants and can be applied on a single-instruction stream parallel machine (Evans, 1978).

We consider the system (158) where A is an (n×n) non-singular matrix with coefficients a_{ij} and $\underline{x}$ and $\underline{b}$ are the two (n×1) vectors,

$$\underline{x} = \begin{pmatrix} x_1 \\ \vdots \\ x_n \end{pmatrix} \qquad \underline{b} = \begin{pmatrix} b_1 \\ \vdots \\ b_n \end{pmatrix} .$$

Consider now two matrices W and Z, where W and Z have the matrix forms for n even:

$$W = \begin{bmatrix} 1 & & & & & & & 0 \\ w_{2,1} & 1 & & & & & 0 & w_{2,n} \\ \vdots & w_{3,2} & \ddots & & 0 & & w_{3,n-1} & \vdots \\ \vdots & \vdots & & 1 & 0 & & \vdots & \vdots \\ \vdots & \vdots & & 0 & 1 & & \vdots & \vdots \\ \vdots & w_{n-2,2} & & & 0 & \ddots & w_{n-2,n-1} & \vdots \\ w_{n-1,1} & 0 & & & & & 1 & w_{n-1,n} \\ 0 & & & & & & & 1 \end{bmatrix} \tag{159}$$

and

$$Z = \begin{bmatrix} z_{1,1} & z_{1,2} & - - - - - - - - - - & z_{1,n-1} & z_{1,n} \\ & z_{2,2} & - - - - - - - - - & z_{2,n-1} & \\ & & \diagdown \qquad \diagup & & \\ & 0 & \times & 0 & \\ & & \diagup \qquad \diagdown & & \\ & z_{n-1,2} & - - - - - - - - - & z_{n-1,n-1} & \\ z_{n,1} & z_{n,2} & - - - - - - - - - - & z_{n,n-1} & z_{n,n} \end{bmatrix} \tag{160}$$

Initially we assume the matrix A to be positive definite and symmetric which thus obviates the need for any interchange of rows.

Let, $$W = [W_1, W_2, \ldots, W_n]$$

and $$Z^T = [Z_1, Z_2, \ldots, Z_n]$$

where

$$W_i = \begin{cases} [\underbrace{0\ldots0}_{i-1}, 1, w_{i+1,i} \ldots w_{n-i,i}, 0, \ldots 0]^T & i = 1(1)\left\lfloor \frac{n-1}{2} \right\rfloor \\ [\underbrace{0\ldots0}_{i-1}, 1, 0, \ldots 0]^T, & i = \frac{n}{2}, \frac{n+2}{2}, \text{ n-even}, \ i = \frac{n+1}{2}, \text{ n odd} \\ [\underbrace{0\ldots0}_{n-i+1}, w_{n-i+2,i} \ldots w_{i-1,i}, 1, 0 \ldots 0]^T & i = \left\lfloor \frac{n+4}{2} \right\rfloor (1) n \end{cases}$$

and

$$Z_i = \begin{cases} [\underbrace{0\ldots0}_{i-1}, z_{i,i} \ldots z_{i,n-i+1}, 0 \ldots 0]^T & i = 1(1)\left\lfloor \frac{n+1}{2} \right\rfloor \\ [\underbrace{0\ldots0}_{n-i}, z_{i,n-i+1}, \ldots z_{i,i}, 0 \ldots 0]^T & i = \left\lfloor \frac{n+3}{2} \right\rfloor (1) n. \end{cases}$$

Hence, we have

$$A = WZ = \sum_{i=1}^{n} W_i Z_i^T \tag{161}$$

To compute the elements of the matrices W and Z, at the beginning of the kth stage we have the column vectors:

$$[w_1, w_2, \ldots, w_{k-1}, w_{n-k+2}, \ldots, w_n]$$

and the matrix

$$A_k = A - \sum_{i=1}^{k-1} W_i Z_i^T - \sum_{i=n-k+2}^{n} W_i Z_i^T \tag{162}$$

whose first and last (k-1) rows and columns are zero.

Thus,

$$\left.\begin{aligned} z_{k,j} &= a_{k,j}^{(k)} \\ z_{n-k+1,j} &= a_{n-k+1,j}^{(k)} \end{aligned}\right\} j=k,n-k+1 \tag{163}$$

and

$$\left.\begin{aligned} z_{k,k}w_{j,k} + z_{n-k+1,k}w_{j,n-k+1} &= a_{j,k}^{(k)} \\ z_{k,n-k+1}w_{j,k} + z_{n-k+1,n-k+1}w_{j,n-k+1} &= a_{j,n-k+1}^{(k)} \end{aligned}\right\} \text{for } j=k+1(1)n-k. \tag{164}$$

We then obtain the matrix

$$A_{k+1} = A_k - W_k Z_k^T - W_{n-k+1} Z_{n-k+1}^T . \tag{165}$$

At least $\left\lfloor\frac{n-2}{2}\right\rfloor$ such stages are required to compute all the elements of the matrices W and Z.

The evaluation of the coefficients $w_{j,k}$ and $w_{j,n-k+1}$ $j=k+1(1)n-k$ can be obtained from (164) in 3 time steps using $2(n-2k+2)+1$ processors.

Similarly, the determination of the matrix A_{k+1} in (165) can be calculated in one multiplication operation using $(2n-2k)^2$ processors and two subtractions using $(n-2k)^2$ processors. Therefore, at the kth stage, 6-time steps are required using $2(n-2k)^2$ processors. Hence the complete matrix factorisation into the W and Z interlocking quadrant factors requires just $3(n-2)$ time steps, using no more than $2(n-2)^2$ processors.

By using (161) we have the linear system (158) in the form

$$W Z \underline{x} = \underline{b} , \tag{166}$$

and the solution vector $\underline{x}$ can be obtained by solving the two alternative systems:

$$W \underline{y} = \underline{b} \tag{167}$$

and

$$Z \underline{x} = \underline{y} . \tag{168}$$

To solve (167) we let

$$W\underline{y} = \underline{b} = \underline{b}^{(1)}$$

$$\rightarrow \sum_{i=1}^{n} y_i \underline{w}_i = \underline{b}^{(1)} ,$$

hence,

$$\left.\begin{aligned} y_k &= b_k^{(k)} \\ y_{n-k+1} &= b_{n-k+1}^{(k)} \end{aligned}\right\} \text{for } k=1,2,\ldots \left\lfloor\frac{n}{2}\right\rfloor \tag{169}$$

where

$$\underline{b}^{(k)} = \underline{b}^{(k-1)} - y_{k-1}w_{k-1} - y_{n-k+2}w_{n-k+2}. \qquad (170)$$

Thus, at the kth stage we compute the quantity (170) in one multiplication using 2(n-2k) processors and two subtractions using (n-2k) processors. Therefore, 3 time steps are required at the kth stage and hence the system (167) requires $3\left\lfloor\frac{n-2}{2}\right\rfloor$ steps to be solved.

The system (168) can be solved in the following manner. At the kth stage we have to solve the (2×2) system:

$$\begin{aligned} z_{\ell-k+1,\ell-k+1}x_{\ell-k+1} + z_{\ell-k+1,n-\ell+k}x_{n-\ell+k} &= y_{\ell-k+1} \\ z_{n-\ell+k,\ell-k+1}x_{\ell-k+1} + z_{n-\ell+k,n-\ell+k}x_{n-\ell+k} &= y_{n-\ell+k} \end{aligned} \qquad (171)$$

where

$$\ell = \left\lfloor\frac{n-1}{2}\right\rfloor.$$

We then set the quantities

$$y_j = y_j - z_{j,\ell-k+1}x_{\ell-k+1} - z_{j,n-\ell+k}x_{n-\ell+k}, \qquad (172)$$

$$\text{for } j=1(1)\ell-1 \text{ and } j=\ell+k+1(1)n.$$

Since, the inverse of the coefficient matrix in (171) is known from the solution (164) derived in an earlier stage the solution of the (2×2) linear systems (171) requires three time steps and it can be verified that the calculation of the intermediate vector quantities (172) can be obtained in one multiplication operation using 2(n-2k) processors and two subtractions using (n-2k) processors. Therefore, the solution of the linear system (168) can be solved in a total of $5\left\lfloor\frac{n-1}{2}\right\rfloor$ time steps, using no more than (n-2) processors.

Finally, we have the result that

$$WZ\underline{x} = \underline{b}$$

requires a total of (4n-8) time steps.

From the above analysis it is obvious that if any of the (2×2) linear systems (164) is singular then the process will break down. However if this happens an evasive pivoting strategy can be completed by interchanging columns (Evans & Hatzopoulos, 1979).

On the computers envisaged for the future where a number of processors are available to carry out a large number of operations in parallel, it is obvious that the operations involved in (164) can be completed in just 3 operations using no more than (2n+1) processors in parallel. In addition, the expressions in (165) can also be done in 3 operations using no more than n(n-1) processors. Finally, the solution of the linear systems (170) and (172) can be evaluated in 3 operations each using no more than 2(n-1) processors and each of the systems given by (171) can be solved in 3 operations.

Finally, it can be readily seen that the new factorisation procedure in the proposed method has a large degree of parallelism whereby the solution process can be obtained in O(n) operations.

8. REFERENCES.

Benson, A. and Evans, D.J. (1977) A normalised algorithm for the solution of symmetric general five-term linear systems of semi-bandwidth m. Computer Studies 15, Department of Computer Studies, Loughborough University of Technology, Loughborough, U.K.

Buzbee, B.L., Golub, G.H. and Nielson, C.W. (1970) On direct methods for solving Poisson's equation. SINUM. 7, pp.627-656.

Chandra, R., Eisenstat, S.C. and Schultz, M.H. (1975) Conjugate gradient methods for partial differential equations, in Advances in Computer Methods for P.D.E.s A.C.I.A., p.60-64.

Cooley, J.W. and Tukey, J.W. (1965) An algorithm for machine calculation of complex Fourier series. Math.Comp.19, pp.297-301.

Crank, J. and Nicolson, P. (1947) A practical method for the numerical evaluation of partial differential equations of the heat conduction type. Proc.Camb.Phil.Soc., 43, pp.50-87.

Evans, D.J. (1967) The use of preconditioning in iterative methods for solving linear equations with symmetric positive definite matrices, J.I.M.A., 4, pp.295-314.

Evans, D.J. (1971) The numerical solution of the fourth boundary value problem for parabolic partial differential equations. J.I.M.A., 7, pp.61-75.

Evans, D.J. (1972A) A new iterative procedure for the solution of sparse systems of linear difference equations, in Sparse Matrices and Their Applications. Eds. D.J. Rose and R.A. Willoughby, Plenum Press, N.Y.

Evans, D.J. (1972B) An algorithm for the solution of certain tridiagonal systems of linear equations. Comp.Jour., 15, pp.356-359.

Evans,D.J. (1973) Comparison of the convergence rates of iterative methods for solving linear equations with pre-conditioning. Caratheodory Symp. Greek Math.Soc., pp.106-135.

Evans, D.J. (1974) Iterative sparse matrix algorithms, in Software for Numerical Mathematics, Academic Press, edit. D.J. Evans, pp.49-84.

Evans, D.J. (1977) On the use of fast methods for solving boundary value problems. Comp.J. 20, pp.181-184.

Evans, D.J. (1978) On the numerical solution of sparse systems of finite element equations, in Proc. of MAFELAP 1978, Ed. J.R. Whiteman, 1979.

Evans, D.J. and Gane, C.R., (1974) Alternating direction preconditioning techniques. Computer Studies 18, Department of Computer Studies, Loughborough University of Technology, Loughborough, U.K.

Evans, D.J. and Hatzopoulos, M. (1976) The solution of certain banded systems of linear equations using the folding algorithm. Comp.J., 19, pp.181-186.

Evans, D.J. and Hadjidimos,A. (1979) On the factorisation of special symmetric periodic and non-periodic quindiagonal matrices. Computing 21, pp.259-266.

Evans, D.J. & Hatzopoulos, M. (1979) A parallel linear equations solver. Int.Jour.Comp.Math. (in press).

Evans, D.J. and Lipitakis, E.A. (1979A) On sparse LU factorisation procedures for the solution of parabolic differential equations in three space dimensions. accepted for pub. Int.Jour.Comp.Maths. 7 (1979).

Evans, D.J. and Lipitakis, E.A. (1979B) A normalised implicit conjugate gradient method for the solution of large sparse systems of linear equations. submitted to Comp.Meth.App.Mech. & Eng.

Evans, D.J. and Missirlis, N.M. (1979) The modified alternating direction preconditioning methods for the solution of elliptic partial differential and biharmonic equations. B.I.T. (in press).

Gustafsson, J. (1977) A class of first order factorisation methods. Report 77.04R Comp.Science, Chalmers Univ. of Technology, and Univ. of Goteborg.

Hadjidimos, A. (1977) On comparing optimum alternating direction preconditioning and extrapolated alternating direction implicit schemes. Jour.Math.Anal.Appls., 59, pp.573-586.

Hageman, L.A. (1975) The solution of linear equations resulting from finite element discretisations of multi-dimensional boundary value problems. W.A.P.D.-TM-1209. Bettis Atomic Power Lab. West Mifflin, Pennsylvania 15122.

Hockney, R.W. (1965) A fast direct solution of Poisson's equation using Fourier analysis. J.A.C.M. 12, pp.95-113.

Kershaw, D.S. (1977) The incomplete Cholesky-conjugate gradient method for the iterative solution of systems of linear equations. J.Comp.Phys. 26, pp.43-65.

Lipitakis, E.A. (1978) Computational and algorithmic techniques for the solution of elliptic and parabolic P.D.E.'s in two and three space dimensions. Doctoral Thesis, Loughborough University of Technology, Loughborough, U.K.

Lipitakis, E.A. and Evans, D.J. (1979A) On the solution of mildly non-linear elliptic partial differential equations. submitted to Int.Jour.Num.Meth.Eng. (1979).

Lipitakis, E.A. and Evans, D.J. (1979B) Normalised factorisation procedures for the solution of self adjoint elliptic partial differential equations in 3 space dimensions. accepted for pub. in Mathematics and Computers in Simulation, (1979).

Meijerink, J.A. and Van der Vorst, H.A. (1977) An iterative solution method of linear systems of which the coefficient matrix is a symmetric M-matrix. Math.Comput. 31, pp.148-162.

Missirlis, N.M. (1979) Preconditioned iterative methods for solving elliptic partial differential equations. Doctoral Thesis, Loughborough University of Technology, Loughborough, U.K.

Nakamura, Shoichiro, (1977) Computational Methods in Engineering Science with applications to fluid dynamics and nuclear systems. J. Wiley and Sons, N.Y.

Peaceman, D.W. and Rachford, H.H. (1955) The numerical solution of parabolic and elliptic differential equations, S.I.A.M., 3, pp.28-41.

Roache, P.J. (1972) Computational Fluid Dynamics, Hermosa Pubs. Albuquerque, N.M.

Stone, H.L. (1968) Iterative solution of implicit approximations of multi-dimensional partial differential equations. S.I.N.U.M. 5, pp.530-558.

Varga, R.S. (1962) Matrix Iterative Analysis, Prentice-Hall Inc., Englewood Cliffs, New Jersey.

Wachspress, E.L. (1966) Iterative Solution of Elliptic Systems and Applications to the Newton Diffusion Equations of Reactor Physics. Prentice Hall Inc.

Young, D.M. (1971) Iterative Solution of Large Linear Systems Acad.Press, N.Y.

Young, D.M. (1975) Notes on the conjugate gradient method. unpublished mss.

CHAPTER 8

NUMERICAL METHODS OF TURBOMACHINERY

Wagdi G. Habashi

Concordia University, Montreal, Canada

INTRODUCTION

The flows within turbomachines are of sufficient complexity to warrant considerable simplifying assumptions before a numerical solution is attempted. The actual flow is unsteady, three-dimensional, compressible and viscous. The solution of the corresponding Navier-Stokes equations would be a formidable task. Even with the advent of three-dimensional solutions, aerodynamic designers are best served, at least in the preliminary stages, by two-dimensional sectional information.

Numerical analysis of the flows in turbomachines has in general followed the Wu (1952) theory. In this theory, three-dimensional flows are decoupled and calculated on two families of intersecting stream surfaces. Flows are calculated independently on the S1 or blade-to-blade surfaces (Figure 1) and on the S2 or hub-to-shroud surfaces (Figure 2). Linkage terms exist between the two surfaces. An iterative procedure would be required to generate a quasi-three-dimensional solution, with the solution on one surface feeding into the other until convergence is achieved.

This work presents a survey of the numerical methods currently utilized in the analysis of turbomachines. A brief review of the traditional solutions is undertaken, while a stronger emphasis is given to more recent developments in the field. Recent reviews such as those of Gostelow (1973), Perkins and Horlock (1975), Japikse (1976) and Scholz and Klein (1977) have admirably covered the merits and limitations of methods appearing before 1976. Less extensive reviews can be found in the articles by Marsh and O'Carroll and Morgan in the Hewitt et al (1976) reference.

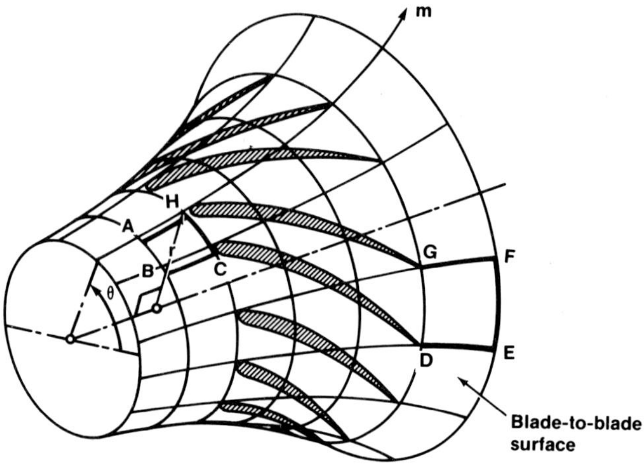

Figure 1 Blade-to-blade or S1 surface of revolution

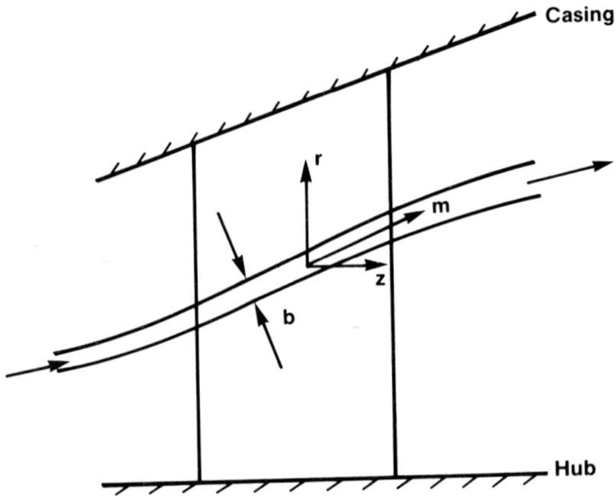

Figure 2 Hub-to-shroud, S2 surface

I. BLADE-TO-BLADE CALCULATIONS (S1 SOLUTIONS)

We shall survey the blade-to-blade solutions that are currently used in the solution of the flow on S1 surfaces. Among the well-tested and established methods, one can cite:

1. Singularity methods.
2. Streamline curvature methods.
3. Matrix or finite difference solutions of the compressible potential equation.

Recently, however, a spate of new blade-to-blade calculation methods have appeared that are invariably based on Finite Difference (FD) or Finite Element (FE) discretizations. Among them one can cite:

4. Time dependent and Relaxation Finite Area and Finite Difference solutions of the inviscid Euler equations.
5. Solutions of the potential flow equation by Finite Difference, Relaxation and Finite Element methods in the transonic regime.
6. Finite Difference and Finite Element Solutions of the Navier-Stokes equations.

The blade-to-blade solution domain generally used in field methods is shown in Figure 3.

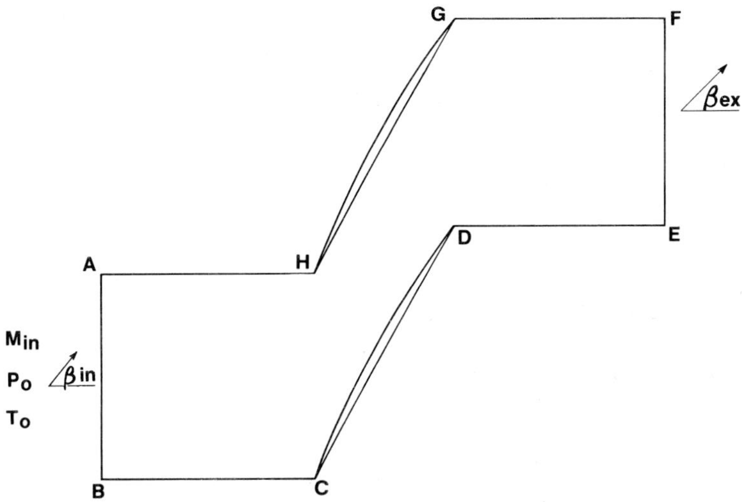

Figure 3 Region of numerical solution in blade-to-blade plane

On the blade surface CD and HG a no-penetration condition is imposed for inviscid solutions, and a condition of zero velocity is enforced for viscous solutions. Boundaries AH and GF are periodic with respect to BC and DE, respectively. Many authors invoke uniform inlet conditions on AH, while others use a more relaxed condition of parallel flow, at the inlet angle βin. On EF the exit angle βex could be specified or, for supersonic exits, the back pressure is specified in general.

1. Singularity methods

Several successful singularity methods exist that replace the blades by a vortex distribution on their contours. When formulated as a Dirichlet problem, as Martensen (1959), a Fredholm integral equation of the second kind with a regular kernel is obtained. The panel method of Hess and Smith (1966), consisting of replacing blade profiles by sources and sinks and satisfying a no-penetration condition at collocation points, has also been successfully applied to cascades (Giesing (1964)). In general, if N unknown singularities are chosen on each blade surface, one obtains a system of 2N linear algebraic equations from the appropriate conditions for no-penetration, or flow tangency at the boundary.

In the panel method this condition can be stated as:

$$2\pi\sigma(p) - \iint \frac{\partial}{\partial n}\left(\frac{1}{r(p,q)}\right)\sigma(q)\,ds = -\bar{n}(p)\cdot\bar{V}_\infty + F \qquad (1)$$

where $\frac{\partial}{\partial n}$ denotes differentiation in the direction of the outward normal n(p), which in turn depends on position; σ is the singularity strength.

The kernel of the integral equation is the outward normal velocity at point p due to a unit point source at point q. After discretization, one obtains the matrix equation

$$\sum_{q=1}^{N} A_{pq}\sigma_q = -\bar{n}_p \cdot \bar{V}_{\infty p} + F_p \qquad (2)$$

This matrix equation is solved by inversion, iterative methods or other appropriate methods.

Further improvements to this theory have been suggested by Argyris and Scharpf (1972), who used Hermitian polynomials to approximate the blade curves, hence achieving slope continuity, and Lagrangian interpolation polynomials to describe the source distribution in each element. This is in contrast to rectilinear approximation of boundaries and the use of constant source strength in the work of Smith and Hess. Furthermore, rather than satisfying a no-penetration condition at collocation points, Argyris suggests an extremization method by means of which the flow across the complete blade surface is minimized.

Singularity methods differ generally in their methods of treating the Kutta condition at the trailing edge. One is referred to the discussion in Gostelow (1975) for the merits of various Kutta condition treatments. Some of the important work in the method of singularities has involved Hess and Smith (1967), Wilkinson (1968), Van den Braembussche (1973), Ribaut and Vainio (1975), Ribaut (1977) among others.

2. Streamline curvature methods

By considering a fluid particle bounded by two streamlines and two normals to the streamlines and under the assumption that the flow is inviscid and steady but quasi-three dimensional, Bindon and Carmichael (1971) show that the governing equations are

$$\frac{\partial\psi}{\partial n} = \rho V b \qquad \text{or} \qquad \frac{\Delta\psi}{\Delta n} = \rho V b \ , \quad \text{in FD form} \tag{3}$$

$$1 = \frac{-R}{V}\frac{\partial V}{\partial n} \qquad \text{or} \qquad 1 = -\frac{R}{V}\frac{\Delta V}{\Delta n} \ , \quad \text{in FD form} \tag{4}$$

where R is the streamline radius of curvature, n the distance normal to the streamline and b is the stream tube height normal to the flow. The two equations may be combined into

$$\Delta V = -\frac{\Delta\psi}{\rho R b} \tag{5}$$

A velocity distribution is assumed along one channel boundary, termed the commencing surface. The surface curvature is calculated at a series of meridional stations along the boundary and equations (3) and (4) are used to find the velocity and pitchwise location of the first chosen stream function value at each of these stations. Thus the streamline curvature can be calculated and the process repeated to locate the next streamline. The process is stopped when the final, or flow bounding streamline, is determined. Unless the correct velocity distribution is selected at the commencing surface, the bounding streamline will not coincide with the second channel surface. An iterative loop is set up whereby the velocity distribution is changed according to the deviation of the final streamline from the terminating surface and the process is repeated until the two geometries coincide. Other streamline curvature efforts include Katsanis (1965), Wilkinson (1970), Frost and Smith (1973). Wilkinson reports solution times of between 20 to 40 seconds on an IBM 360/75 for problems using 30 meridional stations and 7 pitchwise stations.

3. Matrix and finite difference schemes

Denoting the meridional direction as m, the circumferencial direction as θ, and the stream sheet thickness as b, we proceed to define a stream function satisfying the continuity equation identically:

$$\frac{\partial\psi}{\partial m} = -\rho b v_\theta \qquad ; \qquad \frac{\partial\psi}{r\partial\theta} = \rho b v_m \tag{6}$$

By introducing this stream function into the equations of motion

$$\frac{\partial}{\partial m}[p + \rho v_m^2] + \frac{1}{r}\frac{\partial}{\partial\theta}[\rho v_m v_\theta] = 0$$

$$\frac{\partial}{\partial m}[\rho v_m v_\theta] + \frac{1}{r}\frac{\partial}{\partial\theta}[p + \rho v_\theta^2] = 0 \tag{7}$$

one obtains

$$\frac{\partial^2\psi}{\partial m^2} + \frac{\partial^2\psi}{r^2\partial\theta^2} = F[m,\ \theta,\ \rho,\ \psi,\ \psi_m,\ \psi_\theta] \qquad (8)$$

and the density is given by the isentropic relationship

$$\frac{\rho}{\rho_o} = \left[1 - \frac{\gamma-1}{2}\left(\frac{\nabla\psi}{a_o b\rho}\right)^2\right]^{\frac{1}{\gamma-1}} \qquad (9)$$

Katsanis (1968) formulates a conventional five point star finite difference discretization of the above equation. A regular uniform square grid is utilized. The blade profiles are represented by splines, whose intersections with the grid determine the length of the stars' arms near the boundaries. The discretization yields a system of linear algebraic equations, expressible in matrix form as

$$[A]\ \{\psi\}^{(n+1)} = \{F\}^{(n)} \qquad (10)$$

$[A]$ is a square (n x n) matrix, n being the number of mesh points, and $\{F\}$ is the FD nodal representation of $F[m,\theta,\rho,\psi,\psi_m,\psi_\theta]$ in equation (8). The resulting matrix is always banded. An initial stream function distribution is assumed, the right-hand side forcing function $\{F\}$ calculated and a new stream function distribution established from the solution of the above matrix equation. Katsanis chooses to solve this system by an iterative scheme and hence the solution possesses two levels of iteration: an outer one to establish the correct compressible flow field and an inner one to establish the stream function field, at each outer iteration.

To overcome the loss of accuracy of irregular stars near blade boundaries, Smith (1970) and Davis and Millar (1972) use an asymmetric finite difference grid. A curvilinear grid is formed by unevenly spaced vertical lines across the pitch (quasi-orthogonals), each line having the same number of equally spaced grid points in the pitch direction.. To ensure an overall second-order accuracy in their discretization, these authors find it imperative to use a 10 point FD stencil for the Laplacian term and a 12 point stencil for the first-order derivatives included in the right-hand side. These stencils must change shape at the grid boundaries so that no values outside the grid are required. One advantage of the matrix approaches is the fact that the matrix $[A]$ does not vary with iteration and hence could be decomposed, say by an L/U [lower and upper triangular] decomposition, while still preserving its bandedness. The solution, at each inner iteration, becomes a simple substitution process. For example, if

$$[A] = [L]\ [U] \qquad (11)$$

where $[L]$ and $[U]$ are banded lower and upper triangular

matrices, respectively, upon recalculating {F} at each iteration, one obtains

$$[L]\{Q\} = \{F\} \quad \text{, and hence } \{Q\} \qquad (12)$$

$$[U]\{\psi\} = \{Q\} \quad \text{, and hence } \{\psi\} \qquad (13)$$

i.e. the solution is obtainable in two substitution steps. The solutions based on stream function are plagued with the implicitness of the density equation and hence the double valuedness of the density near transonic Mach numbers for a given flow rate. Katsanis (1968) uses a Newton-Raphson method to determine the density corresponding to the subsonic velocity. Katsanis, Smith and Frost (1969) present extensive results for a large range of subsonic Mach numbers, while Davis and Millar illustrate their method with a comparison to an incompressible exact solution of Gostelow (1964) for a compressor blade. In all comparisons against exact analytical incompressible solutions the results of the FD schemes, whether using a regular or asymmetric grid, are invariably slightly inaccurate near the leading and trailing edges.

4. Transonic relaxation methods

Recently researchers have successfully extended the relaxation methods of isolated transonic airfoil flows to the blade-to-blade solutions of arbitrary cascades. Such isolated airfoil methods have been well developed by Murman et al (1971) and Jameson (1974) among others.

Relaxation methods developed for cascades have utilized both the transonic small disturbance equation and the full potential equation. With the latter, a transformation is often needed to convert the solution domain into a shape suitable to standard FD relaxation schemes. If a conformal mapping with modulus h is performed, the potential flow equation in the mapped plane becomes

$$(a^2-u^2)\Phi_{xx} - 2uv\Phi_{xy} + (a^2-v^2)\Phi_{yy} + (u^2+v^2)(uh_x + uh_y) = 0 \qquad (14)$$

where $u = \Phi_x/h$; $v = \Phi_y/h$

Dodge (1973) initiated such relaxation methods and obtained excellent results for supercritical cascades. For more general transonic conditions, Dodge (1976) constructed a mesh in the blade-to-blade plane closely aligned to characteristic lines. Local derivatives of the full potential equations are transformed to derivatives in the direction of the resulting skewed mesh. Second-order differences are used for elliptic operators, first-order for hyperbolic ones and similarly for first-order derivatives. Run times range around 5 minutes per case on a CDC 6400 computer.

Luu and Coulmy (1975) solve the full potential equation in a conformal coordinate system formed by the net of stream (ψ) and potential (Φ) functions of the corresponding incompressible flow through the given cascade. To obtain this net, a method of singularities is used.

The blade is first replaced by a polygonal boundary and collocation points are chosen at the center of each of the constituting segments. A constant strength source and vortex are assumed, for example, on each segment and the noncirculatory part of the stream function is obtained; the circulatory part is obtained from the superposition of vortices. The positioning in the physical plane, of the grid points where Φ and ψ are prescribed could be obtained in an iterative fashion. Rather than resort to such a time-consuming process, these authors choose to determine the grid points at the blade boundaries (ψ = constant) by interpolation from the numerical sets $x(\phi)$ and $y(\phi)$ defined at the collocation points. This, however, would result in non-vertical lines in the transformed plane due to the difference in progression of ϕ on the suction and pressure sides of the blade forming the channel. A further local transformation defined by $\phi = F(\overline{\Phi})$ is introduced that ensures the production of a regular mesh in the $(\overline{\Phi},\psi)$ plane.

The transonic potential field, in the mapped plane, is computed using central differences in the subsonic region and Jameson's (1974) rotated upwind difference scheme in the supersonic region. Flows over compressor and turbine cascades are presented for a range of Mach numbers, but no comparisons to other solutions are made. Solution times of 15 minutes on a Univac 1110 are quoted. Inclusion of viscous effects, through boundary layer calculations, although not implemented, is also possible with this method.

A cascade design procedure based on the prescription of required velocity profiles in a potential stream function plane has recently been developed by Schmidt (1979). The scheme is more palatable to design engineers, who do not easily accept the hodograph plane due to the use of velocity coordinates. The boundary conditions are prescribed as velocity magnitudes on the stagnation streamlines of the suction and pressure sides of a flow channel. Since this prescription is given over the arc length, pressure gradients determined from boundary layer calculations can be considered exactly. The boundary conditions are transformed into the computational plane by integration. Solution in the mapped plane follows a scheme very similar to Luu and Coulmy.

In a variant of Luu and Coulmy's method, Ives and Liutermoza (1976, 1978) conformally map an arbitrary cascade of aerofoils into a bounded rectangular computational domain. This is done through a series of conformal mappings that first see the cascade mapped into the region interior to a near circle. The interior of the near circle is mapped into the interior of a unit circle by a similar method to that for isolated airfoils (Sells, 1968). Further mappings then transform the region in question to a rectangular domain. Several advantages ensue, since the solution can now be obtained using the Finite Difference transonic relaxation methods developed for isolated airfoils and employing uniform grids. The transformation also magnifies high gradient regions at leading and trailing edges. Ives and Liutermoza then proceed to solve the full transonic potential flow equation by a variant of the rotated upwind difference scheme of Jameson (1974). Further improvements have subsequently been brought by utilizing a second-order scheme throughout (Ives and Liutermoza, 1978). Line relaxation is used to solve the FD equations without the need for explicit damping terms.

The method's reproduction of the incompressible flow over a Gostelow (1964) blade at no incidence is indeed excellent, using 35 points on each blade surface. Further successful comparisons are made to experimental data. For supercritical cascades in particular, two blades designed by the Korn (1975) hodograph method are analyzed by this relaxation method. While giving excellent results for a case with a sharp trailing edge, the method does not seem accurate at blunt trailing edges.

In this calculation these authors found it necessary to introduce an arbitrary stream tube contraction (Ives, 1978; Stephens, 1978) in order to obtain the best possible reproduction of Korn's (1975) hodograph method design conditions for the blade in question. The need for this contraction is unclear and, in the following section, good Finite Element results will be demonstrated for this case without resort to this artifice. Published results have been limited to low solidity cascades [chord/pitch<1], as the success of the mapping seems to be strongly influenced by cascade solidity.

A refreshingly novel relaxation solution of the full potential equation, using a rectangular mesh, is proposed by Rae and Homicz (1978). The use of complicated coordinates transformations severely restricts the applicability of most of the previously described relaxation schemes. Their use in an iterative fashion requires a user well schooled in conformal transformations. Recognizing the large success, availability and good documentation of the NASA codes (Katsanis and McNally), Rae and Homicz adapt a transonic relaxation scheme to the rectangular meshes used in these programs. While central differences are used at subsonic points, Jameson's (1974) rotated upwind difference scheme is used for supersonic points. The no-penetration condition is satisfied at blade locations, which while not being nodal points, are the location of the normal projected from the nearest flow field mesh point to the surface. This necessitates the introduction of "dummy points" inside the blade, a fact that should not detract from the other obvious advantages of this simple approach. The method, being in its developmental stages, is hampered by slow convergence for cases with high solidity and very thick blades. If future improvements could relax these limitations, this technique may become popular because of its innate simplicity and elegance.

Another relaxation solution method is presented by Sator (1976, 1977) using the transonic small disturbance equation to solve the flow over a cascade of thin airfoils. Due to the small disturbance approximation the geometric complexity of the solution domain is considerably reduced. Consider the equation

$$[K - (\gamma + 1)\Phi_x]\,\Phi_{xx} + \Phi_{\eta\eta} = 0 \qquad (15)$$

where $K = \dfrac{1-M_\infty^2}{(\gamma M_\infty^2)^{2/3}}$

$\eta = (M_\infty^2\,\delta)^{1/3}y$

δ = profile thickness

Murman, Cole and Krupp's (1971) well known relaxation methods are utilized to obtain a solution in a non-uniform grid, i.e. second-order central differences are used in elliptic regions, first-order backward differences in hyperbolic regions, together with an appropriate shock operator. The integral boundary layer method of Walz (1969) is used to account for viscosity effects.

While the approach is limited to thin blades [thickness/chord $\leq$ 10%; camber $\leq$ 10%] and to small incidence angles [$-3 \leq i \leq 3$], the extensive results presented in Sator (1976) compare stunningly well against experimental results. Solutions are obtained both in the lower transonic region, where shocks only occur in the blade channel and in the upper transonic region, with detached bow shocks. The inverse method has also proven very successful in determining blade geometries, for given transonic distributions.

5. Finite element methods

The application of Finite Element methods to fluid flow problems has enjoyed a phenomenal growth in the last decade. It would be difficult to attempt a review of such progress here. The reader is referred to the review by Shen (1977) which concisely introduces the Finite Element method, contrasts it to finite differences and proceeds to illustrate its main areas of progress. While several textbooks introduce Finite Element applications in Fluids, Chung's (1978) book admirably covers many of the topical areas in great detail.

Application of the Finite Element method to cascade blade-to-blade problems was only recently attempted. Thompson (1973) applies the finite element approach to two-dimensional incompressible cascades using rectilinear and isoparametric (curvilinear) triangular elements. The isoparametric concept of Finite Elements affords the user the same level of interpolation polynomials for functional variation, as well as geometric variation. By solving the flow over a cascade of ovals, this author shows the results using rectilinear elements to be very discouraging, while isoparametric elements bring a dramatic improvement. Habashi (1975), however, demonstrates that the rectilinear elements can achieve good accuracy. By using a simple inverse Joukowsky transformation, the blade is mapped into a near circle, around which an appropriate cylindrical coordinates mesh can be generated. Since the Laplace equation is unchanged by a conformal mapping, the discretization of this inner patch is carried out in the mapped plane. Outside this patch, where gradients are low, a rectilinear grid to the outer boundary is used. Results were presented for the incompressible flow over a cascade of Joukowsky airfoils by the stream function approach. To limit the domain size in the axial direction the method assumes the effect of the cascade at inlet and exit to be represented asymptotically by a Laurent series of singularities. The strength of the singularities is obtained along with the finite element nodal unknowns solution. This amounts to a "patched asymptotic expansion" method between the numerical solution domain and the free stream. Compressible calculations around isolated airfoils are also presented and the extension to cascade flows is straightforward.

Hirsch and Warzée (1977, 1978) present subsonic results using the stream function for compressor and turbine blades up to a suction surface Mach number of 0.75. In a highly sophisticated approach they link the blade-to-blade and throughflow programs into a quasi three dimensional program of slightly larger core requirements than each of the individual ones. This is possible because of the strong similarity in the form of the two governing equations and because of the modular structure of many finite element programs. Excellent results are obtained in this work.

Implementation of finite element solutions on a daily basis, as would be required in an industrial setting, is no easy task. The mesh generation, even if semi-automated, can be laborious and cumbersome enough to discourage even the most enterprising solver. Industrial designs often require a trial and error approach in which constant modifications to the blade profile are made and the effects assessed. For an analysis routine to be practical requires both a flexible mesh generation method and an analysis program efficient enough to be run in an interactive manner,say at a video terminal. In view of these difficulties and in order to extend the Mach number range of the subsonic solutions, Habashi, Dueck and Kenny (1978, 1979) developed a fully-automated interactive finite element mesh generator, appropriate to general cascades. The process is illustrated in Figures 4 to 7.

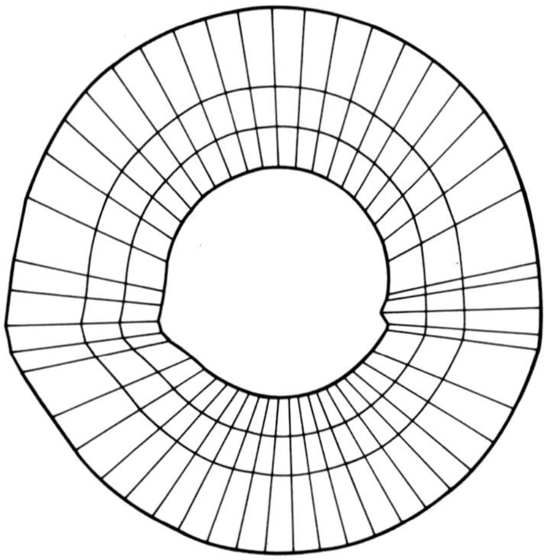

Figure 4 Finite element layers around mapped blade

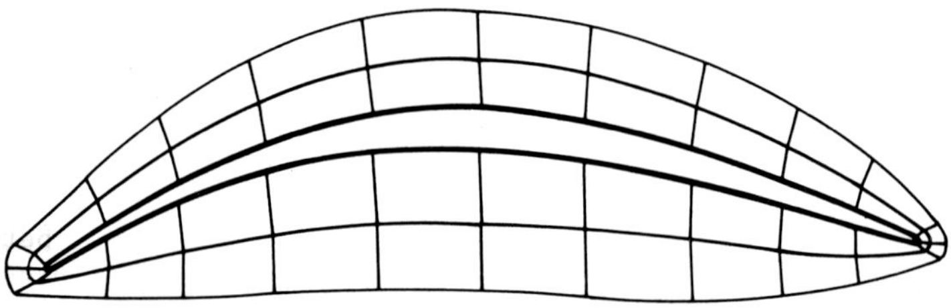

Figure 5 Finite element layers in the vicinity of the blade, in physical plane

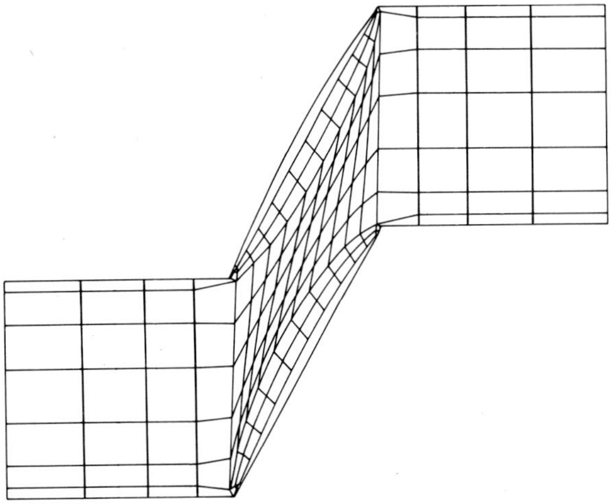

Figure 6 Finite Element mesh for blade-to-blade cascade analysis

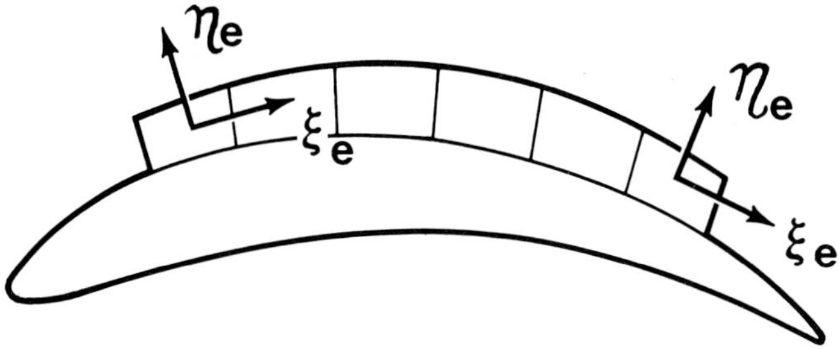

Figure 7 Isoparametric quadrilaterals and local linearization coordinates

A blade profile is first splined and then fitted with appropriate leading and trailing edge circles. Several points are chosen on these circles to ensure a faithful representation of these crucial zones. The remaining subdivisions are then chosen along the blade chord in an expanding/contracting fashion according to relative curvature of the blade surface or according to requirements defined by the user. The airfoil is then mapped by a one-step explicit conformal transformation into a near circle around which contiguous layers reproducing the airfoil shape can be generated (Figure 4). The number of layers and their thickness are parameters of the program. The generated mesh is then mapped back to the physical plane (Figure 5). The mapping has the advantage of magnifying leading and trailing edge regions. Since the solution is carried out in the physical plane,no limitations on solidity exist. To complete the mesh generation a number of subdivisions are chosen across the pitch, outside the generated layers, where eight-node quadrilateral elements are constructed in an isoparametric fashion (Figure 6). The density of these elements across the pitch and in the upstream and downstream sections is defined as an interactive input. Furthermore, element sizes in the upstream and downstream regions can be tailored (by contraction or expansion) to accommodate the transition, from uniform flow far upstream or downstream, to the regions of high flow gradients within the cascade. It may also be worthwhile to mention that blade splining permits the selection of exact mid-points for the isoparametric elements, a fact of considerable importance when such elements are used. The only input to the mesh generation program consists of the upper and lower blade surface coordinates, together with an additional point on the leading and trailing edges to define the circles.

A further advantage of such automated mesh generation is the flexibility of choice of the type of element. For example, each eight-node quadrilateral element can be broken down during the analysis into six triangular elements without generating additional nodes. The analysis program contains an option to carry out solutions using triangles or quadrilaterals. In a well-structured finite element program, such an option may be included with only the addition of one subroutine.

To permit the extension of the finite element solution of the potential equation to high subsonic and possibly transonic Mach numbers, Habashi (1975), Shen and Habashi (1976) have devised a local linearization scheme. The scheme takes advantage of the fact that since finite element discretizations only require a locally valid equation, it is hence permissible to perform a different linearization in each element with respect to the local Mach number in the element. Having established an initial velocity field for a particular flow, say V_e in each element, one defines the local velocity direction as ξ_e and its normal as η_e (Figure 7). Assuming the next iterative step to consist of a perturbation within each element, around V_e, gives:

$$V_e^{(n+1)} = [V_e^{(n)} + u', v'] \text{ where } \frac{u'}{V_e}, \frac{v'}{V_e} << 1 \tag{16}$$

and $$\Phi = V_e \xi_e + \phi(\xi_e, \eta_e)$$

The perturbation potential (ϕ) is governed by the Prandtl-Glauert equation:

$$\rho_e[(1-M_e^2) \frac{\partial^2\phi}{\partial \xi_e^2} + \frac{\partial^2\phi}{\partial\eta_e^2}] = 0 \tag{17}$$

for which the stretching $\xi_e' = \dfrac{\xi_e}{(1-M_e^2)^{\frac{1}{2}}}$ $\quad \eta_e' = \eta_e$ gives the

Laplace equation with respect to ξ'_e , η'_e
An applicable variational principle can easily be formulated. Reconverting into global variables, the derived variational integral becomes (for N elements)

$$\delta I(\Phi) = \frac{1}{2}\delta \sum_{e=1}^{N} \rho_e \iint_{A_e} [(1-M_e^2)\Phi_{\xi_e}^2 + \Phi_{\eta_e}^2 + 2V_e M_e^2 \Phi_{\xi_e}] d\xi_e d\eta_e - \oint \rho \delta\Phi (\frac{\partial \Phi}{\partial n}) ds \quad (18)$$

$$\delta I(\Phi) = [\frac{\delta}{2} \sum_{e=1}^{N} I_1(\Phi)] + \delta I_2(\Phi) \quad (19)$$

where $I_2(\Phi)$ applies only to elements on the outer contour of the domain analyzed.

This variational integral provides the basis for an iterative procedure where at each step new values of V_e, M_e, ρ_e as well as a new ξ_e direction are obtained.

Within each element one assumes:

$$\Phi(x,y) = \sum_{i=1}^{8} N_i(\zeta,s)\Phi_i \quad (20)$$

ζ and s being normalized local coordinates within an element and N_i are the shape functions, whose explicit form can be found in any finite element textbook (e.g. Chung, 1978). Furthermore, in the element, the isoparametric concept is used

$$m = \sum_{i=1}^{8} N_i(\zeta,s)m_i \quad ; \quad \theta = \sum_{i=1}^{8} N_i(\zeta,s)\theta_i \quad (21)$$

In other words the geometry of each element can be exactly approximated up to a parabolic variation between points, straight lines being special cases of parabolas. Such a description is useful in the representation of highly cambered turbomachinery blades, especially in leading and trailing edge regions.

From equation (18) and upon minimization with respect to the nodal values Φ_i, (i=1,8) within each element, one obtains an 8x8 element influence matrix whose coefficients are given by

$$k_{ij} = \rho_e \iint_{A(e)} [(1-M_e^2) \frac{\partial Ni}{\partial \xi_e} \frac{\partial Ni}{\partial \xi_e} + \frac{\partial Ni}{\partial \eta_e} \frac{\partial Ni}{\partial \eta_e}] d\xi_e d\eta_e \quad (22a)$$

and the element contribution to the right hand side

$$f_i = -\rho_e V_e M_e{}^2 \iint_{A(e)} \frac{\partial Ni}{\partial \xi_e} d\xi_e d\eta_e \tag{22b}$$

Since all shape functions (Ni) are in terms of the local coordinates (ζ, s), simple Jacobian transformations are made from (ζ, s) to (m, θ) and finally to (ξ_e,η_e). The angle between ξ_e and m in each element is given at each iteration by

$$\sigma^{(e)} = \tan^{-1}\left\lfloor \frac{\frac{\partial \Phi}{r d\Theta}}{\frac{\partial \Phi}{\partial m}} \right\rfloor^{(e)} = \tan^{-1}\left\lfloor \frac{\frac{1}{r}\frac{\partial Ni}{\partial \Theta}\{\Phi i\}}{\frac{\partial Ni}{\partial m}\{\Phi i\}} \right\rfloor^{(e)} \tag{23}$$

The integral I_2 of equation (19) vanishes on all surfaces where $\partial\Phi/\partial n=0$, implicitly accounting for the no-penetration condition on the blade. It also cancels out on all periodic boundaries. Its minimization only adds a contribution to equation (22b) in elements on EF (Figure 3)

$$\{Fi\} = \rho_{ex}\cdot R\cdot M_{ex}\cdot a_o\cdot p\cdot \cos\beta_{ex} \sum_{e=1}^{n_1} \left(\int_{-1}^{1} Nids\right)\cdot \frac{L^{(e)}}{2} \tag{24}$$

where $L^{(e)}$ is the boundary length of an element on EF
p is the cascade pitch
n_1 is the number of elements on EF

All integrations in Equation (22) are carried out by a 3x3 numerical Gaussian integration scheme within each element. After assembly the resulting global matrix equation is of the form:

$$[A(\Phi)]\{\Phi\} = \{F\} \tag{25}$$

and is linearized at each iterative step (n) as:

$$[A(\Phi)^{n-1}]\ \{\Phi\}^{(n)} = \{F\}^{(n-1)} \tag{26}$$

The matrix [A] is formed and solved at each iterative step. It is a narrowly banded sparse symmetric matrix. One immediately notes the ensuing computer storage advantage over finite difference methods where [A] is not symmetric. Storage requirements are dynamically adjusted within the program and the matrix stored in vector form by the skyline variable bandwidth method (Bathe and Wilson, 1976). Solution is carried out by an L/U symmetric decomposition scheme. Convergence for a density error of 0.1% is achieved for transonic Mach numbers in less than 10 iterations, normally 5. For 21 points on each blade surface, including element mid-points, solution times are of the order of 5-10 seconds on a Cyber 170/Model 175. It is thus practical to carry out the analysis in an interactive manner at video terminals. The interactive mesh generation requires only between 1 and 2 seconds of execution time on the same computer.

Several test cases have been analysed to verify the application of this method to a spectrum of conditions from incompressible to transonic flows. Comparisons are made with both exact and analytical results. Figure 8 shows the comparison to an analytical solution by Gostelow (1973) for the incompressible flow over a compressor blade at a very severe negative incidence of 70°. The results obtained with 25 points on each blade surface demonstrate smoothness and uniformity of flow over the complete blade profile, even in the high gradient leading and trailing edge regions, and provides excellent agreement with Gostelow's results.

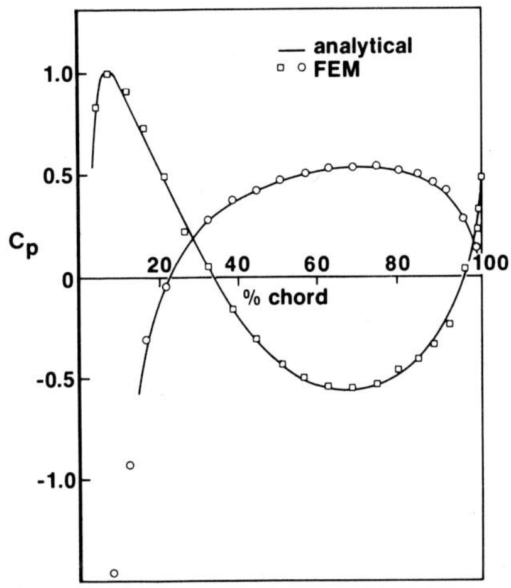

Figure 8 Incompressible flow by Finite Element method (-70° incidence)

To verify the applicability to shockless transonic flow and test the accuracy of the method, two test cases are presented. Both blade profiles were designed using the hodograph method of Korn et al (Korn, 1975) to provide a specified pressure distribution. Second-order accurate finite difference solutions are also available for both blades (Ives and Liutermoza, 1977, 1978). The first test case is a supercritical blade profile designed for shockless performance at an inlet Mach number of 0.78. Results of the finite element analysis are provided in Figure 9 with 35 points on the blade.

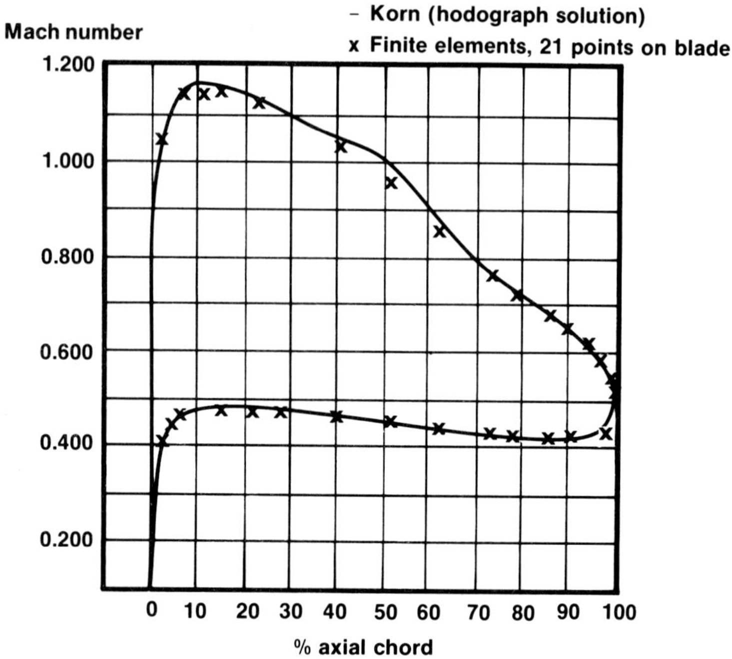

Figure 9 Finite Element results for Korn supercritical cascade I

The results had a strong convergence with mesh refinement, with anticipated behaviour at leading and trailing edge precisely reproduced. A second-order finite difference solution by Ives and Liutermoza (1978) with 54 points on each blade surface is available for comparison. The improvement brought by finite elements over the last 25% of the blade, on both surfaces, is good. No artificial stream tube contraction was included in the finite element solution.

The second test case is a hodograph designed blade of Korn and Garabedian (Ives, 1978). The finite element results are shown in Figure 10. They match the design Mach No. profiles to a good degree of accuracy with only 21 points on the blade.

6. Finite area and finite difference solutions of the inviscid Euler equations

This, undoubtedly, is one of the computational methods in blade-to-blade solutions to enjoy the largest growth in the last few years. Time dependent methods have emerged, providing good and reasonably accurate solutions into the transonic regime. In the time dependent approach two distinct routes have been followed, with more or less the same degree of success. In the first approach, which we shall label the differential approach, time-splitting or predictor-corrector methods are

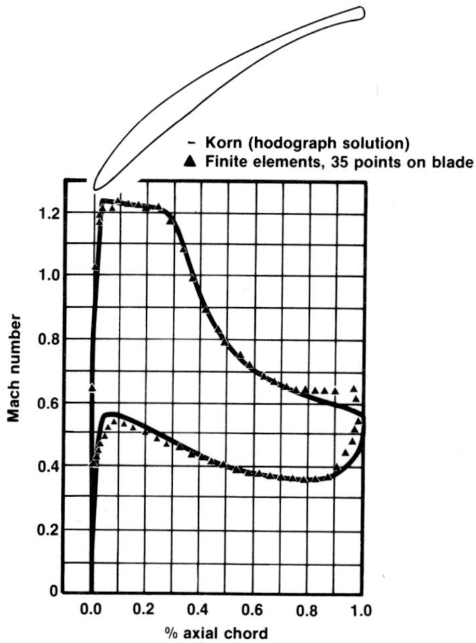

Figure 10 Finite Element results for Korn supercritical cascade II

used. This is illustrated in the method of Gopalakrishnan and Bozzola (1971). These authors start by mapping the blade-to-blade region into a rectangular field, which can easily be discretized using a uniform finite difference grid. The coefficients of the point by point transformation are absorbed into the governing equations.

The time dependent Euler equations can be written as

$$\overline{W}_t - \overline{F}_x - \overline{G}_y = 0 \tag{27}$$

$$\overline{W} = \begin{bmatrix} \rho \\ \rho u \\ \rho v \\ E \end{bmatrix}, \quad \overline{F} = \begin{bmatrix} \rho u \\ p+\rho u^2 \\ \rho uv \\ u(p+E) \end{bmatrix} \text{ and } \overline{G} = \begin{bmatrix} \rho v \\ \rho uv \\ p+\rho v^2 \\ v(p+E) \end{bmatrix}$$

where the internal energy $E = \frac{p}{\gamma-1} + \frac{1}{2}\rho(u^2+v^2)$ (28)

These authors assume an initial solution that specifies a linear variation of all properties between the leading and trailing edges in the flow direction and constant properties across the pitch. In their first paper (1971) they analyze turbine cascades using a time-splitting approach. In their analysis of compressor cascades (1972), however, they adopt the MacCormack (1971, 1972) predictor-corrector scheme which has the same second-order accuracy and stability limit as the Lax-Wendroff procedure, but is easier to program than their previous scheme. The MacCormack scheme consists of two steps: A predictor step [first-order forward, to approximate the first-order spacial derivatives]

$$\tilde{W}(i,j,k) = W(i,j,k) + \Delta t\left[\frac{F(i,j+1,k)-F(i,j,k)}{\Delta x} + \frac{G(i+1,j,k)-G(i,j,k)}{\Delta y}\right] \qquad (29)$$

A corrector step [second-order backward, using the predictor values]

$$W(i,j,k+1) = \tfrac{1}{2}\left[W(i,j,k)+\tilde{W}(i,j,k)+\Delta t\left\{\frac{\tilde{F}(i,j,k)-\tilde{F}(i,j-1,k)}{\Delta x} + \frac{\tilde{G}(i,j,k)-\tilde{G}(i-1,j,k)}{\Delta y}\right\}\right] \qquad (30)$$

where W(i,j,k) represents the value of W at node ($x=i\Delta x$, $y=j\Delta y$) at time ($t=k\Delta t$). A final smoothing process is added to the MacCormack scheme:

$$W_s(i,j,k+1) = \frac{1}{\phi+4}\left[\phi W(i,j,k+1) + W(i,j+1) + W(i+1,j,k+1) + W(i,j-1,k+1) + W(i-1,j,k+1)\right] \qquad (31)$$

where ϕ is the smoothing operator lying usually between 200 and 1000. The procedure is conditionally stable with the stability criterion given by

$$\frac{\Delta t}{\min(\Delta x,\Delta y)} \leq \frac{2}{3a(1+M)} \qquad (32)$$

As in other FD schemes, flow variables on the blade surface are computed by first-order analogs of the differential equation using one-sided derivatives. About 1200 cycles are needed to settle the solution, and users have found the method excessively slow in convergence. It is also a controversy how much of an error can be introduced by the geometric transformation, especially in turbine applications where radii of curvature can be very important. The method has compared unfavorably to the Finite Area method of McDonald in the VKI (1973) lecture series 59 (Couston, 1976). Although good comparisons to experimental data are shown in their first paper, no comparisons are made to compressor blade data in the second. In the only comparison to other numerical methods the authors have compared one of their computations to Katsanis' (1968) method for a turbine rotor tip section at subsonic speed. The results, although agreeing qualitatively, differ substantially near leading and trailing edges.

On the other hand, McDonald (1971) expresses the transient conservation Euler equations in integral form. These conservation equations are then applied to "finite area elements" where each interior point is surrounded by six sides. An isentropic flow assumption was originally made. More recent extensions (Denton, 1974) assume the stagnation enthalpy to be conserved through shocks, which while not being true in the transient state, is exact when the flow becomes steady.

For a control volume of area ΔA and a time step Δt, one can write (following Denton, 1974):

$$\begin{aligned}
&\text{Mass} && \Delta A \frac{\Delta\rho}{\Delta t} = \Sigma\,(\rho u dA_m + \rho v dA_\theta) \\
&\text{m-momentum} && \Delta A \frac{\Delta(\rho u)}{\Delta t} = \Sigma((p + \rho u^2)\, dA_m + \rho uv dA_\theta) \\
&\theta\text{-momentum} && \Delta A \frac{\Delta(\rho v)}{\Delta t} = \Sigma((p + \rho v^2)\, dA_\theta + \rho uv dA_m) \\
&\text{Energy} && \Delta A \frac{\Delta(\rho E)}{\Delta t} = \Sigma(\rho u H_o dA_m + \rho v H_o dA_\theta)
\end{aligned} \tag{33a}$$

where E = specific internal energy $= c_v T + \dfrac{(u^2 + v^2)}{2}$

H_o = specific stagnation enthalpy $= c_p T + \dfrac{(u^2 + v^2)}{2}$

In addition one imposes an equation of state, say

$$p = \rho RT = \frac{\rho R}{2c_v}\,[2E-(u^2+v^2)] \tag{33b}$$

In this method the non-rectangular grid is formed by the intersection of vertical unevenly spaced axial stations (quasi-orthogonals) and evenly spaced quasi-streamlines in the pitchwise direction. Since rounded leading and trailing edges cannot be handled by such grids, it is customary to replace them with cusps. In Denton's work, the six-sided control volume polygon is replaced by the simpler quadrilaterals formed by the intersection of the vertical lines and the quasi-streamlines. The calculating points become the mid-points of each quasi-streamline segment (Figure 11).

To ensure solution stability Denton solves the equation in the order continuity-pressure-momentum, the order in which the momentum equations are solved being unimportant. While calculating points at periodic boundaries requires no special treatment, those on the blade prove the most difficult to treat. Equations can be derived in the customary Finite Difference way of imagining the grid to extend into the blade and obtaining the required fluxes at the boundary by some extrapolation of the properties inside the boundary.

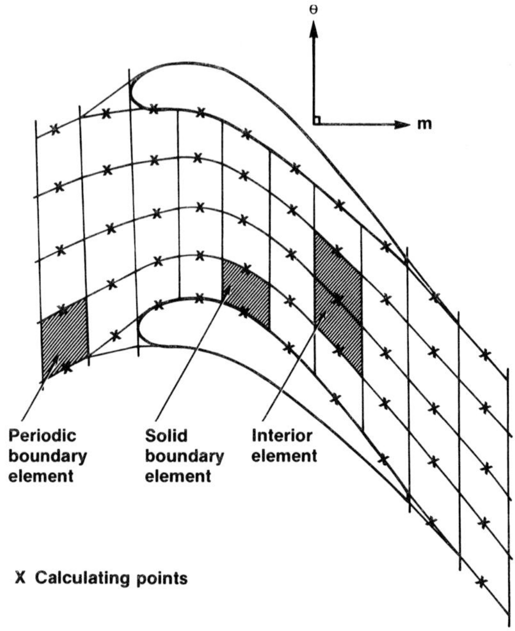

Figure 11 Grid used in Finite Area method

Although McDonald's original method is only first-order in accuracy, Denton, while keeping the basic transient scheme unchanged, uses more accurate differencing methods at steady state. This is achieved by adding correction factors, which when added to the fluxes and pressures assumed to act on the vertical faces of the elements make the resulting value equal to that obtained from a curve fit along the quasi-streamline. These correction factors also need to be damped. The method has been extensively tested for turbine blades by McDonald and by Bryce and Lichtfield (1977) and for both compressor and turbine blades by Denton (1974, 1976). Leading and trailing edge inaccuracy is amply demonstrated in the comparison against an exact solution by Gostelow (1964) for a compressor blade.

Another recent time dependent Finite Difference contribution is that of Nagayama and Adachi (1977), who label their method "Modified FLIC" (Fluid-in-Cell). To overcome the necessity of a domain transformation and to improve FD boundary condition handling, these authors use a triangular mesh system in the blade-to-blade plane. The mesh is formed in the same manner as the McDonald scheme, except that each quadrilateral can be broken down into two triangles. It is not quite clear why these triangular meshes would be superior in boundary condition approximation to McDonald's scheme. The only advantage that can be seen is the fact that rectilinear segments using the true blade coordinates would be used, rather than the cusps used in the Finite Area methods. A time-splitting scheme is utilized (Gentry et al, 1966), with suitable modifications for the triangular net. Intermediate values of the variables are calculated in a first step by neglecting convective terms. In a correcting second step the new values are calculated by applying the conservation laws to these intermediate values. Good agreement with experimental results is shown for the hub, tip and mean sections of a transonic turbine cascade. For a mesh of about 500 triangular elements, solution times of the order of 30 min. on a Univac 1108 are quoted. Other Finite Area results are presented by Veuillot (1976), who also presents some results using the Finite Difference method of Gopalakrishnan and Bozzola.

Time marching Finite Area solutions offer some considerable advantages; not least among them is the fact that shock waves appear naturally as the solution evolves and no special artifices are required. Moreover, the constant hyperbolic nature of the equations make special operators unnecessary in mixed flow situations. The extension of such methods to three-dimensional flows, while not simple, is straightforward. However, the disadvantages of the method can sometimes be quite serious. Computational times, even for shockless flows, can be excessive. Solutions are strongly a function of the zeroth order or guessed basic flow. Present iterative accuracy is limited to first order, and refinement through mesh size reduction can significantly affect solutions cost. In representing rounded leading and trailing edges the method arbitrarily replaces them by equivalent wedges. While introducing a tolerated approximation for thick leading edges as in turbines, the method introduces a significant approximation in compressor blades. The resultant velocity distribution is often jagged or sawtooth in character requiring considerable interpretation or artificial smoothing (Bryce and Lichtfield, 1977). These authors also state that even with a stipulated convergence criterion on axial velocity of 0.01%, fairly significant changes of total pressure still exist after the solution has converged. This is most noticeable at leading and trailing edges but even in the channel fluctuations of $\pm$ 5% still exist. Stability often can only be achieved by the addition of artificial viscosity terms (Denton, 1976)with the result that shock waves appear smeared, typically over 5 grid points or more.

Due to the slowness of convergence of hyperbolic time-marching equations Singhal and Spalding (1978) "partially-parabolize" the governing equations and essentially introduce a time behaviour governed asymptotically by a diffusion process. They consider a non-orthogonal grid system formed by quasi-orthogonals and actual streamlines. Several advantages are claimed for the scheme, among them:

a) Blade surfaces are easy to handle.

b) Governing flow equations are simplified because the mass fluxes across the control volume faces are known.

The variables used are u, v, p, ρ where u and v are not orthogonal to each other, but are aligned with the streamline and circumferential directions, respectively. This permits each momentum equation to contain only one pressure term. Velocities and pressure are stored at staggered positions in the grid.

An initial guess for the pressure field is made. With an inlet value of ρ, u, v known, the method obtains these values at the next downstream section from the two simple momentum equations and the equation of state. A set of pressure correction equations is solved at this station using a tri-diagonal matrix algorithm and the pressure, velocities and densities corrected. Several successful examples are demonstrated for subsonic and transonic turbine cascades. Solution times are of the order of 50 seconds for a 20x38 mesh, on a CDC 6600 computer.

7. Finite difference and finite element solutions of the Navier-Stokes equations

a. Compressible time-dependent approaches with primitive variables (u,v,p) We shall present in this section the finite difference approach of Kurzrock and Novick (1975), the finite element approach of Bowley and Prince (1971), and the method of characteristics of Delaney and Kavanagh (1976). The first two approaches are time-marching solutions of the Navier-Stokes equations written in conservation form. Kurzrock and Novick first map the domain between two blades into a rectangular bounded region, in which the blade surfaces appear as a slit. Their equations contain rotation, w, stream tube contraction and radius change. These are written as:

$$\frac{\partial \overline{U}}{\partial t} + \frac{\partial \overline{E}}{\partial m} + \frac{\partial \overline{G}}{\partial \theta} + \overline{J} = \overline{L} \tag{34}$$

where the column vectors are defined as

$$U = \begin{bmatrix} Rb\rho \\ Rb\rho W_m \\ Rb\rho W_\theta \\ Rb\left(\rho H_R - \dfrac{\gamma-1}{\gamma} P\right) \end{bmatrix} \qquad E = \begin{bmatrix} Rb\rho W_m \\ Rb\left(\rho W_m^2 + \dfrac{1}{\gamma M_r^2} P\right) \\ Rb\rho W_m W_\theta \\ Rb\rho W_m H_R \end{bmatrix}$$

$$G = \begin{bmatrix} \rho W_\theta b \\ \rho W_m W_\theta b \\ (\rho W_\theta^2 + \frac{1}{\gamma M_r^2} P) b \\ \rho W_\theta H_R b \end{bmatrix} \quad J = \begin{bmatrix} 0 \\ -b\rho (W_\theta - R\omega)^2 \frac{\partial R}{\partial m} - \frac{P}{\gamma M_r^2} \frac{\partial (Rb)}{\partial m} \\ b\rho W_m (W_\theta - 2R\omega) \frac{\partial R}{\partial m} \\ 0 \end{bmatrix}$$

The reader is referred to Kurzrock for the viscous term L.

MacCormack's (1971) second-order explicit finite difference predictor-corrector method is utilized. The approach has many similarities to the method of Gopalakrishnan and Bozzola (1971) and similar comments hence apply. By carrying out numerical experimentation for the one-dimensional shock transition problem, these authors deduce that the coefficient of viscosity must be adjusted to ensure that the numerical Reynolds number, based on mesh size, be 10 or less to adequately describe shock wave transition. Only inviscid boundary conditions are specified along the blade surface and the authors state that a viscous boundary condition specification would just increase the total pressure losses by adding a viscous profile loss. A discretized method of characteristics is used to determine the inviscid boundary conditions along the blade surface as well as those at the exit plane, for constant back pressure. The static pressure distribution over an MCA (multiple circular arc) blade cascade is obtained numerically and experimentally. Results seem excellent, with only a 22 x 10 mesh (253 grid points) used to discretize the domain for numerical solution. Solution times are reported to be less than 3 min. on an IBM 370/165. Numerical rotor flows are presented with no comparison to measured data.

On the finite element front Bowley and Prince (1971) develop a time dependent discretization procedure for two-dimensional viscous flows in cascades. The quadrilateral mesh used is formed by the intersection of quasi-streamlines with unevenly spaced quasi-orthogonals. The Navier-Stokes equations are written in the form:

$$\int_v \frac{\partial U}{\partial t} dV - \int_s F ds = 0 \tag{35}$$

U being the same as in Equation 34. The reader is referred to the original paper for the term F.

The time derivative is discretized in a finite difference manner, a time-splitting one specifically. The field variables at each time step are approximated by a subdomain collocation finite element discretization. This is basically a weighted residual method, for which the weighting function is unity over the element and zero outside it. The unknown variables are represented by a bilinear distribution, in each element. Qualitatively good results are obtained for the viscous flow in the inlet between two parallel plates. Successful comparisons to experimental results were obtained for transonic turbine blades. The viscous solution is further compared to a corresponding inviscid one using the same approach. These authors suggest further refinements to their method could include a finer breakup at leading and trailing edges and in the boundary layer. Perhaps both objectives could be achieved by the use of higher order elements.

A third time-dependent approach is the method of characteristics of Delaney and Kavanagh (1976). They apply a second-order time-dependent method of characteristics using bicharacteristics. Their inverted scheme fixes the solution point in the new time plane determined by the time increment Δt, and the bicharacteristics and particle path are projected back from the solution point into the initial data plane. This permits the use of a regular finite difference grid formed by equally spaced quasi-orthogonals and quasi-streamlines. Two turbine test cases are successfully demonstrated in this work but no indication is given of the total number of iterations required for a steady-state solution.

b. Incompressible stream function-vorticity formulations Two such formulations have recently been proposed by Bosman, Chan and Halton (1978), Khalil, Tabakoff and Hamed (1978). By introducing a stream function satisfying continuity identically and by eliminating the pressure between the two momentum equations, one obtains the general forms:

$$\frac{1}{r}\left[\frac{\partial}{\partial m}\left(\frac{\dot{m}r}{\rho b}\frac{\partial\psi}{\partial m}\right)+\frac{\partial}{r\partial\theta}\left(\frac{\dot{m}}{\rho br}\frac{\partial\psi}{\partial\theta}\right)\right] = -\zeta \tag{36}$$

$$\frac{\partial}{\partial m}\left[\frac{\dot{m}}{b}\frac{\partial\psi}{\partial\theta}\zeta\right] - \frac{\partial}{\partial\theta}\left[\frac{\dot{m}}{b}\frac{\partial\psi}{\partial m}\zeta\right] - \frac{\partial}{\partial m}\left[r\frac{\partial}{\partial m}(\mu_e\zeta)\right] + \frac{\partial}{\partial\theta}\left[\frac{1}{r}\frac{\partial}{\partial\theta}(\mu_e\zeta)\right] + G_1 = 0 \tag{37}$$

where G_1 is a source term.

Bosman et al solve iteratively for ζ and ψ from the stream function and vorticity transport equations by a relaxation technique with a finite difference approximation using central differences for diffusion terms and upwind differences for convection terms. Khalil et al parabolize the equations by introducing a fictitious time derivative, then use an Alternating Direction Implicit (ADI) solution method. This well known method is a time-splitting procedure by means of which the equations are solved implicitly in one direction in each half step, yielding successive tridiagonal systems of linear algebraic equations. In the first approach a slip model is used at the blade surface, while in the latter method the wall vorticity is updated at each iterative step from the no-slip condition. Both schemes use an enhanced effective viscosity to model the stream turbulence level.

The downstream boundary condition poses a particular problem with such schemes since equations 36 and 37 are satisfied for arbitrarily imposed values of ψ and ζ on this boundary. However, to satisfy the original lower order dynamical equations of motion, unique values of these two parameters are necessary. A trial and error procedure must be adopted whereby the exit angle is guessed,and an evaluation of the torque developed by the channel is obtained through the integration of the difference in pressure and shear forces acting on the blade surfaces. This torque is compared to the imposed change of angular momentum between exit and inlet. If the value of the predicted torque was not equal to the rate of change of the angular momentum, then the direction of the exit flow velocity is altered. A complete re-solution is required for each new selection of exit angle.

II. HUB-TO-SHROUD CALCULATIONS (S2 SOLUTIONS)

In this section the three main approaches to the throughflow problems will be presented, namely: <u>1. Streamline curvature method</u> <u>2. Matrix or finite difference methods</u> <u>3. Finite element methods</u>.

Extensive calculations have been made using the first two methods and comparisons between them have been carried out by several authors. The finite element method is a recent newcomer to this field, but is quickly establishing itself as particularly well-suited to tackling the throughflow problem.

All three methods start from the assumption of steady, inviscid flow and numerically solve the radial component of the Navier-Stokes equation to establish a velocity-density field satisfying the continuity equation. Passage-averaged continuity momentum and energy equations are derived and are assumed to hold in a plane θ = constant. Alternatively the solution can be viewed as that on a mean blade-to-blade stream surface whose shape is given by the geometry of the blade rows. Methods, however, differ in their choice of the mean S_2 streamsheet. Some authors have chosen the $\psi = 0.5$ line, others the mean camber line shape or a mass-averaged mean streamsurface. Once the streamsheet shape has been selected the flow field is broken down into a network which depends on the solution method and the unknowns are determined at each of these nodal points. With blade geometry given, a cascade model is used to determine the air deflection and the pressure loss. Boundary layer growth along the hub and shroud is calculated and specified as a blockage factor, whose effect is included by recalculating new hub and shroud displaced geometries.

1. Streamline curvature method

The streamline curvature method has been widely used in calculating throughflow behaviour. While several restrictions were originally placed on both the grid and the positions of calculating stations, modern versions of this technique have eliminated such restrictions. Among workers in this method we can cite Katsanis (1966), Smith (1966), Novak (1967), Frost (1970), Davis and Millar (1971), Katsanis and McNally (1973) and Denton (1978).

The notation of Davis and Millar will be adopted in this section. Selecting axial stations and defining arbitrary quasi-orthogonals in a direction s, the radial equilibrium equation can be written in the non-linear partial differential equation form:

$$\frac{dW_m}{ds} = [C_1(s\ ,W)\ ]\ \frac{1}{W_m} - [C_2(s,W)]W_m \tag{38}$$

where W_m is the meridional velocity. This equation can be written in a relative or absolute coordinate system. This is coupled to the continuity requirement

$$m = 2\pi \int_h^t \rho W_m\ r\ \cos\ (\psi + \phi)\,ds \tag{39}$$

where ψ = angle between quasi-orthogonal and radial direction and ϕ = angle between meridional direction and engine axis.

The solution of these equations is obtained by an iterative technique at a discrete number of points along the quasi-orthogonals. An initial guess is made of streamline location in the field, for example by postulating that equal flows pass through annuli of equal area. The functions C_1 and C_2 can hence be estimated based on this streamline distribution and its curvature. With C_1 and C_2 known from a previous iteration, equation 38 becomes an ordinary differential equation along each quasi-orthogonal (s):

$$\frac{dW_m^{(n+1)}}{ds} = C_1^{\,n} \cdot \frac{1}{W_m} - C_2^{\,n} \cdot W_m \tag{40}$$

Selecting a value of W_m at an arbitrary position along (s), the above equation is integrated along s and complete W_m and density distributions are obtained. With these distributions the mass flow rate m is calculated by fitting a polynomial to $(\rho W_m r)$. Since this mass flow rate will generally not agree with the inlet m, another W_m guess is taken and an appropriate iteration monitoring followed until convergence at that station is achieved. The new streamline pattern at each iteration will yield the new required values of the C_1 and C_2 functions.

The second-order derivatives contained in C_1 and C_2 require accurate evaluation along the streamlines. The quality and accuracy of streamline curvature techniques are hence strongly dependent on the curve fit or spline methods used. Such splines have been assessed by Wilkinson (1970), who has also applied the method to blade-to-blade calculations.

2. Matrix or finite difference methods

To enable solvers to utilize a field method that avoids some of the difficulties of streamline curvature methods in highly curved channels, Marsh (1966) formulates a finite difference approach to the throughflow problem. By defining a stream function as:

$$\frac{\partial\psi}{\partial r} = \rho r b W_z \quad ; \quad \frac{\partial\psi}{\partial z} = -\rho r b W_r \tag{41}$$

the equation of motion becomes:

$$\frac{\partial^2\psi}{\partial r^2} + \frac{\partial^2\psi}{\partial z^2} = F[r,z,\rho,\psi,\psi_r,\psi_z] \tag{42}$$

The nonlinear function F can be explicitly written as:

$$F = \frac{\partial\psi}{\partial r}\cdot\frac{\partial}{\partial r}[\ell n(\rho r b)] + \frac{\partial\psi}{\partial z}\frac{\partial}{\partial z}[\ell n(\rho r b)] + \frac{\rho r b}{W_z}\Big[\frac{\partial I}{\partial r} - T\frac{\partial s}{\partial r} - \frac{W_\theta}{r}\frac{\partial}{\partial r}(rV_\theta) - F_r\Big] \tag{43}$$

where I = rothalpy.

The finite difference discretization of the above Poisson equation would be very simple, were it not for the complexity of the boundary conditions. While some methods have attempted to use a rectangular mesh (Biniaris, 1975), the shape of hub and shroud generally makes preferable the use of a distorted grid formed by the intersection of quasi-orthogonals and quasi-streamlines (Figure 12).

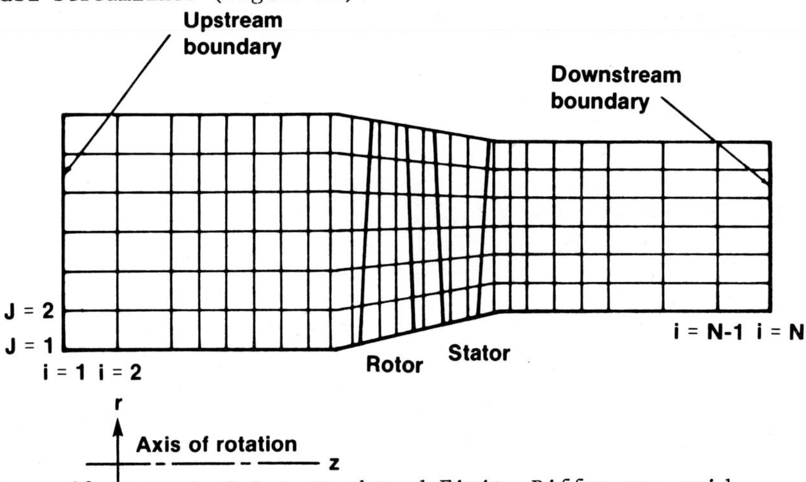

Figure 12 A typical hub-to-shroud Finite Difference grid

The FD stars and the iterative solution method are very similar to the blade-to-blade solution. Other works concerning matrix throughflow analysis include Smith (1970), Davis and Millar (1972, 1974) and Katsanis and McNally (1974). The latter authors use a fully orthogonal mesh generated within the program. In the presence of transonic flow, the finite difference method is used to obtain a reduced mass flow subsonic solution for flow angles and streamline curvatures.This information is then used with the streamline curvature method to obtain the transonic solution.

Detailed comparisons between matrix and streamline curvature methods have been carried out by Frost (1970) and Davis and Millar (1974). Frost further compares streamline curvature duct methods to methods discretizing within the blade rows. Two turbomachines are selected as test cases: A two-stage turbine and single-stage compressor. Agreement is indeed good between matrix and streamline curvature throughflow methods. In a streamline curvature duct model the omission of blade blockage and blade radial tilt effects significantly altered the results. David and Millar's conclusions are essentially the same, further concluding that input preparation was comparable for both methods. With present day computers the storage requirements of the matrix method are not significant, and cannot be used as an argument in favour of streamline curvature methods.

3. Finite element solutions

Finite element solutions of the throughflow problem are presented by Hirsch and Warzée (1974, 1976), Oates, Knight and Carey (1974, 1975). Using an actuator disc approach, Oates et al formulate the throughflow problem by finite elements to yield a system of nonlinear algebraic equations solved by a Newton-Raphson iteration method. We present here Hirsch and Warzée's approach because it is readily adaptable to many finite element programs written for quasi-harmonic equations. These authors rewrite equation (42) in a form suggested by Gelder (1971) as:

$$\frac{\partial}{\partial r}\left(\frac{1}{\rho rb}\frac{\partial\psi}{\partial r}\right) + \frac{\partial}{\partial z}\left(\frac{1}{\rho rb}\frac{\partial\psi}{\partial z}\right) + F[r,z,\rho,\psi,\psi_r,\psi_z] = 0 \tag{44}$$

for which a solution is obtained by the extremization of the quasi-variational integral

$$\delta I(\psi)^{(n+1)} = \delta\tfrac{1}{2}\iint\frac{1}{r}\left[\left(\frac{1}{\rho rb}\right)^{(n)}\left(\frac{\partial\psi}{\partial r}\right)^2 + \left(\frac{1}{\rho rb}\right)^{(n)}\left(\frac{\partial\psi}{\partial z}\right)^2 - 2F\psi\right]dA - \oint_c \frac{\delta\psi}{r}\left(\frac{\partial\psi}{\partial n}\right)ds = 0 \tag{45}$$

after multiplying equation (44) by 1/r for axisymmetric problems. The contour integral in equation (45) vanishes since on all boundaries ψ is either known (hub, shroud) or $\frac{\partial\psi}{\partial n} = 0$ (parallel inlet/exit, not necessarily uniform). It is evident that one great advantage of the Finite Element method is the implicitness of such normal derivative boundary conditions.

Hirsch and Warzée select an 8-node isoparametric quadrilateral element and carry out the integration of equation (45) using a 2x2 Gaussian integration scheme. The procedures for forming the matrix and for iteration follow closely those outlined in the FE section concerning blade-to-blade solutions (section I.5, equations (20) to (22)). A typical Finite Element mesh is shown in Figure 13.

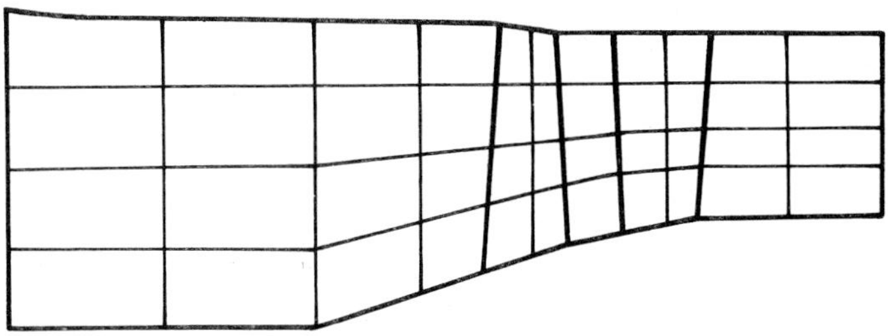

Figure 13 A typical hub-to-shroud Finite Element mesh

A possible solution procedure consists of the following steps: a) An initial stream function distribution is determined by taking $\rho=\rho_o$ throughout the field and setting F=0, in essence solving Laplace's equation in the field. For high Mach numbers, weighted values obtained from a smaller mass flow rate could also be used. b) The normalized stream function $\frac{\partial\psi}{\partial z} = -\frac{\rho b r}{\dot{m}} V_r$; $\frac{\partial\psi}{\partial r} = \frac{\rho b r}{\dot{m}} V_z$ is set to zero at the hub, and to 1 on the shroud. At inlet all the thermodynamic variables are computed at the grid points of the inlet station. c) A marching procedure is carried out to trace the streamlines at each of the nodes of a station (i) from those at the preceding station (i-1). This is done by first locating the element where the streamline originates. A simple interval-halving method can then be used to determine the origin of the streamline, within that element. d) Conservation principles are then applied along streamlines

$$\begin{aligned} &\text{In a duct region} \quad rV_\theta && = \text{constant} \\ &\text{In a stator} \quad H = c_p T_o && = \text{constant} \\ &\text{In a rotor} \quad I = c_p T_{oR} && = \text{constant} \end{aligned} \tag{46}$$

and hence all thermodynamic properties at station (i) would have been determined. e) Hirsch and Warzée use NASA correlations to account for deviations and losses. f) The density is solved iteratively, at each point, from the mass flow function. For example in a stator

$$\frac{\rho}{\rho o} = \left[1 - \frac{\gamma-1}{2a_o^2\rho^2 b^2 r^2}\left(\left(\frac{\partial\psi}{\partial z}\right)^2 + \left(\frac{\partial\psi}{\partial r}\right)^2(1+\tan^2\alpha)\right)\right]^{\frac{1}{\gamma-1}} \tag{47}$$

g) With an automatic mesh generator available it would be a

simple matter to adjust the FE mesh before the next iteration to conform to the streamline shapes obtained in (c). h) The process is repeated from (a) with the new density and stream function distribution and the process is judged to have converged when the stream function is lower than a specified percentage error in the entire field.

Impressive results are presented by these authors for the NASA-TASK-1 single stage transonic compressor as well as the NGTE-106-6 stage compressor. Selected results are shown in Figure 14.

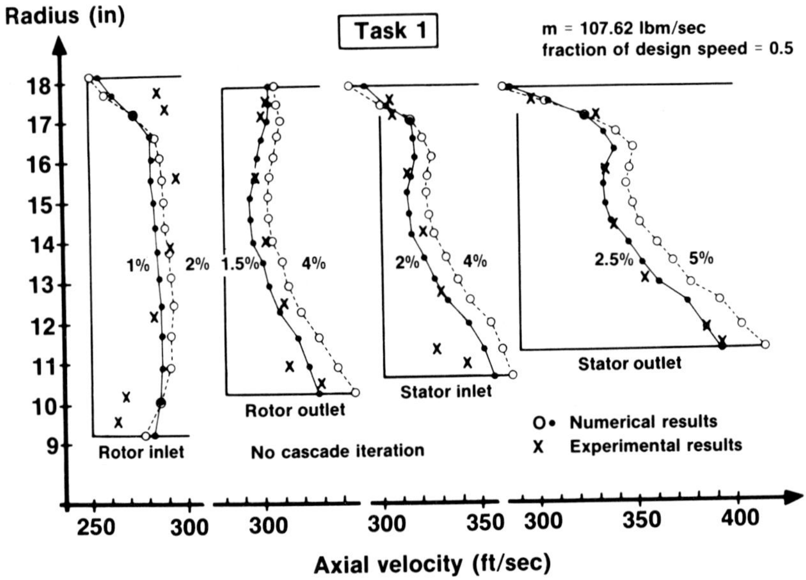

Figure 14 NASA TASK 1 transonic compressor FE results after Hirsch and Warzée

An interesting discussion by Japikse, at the end of the Hirsch and Warzée (1976) article, provides the impetus for a streamline curvature/finite element comparison.

III THREE-DIMENSIONAL SOLUTIONS

We shall attempt to limit this section to a brief review of three-dimensional calculations that have found an application in turbomachinery passages. Such schemes could be divided into two broad categories: 1. Quasi-three-dimensional solutions. 2. Fully three-dimensional solutions. Each of these approaches has received considerable attention in the last few years and several methods have emerged that should prove very useful in assessing designs where three-dimensional effects may not be negligible.

1. Quasi-three-dimensional solutions

In the quasi-three-dimensional approach an iterative procedure is adopted whereby S_2 solutions are used along with S_1 solutions in an interactive manner. The S_2 solutions define the shape and depth distribution of the streamtubes as well as inlet and exit angles used in the S_1 solutions. The S_1 solutions in turn help to define the shape of the meridional streamsurface. The iteration between the two surfaces, while not trivial, introduces no new numerical concepts. While the S_1 blade-to-blade streamsurfaces are invariably taken as surfaces of revolution, each method differs in the selection of the representative meridional surface. Table I summarizes some of the well known methods.

TABLE I QUASI-THREE-DIMENSIONAL SOLUTIONS

AUTHORS	S1-SOLUTION	S2-SURFACE	S2-SOLUTION
Senoo & Nakase 1972	Singularity method	Mean camberline	Streamline curvature
Bosman & El Shaarawi, 1977	Matrix method with matrix inversion	Mass-averaged	Matrix method with matrix inversion
Novak & Hearsey 1977	Streamline curvature	ψ=0.5 streamline	Streamline curvature
Hirsch & Warzée	Finite Element 8-node isoparametric 2x2 Gaussian integration	True meridional (r-z) plane. Exact pitch-averaged equations	Finite Element 8-node isoparametric 2x2 Gaussian integration

2. Fully three-dimensional solutions

Recently, inviscid 3-D solutions using finite difference, finite volume, finite element and characteristics methods have been formulated with success. Such solutions presently cover the subsonic to supersonic regimes. While for low aspect ratio ducts it is found that such 3-D solutions vary very little from the quasi 3-D ones, the scarcity of experimental and theoretical data for 3-D flows makes extensive comparisons difficult in more general cases.

a. Subsonic 3-D solutions The Finite Element method has recently made inroads into 3-D inviscid solutions. In the absence of a variational formulation, Laskaris (1978) applies a Galerkin weighted-residual method to the discretization of the steady-state hydrodynamic equations of 3-D compressible potential flows. Second-order Lagrangian isoparametric cubic elements are used, hence representing the potential, as well as the geometry, quadratically in the undistorted grid. One of the method's advantages is its good approximation of blade boundaries, with no resort to cusps. To increase the accuracy of representation on blade-to-blade surfaces, the Lagrangian element permits the addition of midpoint nodes on the corresponding faces.

The resulting matrix equations are nonlinear and are solved by a Newton-Raphson method. Two mesh sizes were tested, namely 5x15x5 and 5x29x9 with less than 1% difference in accuracy. Solution times are typically 20 min. for the coarse grid and 160 min. for the finer grid on a GE 605 computer. Qualitative results are presented. Future comparisons to experiments and other numerical schemes will help assess the merits of this approach. It would be interesting to determine what advantages this approach would enjoy, if any, over time-dependent finite volume approaches, in the subsonic regime.

b. Transonic and supersonic 3-D solutions Denton (1975,76) formulates a time-dependent Finite Volume approach for inviscid 3-D calculations. The extension from Finite Area methods in 2-D is straightforward in principle, involving only the inclusion of additional flux terms in each conservation equation and the consideration of a third momentum equation. In addition, for a rotating blade row, the energy equation is modified to account for the non-uniformity of the rothalpy at entry to the blade row. The use of a Finite Volume type grid, however, makes the geometrical complications formidable. The method, as it appears in the open literature, is presently restricted to annular blade rows, with leading and trailing edges lying in planes perpendicular to the axis. Arbitrary cusps have to be inserted along the leading and trailing edge planes. The method was programmed for up to 2560 grid points (8x40x8). Computing time per point per time step is about 2.1×10^{-4} seconds on an IBM370-165 and indications are that convergence occurs in about 1000 time steps. Comparisons between fully 3-D and quasi-3-D solutions show that discrepancies occur because of the inability of the latter to account for asymmetry across the pitch. A large number of test cases are demonstrated by Denton (1976) and Bryce and Lichtfield (1977).

Another interesting transonic solution is proposed by Rae (1978). It is essentially an extension of the well-developed relaxation external aerodynamics methods to internal flows. Because the transonic small disturbance equation is adopted, the scheme is limited to thin, lightly loaded blades. To facilitate the use of type dependent finite difference formulae, the skewed grid used in the blade-to-blade planes consists of quasi-orthogonals and straight quasi-streamlines inclined at the stagger angle. It is found, however, that such skewed grids tend to smear out shock transitions into very gradual compressions. To remedy this, Rae adopts a seven-point formula for the mixed derivative in the skewed grid. A 10x60x20 grid is used to analyze a blade row of hub/tip radius ratio of 0.5 and a solidity of 0.5. The solution required 1980 iterations to convergence (the potential at each grid point was updated 1980 times) and solution times are of the order of 18 minutes on an IBM370/168 ($.47 \times 10^{-4}$seconds/grid-point/updating). Several interesting conclusions emerge from this study such as the ability of 3-D calculations to redistribute the mass flow by means of radial motion, while a quasi-3-D calculation may have predicted choking.

Other three-dimensional solutions include the singularity method of Ribaut (1968), the lifting surface approach of Howells and Lakshiminarayana (1977) and the Characteristics Method of Martinon (1979).

Of several schemes that show promise for the solution of 3-D viscous flows, two schemes will be reported. Dodge (1977) separates the velocity vector into viscous and potential components. By equating the purely potential terms of the equation of motion to the pressure gradient, the pressure variable is eliminated. The equations to be solved are

$$\text{CONTINUITY:} \quad \nabla^2\Phi = \nabla\cdot U \; ; \quad (W = \nabla\Phi + U) \tag{48}$$

$$\text{MOMENTUM:} \quad (W\cdot\nabla)U + (U\cdot\nabla)\nabla\Phi = \nu_e\nabla^2 U + \nu_e\nabla(\nabla^2\Phi) + (\text{def } U)\nabla\nu_e + 2[\nabla(\nabla\Phi)]^*\nabla\nu_e \tag{49}$$

The continuity equation is solved by relaxation with a relatively sparse grid near walls. The parabolic momentum equation is solved in an overlayed much finer subgrid by a marching type solution similar to boundary layer analyses. Since equations (48) and (49) contain coupling terms, iteration must be carried out between them.

Walitt, Liu and Harp (1978) attempt a successive approximation solution of the steady, Reynolds averaged, 3-D Navier-Stokes equations. An Alternating-Direction-Explicit FD solution is carried out in three mutually orthogonal planes: 1. Blade-to-blade surfaces. 2. Cross-sectional surfaces from inlet to exit. 3. Meridional planes in mid-channel of blade surfaces. Solution times on a CDC7600 are measured in hours for each surface.

CONCLUSIONS

This chapter presents an "aperçu" of the status of numerical methods in turbomachinery analysis. The progress achieved in the last few years has been phenomenal. Modern approaches are capable of improved geometry definition, shorter solution times and more realistic approximations of the exact governing equations.

Some comparisons between the various methods have been made in the relevant sections. However, conclusive comparisons among them are, to borrow a frequently used numerical turbomachinery terminology, "quasi"-impossible. Understandably, publications tend to show each method in the best possible light, making the task of gauging the validity, accuracy and ease of implementation of the various schemes quite difficult. Although a standardized series of experimental and analytical test cases have been suggested by Gostelow and Smith (1973), they have rarely been used in recent publications. It becomes the users' task to sift through the maze of claims made about each method. Japikse (1977) addresses himself admirably to this problem.

While accuracy of methods is certainly an important criterion, it is felt that the following criteria are equally as important for a method to be labelled practical: 1. Simplicity of input data required to run a case. 2. Fast solution times coupled with minimum computer storage requirements. 3. Minimum interference required from users in interactive type approaches. Such criteria take more importance at a time when numerical analysis is gaining wide acceptance in day-day operations in industry and is reducing reliance on empirical techniques. These methods are used to reduce component development time, while providing performance improvements that derive from improved design and modelling techniques. Numerical programs are often devised with specific configurations in mind and encounter difficulties under conditions however slightly different from the ones for which they were developed. It may be found, to the user's chagrin, that it might have been more economical to construct a new program rather than modify existing ones to conform to a new situation. It becomes imperative, therefore, in the near future to seriously consider tests and criteria by means of which various methods can be judged uniformly. Such comparisons will permit users to make proper choices and recognize the phenomena, within each technique, that may require further study.

REFERENCES

Argyris, J.H. and Scharpf, D.W. (1969) Two and Three-Dimensional Potential Flow by the Method of Singularity. The Aeronautical J. of the Royal Aeronautical Society, 73 :959-961.

Bathe, K.J. and Wilson, E.L. (1976) Numerical Methods in Finite Element Analysis, Prentice Hall.

Bindon, J.P. and Carmichael, A.D. (1971) Streamline Curvature Analysis of Compressible and High Mach Number Cascade Flows. The J. of Mechanical Engg. Science, 13, 5:344-357.

Biniaris, S. (1975) The Calculation of the Quasi-Three-Dimensional Flow in an Axial Gas Turbine. J. of Engg. for Power, Trans. ASME, 97, 2:283-294.

Bosman, C. and El-Shaarawi, M.A.I. (1977) Quasi Three-Dimensional Numerical Solution of Flow in Turbomachines. J. of Fluids Engg., Trans ASME, 99, 1:132-140.

Bosman, C., Chan, K.C. and Halton, A.P. (1978) A Calculation Method for Incompressible Viscous, Blade-to-Blade Flow Through Radial Turbomachines With Log-Spiral Blade Surfaces. ASME paper 78-GT-195. Also J. of Engg. for Power, Trans. ASME (in press).

Bowley, W.W. and Prince, J.F. (1971) Finite Element Analysis of General Fluid Flow Problems. AIAA paper 71-602.

Bryce, J.D. and Lichtfield, M.R. (1977) Experience of the Denton Blade-to-Blade Time Marching Computer Programs. ARC C.P. 1373.

Chung, T.J. (1978) Finite Element Analysis in Fluid Dynamics. McGraw-Hill.

Couston, M. (1976) Transonic Flows in Axial Turbomachinery. Von Karman Institute for Fluid Dynamics, Lecture Series 84.

Davis, W.R. and Millar, D.A.J. (1972) A Discussion of the Marsh Matrix Technique Applied to Fluid Flow Problems. Canadian Aeronautics and Space Journal, 5, 2:64-70.

Davis, W.R. and Millar, D.A.J. (1974) A Comparison of the Matrix and Streamline Curvature Methods of Axial Flow Turbomachinery Analysis, From a User's Point of View. J. of Engg. for Power, Trans. ASME, 97, 4:549-560.

Delaney, R.A. and Kavanagh, P. (1976) Transonic Flow Analysis in Axial-Flow Turbomachinery Cascades by a Time-Dependent Method of Characteristics. J. of Engg. for Power, Trans. ASME, 98, 3:356-364.

Denton, D.J. (1974) A Time Marching Method for Two- and Three-Dimensional Blade to Blade Flows. ARC R&M No. 3775.

Denton, D.J. (1976) Extension of the Finite Area Time Marching Method to Three Dimensions. In "Transonic Flows in Axial Turbomachinery". Von Karman Institute for Fluid Dynamics, Lecture Series 84.

Denton, D.J. (1978) Throughflow Calculations for Transonic Axial Flow Turbines. J. of Engg. for Power, Trans. ASME, 100, 2:212-218.

Dodge, P.R. (1976) A Non-Orthogonal Numerical Method for Solving Transonic Cascade Flow. ASME paper 71-GT-63.

Dodge, P.R. (1977) Numerical Method for 2D and 3D Viscous Flows. AIAA J., 15:961-965.

Frost, D.H. (1970) A Streamline Curvature Through-Flow Computer Program for Analysing the Flow through Axial Flow Turbomachines. N.G.T.E. Report No. R.312.

Frost, D.H. and Smith, D.J.L. (1973) A Streamline Curvature Computer Program for Analysing the Subsonic Compressible Flow Past a Cascade of Aerofoils. N.G.T.E. Report No. R.327.

Gelder, D. (1971) Solution of the Compressible Flow Equations. Int. J. for Numerical Methods in Engg, 3 :35-43.

Gentry, R.A., Martin, R.E. and Daly, B.J. (1966) An Eulerian Differencing Method for Unsteady Compressible Flow Problems. J. of Computational Physics, 1 :87-118.

Giesing, J.P. (1964) Extension of the Douglas Neumann Program to Problems of Lifting Infinite Cascades. Douglas Aircraft

Company Report No. LB 3176.

Gopalakrishnan, S. and Bozzola, R. (1971) A Numerical Technique for the Calculation of Transonic Flow in Turbomachinery. ASME paper 71-GT-42.

Gopalakrishnan, S. and Bozzola, R. (1972) Computation of Shocked Flows in Compressor Cascades. ASME paper 72-GT-31.

Gostelow, J.P. (1964) Potential Flow through Cascades. A Comparison between Exact and Approximate Solutions. ARC 25, 829.

Gostelow, J.P. (1964) Potential Flow through Cascades, Extension to an Exact Theory. ARC 26, 168.

Gostelow, J.P. (1973) Review of Compressible Flow Theories for Airfoil Cascades. J. of Engg. for Power, Trans. ASME, 95, 4:281-292.

Gostelow, J.P. and Smith, D.J.L. (1973) Test Cases for Turbomachinery Flow Field Computation. CUED/A Turbo/TR 48.

Gostelow, J.P. (1975) Trailing-Edge Flows over Turbomachinery Blades and the Kutta-Joukowsky Condition. ASME Paper 75-GT-94.

Habashi, W.G. (1975) A Study of the Finite Element Method for Aerodynamic Applications. Ph.D. Thesis, Aerospace Engineering Department, Cornell University, Ithaca, N.Y., U.S.A.

Habashi, W.G. (1976) The Finite Element Method in Subsonic Aerodynamics. Proceedings of the 1976 Heat Transfer and Fluid Mechanics Institute, McKillop et al Editors, Stanford University Press :374-389.

Habashi, W.G. (1976) Compressible Potential Flows by the Finite Element Method. Proceedings of the Second Int. Symposium on Finite Element Methods in Flow Problems, Santa Margherita, Italy :83-96.

Habashi, W.G. and Dueck, E.G. (1978) The Finite Element Method for Turbomachinery Calculations. Proceedings of the Int. Conf. on Numerical Methods in Laminar and Turbulent Flow, C. Taylor et al, Editors, Halsted Press: 689-695.

Habashi, W.G., Dueck, E.G. and Kenny, D.P. (1979) A Finite Element Approach to Compressor Blade-to-Blade Cascades. AIAA J. (in Press).

Hess, J.L. and Smith, A.M.O. (1966) Calculation of Potential Flow about Arbitrary Bodies. Progress in Aeronautical Sciences, Pergamon Press, 8 :1-139.

Hewitt, B.L. et al, Editors (1976) Methods and Problems in Aeronautical Fluid Dynamics. Academic Press :117-157.

Hirsch, Ch. and Warzée, G. (1974) A Finite Element Method for Flow Calculations in Turbomachines. Vrije Universiteit Brussel, Department of Fluid Mechanics, Report V.U.B.-STR-5.

Hirsch, Ch. and Warzée, G. (1976) A Finite Element Method for the Axisymmetric Flow Computation in a Turbomachine. Int. J. for Numerical Methods in Engg, 10 :93-113.

Hirsch, Ch. and Warzée, G. (1976) A Finite Element Method for Through-Flow Calculations in Turbomachines. J. of Fluids Engg., Trans. ASME, 98 :403-421

Hirsch, Ch. and Warzée, G. (1977) Finite Element Computation of Subsonic Cascade Flows. Canadian Congress of Applied Mechanics (CANCAM), University of British Columbia, Vancouver, 2 :577-578.

Hirsch, Ch. and Warzée, G. (1978) An Integrated Quasi-3D Finite Element Calculation Program for Turbomachinery Flows. ASME paper 78-GT-56.

Howells, R. and Lakshiminarayana (1977) Three-Dimensional Potential Flow and Effects of Blade Dihedral in Axial Flow Propeller Pumps. J. of Fluids Engg.,Trans. ASME, 99,1:167-186.

Ives, D.C. and Liutermoza, J.L. (1977) Analysis of Transonic Cascade Flow Using Conformal Mapping and Relaxation Techniques. AIAA J., 15 5:647-652.

Ives, D.C. and Liutermoza, J.L. (1978) Second Order Accurate Calculation of Transonic Flow over Turbomachinery Cascades. AIAA paper 78-1149.

Jameson, A. (1974) Iterative Solution of Transonic Flows over Airfoil and Wings, Including Flows at Mach 1. Communications on Pure and Applied Mathematics, 27 :283-309.

Japikse, D. (1976) Progress in Numerical Turbomachinery Analysis. J. of Fluids Engg.,Trans. ASME, 98, 4:592-606.

Japikse, D. (1977) Guest Editorial. J. of Fluids Engg., Trans. ASME, 99, 1:2.

Katsanis, T. (1964) Use of Arbitrary Quasi-Orthogonals for Calculating Flow Distribution in the Meridional Plane of a Turbomachine. NASA-TN-D-2546.

Katsanis, T. (1965) Use of Arbitrary Quasi-Orthogonals for Calculating Flow Distribution on a Blade to Blade Surface in a Turbomachine. NASA-TN-D-2809.

Katsanis, T. (1968) A Computer Program for Calculating Velocities and Streamlines on a Blade-to-Blade Stream Surface of a Turbomachine. NASA-TN-D-4525.

Katsanis, T. and McNally, W.D. (1974) Quasi-Three-Dimensional Flow Solution by Meridional Plane Analysis. Society of Automotive Engineers, Paper 740850.

Khalil, I., Tabakoff, W. and Hamed, A. (1978) Viscous Flow Analysis in Mixed Flow Rotors. ASME Paper 78-WA/GT-3.

Korn, D.G. (1975) Numerical Design of Transonic Cascades. ERDA R&D Report COO-3077072, Courant Institute of Mathematical Sciences, New York University.

Kurzrock, J.W. and Novick, A.S. (1975) Transonic Flow around Rotor Blade Elements. J. of Fluids Engg., Trans. ASME, 97 4:598-607.

Laskaris, T.E. (1978) Finite Element Analysis of Three-Dimensional Potential Flow in Turbomachines. AIAA J., 16, 7 :717-722.

Luu, T.S. and Coulmy, G. (1975) Calcul de l'Ecoulement Transsonique avec Choc à Travers une Grille d'Aubes. Association Technique Maritime et Aéronautique :1-18.

Luu, T.S. and Coulmy, G. (1975) Relaxation Solution for the Transonic Flow through a Cascade. Symposium Transsonicum II, Springer Verlag :331-339.

MacCormack, R.W. (1971) Numerical Solution of the Interaction of a Shock Wave with a Laminar Boundary Layer. Proceedings of the Second International Conference on Numerical Methods in Fluid Dynamics, Springer-Verlag :151-163.

MacCormack, R.W. and Paullay, A.J. (1972) Computational Efficiency Achieved by Time Splitting of Finite Difference Operators. AIAA paper 72-154.

Marsh, H. (1966) A Digital Computer Program for the Through-flow Fluid Mechanics in an Arbitrary Turbomachine using a Matrix Method. ARC R&M 3509.

Martensen, E. (1959) The Calculation of the Pressure Distribution on a Cascade of Thick Airfoils by Means of Fredholm Integral Equations of the Second Kind. NASA-TT-F-702.

Martinon, J. (1979) Use of the Characteristic Method for the Prediction of the Three-Dimensional Flow Field in High Transonic Compressors. ASME paper 79-GT-34. Also J. of Engg. for Power, Trans. ASME (in press).

McDonald, P.W. (1971) The Computation of Transonic Flow through Two-Dimensional Gas Turbine Cascades. ASME paper 71-GT-89.

Murman, E.M. and Cole, J.D. (1971) Calculation of Plane Steady Transonic Flows. AIAA J., 8 1:114-121.

Murman, E.M. and Krupp, J.A. (1971) Solution of the Transonic Potential Equation using a Mixed Finite Difference System. In "Lecture Notes in Physics", 8, Springer Verlag :199-206.

Nagayama, T. and Adachi, T. (1977) Numerical Analysis of Flow through Turbine Cascade by the Modified FLIC Method. 1977 Tokyo Joint Gas Turbine Conference, Paper no. 36.

Novak, R.A. (1967) Streamline Curvature Computing Procedures for Fluid Flow Problems. J. of Engg. for Power, Trans. ASME, 89 :478-490.

Novak, R.A. and Hearsey, R.M. (1977) A Nearly Three-Dimensional Intrablade Computing System for Turbomachinery. J. of Fluids Engg., Trans. ASME, 99, 1:154-166.

Oates, G.C. and Carey, G.F. (1974) A Variational Formulation of the Compressible Throughflow Problem. Technical Report AFAPL-TR-74-78.

Oates, G.C., Knight, C.J. and Carey, G.F. (1976) A Variational Formulation of the Compressible Throughflow Problem. J. of Engg. for Power, 98, 1:1-8.

Perkins, H.J. and Horlock, J.H. (1975) Computation of Flows in Turbomachines, Finite Elements in Fluids, 2, Chapter 8, John Wiley (Gallagher, R.H., Oden, J.T., Taylor, C. and Zienkiewicz, O.C. eds).

Rae, W.J. (1977) Calculations of Three-Dimensional Transonic Compressor Flowfields by a Relaxation Method. J. of Energy, 1, 5:284-296; also AIAA paper 77-199.

Rae, W.J. and Homicz, G.F. (1978) A Rectangular-Coordinate Method for Calculating Nonlinear Transonic Potential Flowfields in Compressor Cascades. AIAA paper 78-248.

Ribaut, M. (1968) Three-Dimensional Calculation of Flow in Turbomachines with the Aid of Singularities. J. of Engg. for Power, Trans. ASME, 90, 3:258-264.

Ribaut, M. and Vainio, R.(1975) On the Calculation of Two-Dimensional Subsonic and Shock-Free Transonic Flow. J. of Engg. for Power, Trans. ASME, 97, 4:603-609.

Ribaut, M. (1977) On the Calculation of Three-Dimensional Divergent and Rotational Flow in Turbomachines. J. of Fluids Engg., Trans. ASME, 99, 1:187-196.

Sator, F.G. (1976) Détermination de Profile de Grilles d'Aubes pour Compresseurs Axiaux Transsoniques. Thèse no.254, Département de Mécanique, Ecole Polytechnique Fédérale de Lausanne, Suisse.

Sator, F.G. (1977) Inverse Computation of Profile Shapes for Given Transonic Flow-Configurations with and without Detached Bow-Shocks in Two-Dimensional Turbomachinery Cascades. J. of Engg. for Power, Trans. ASME, 99, 4:601-607.

Schmidt, E. (1979) Computation of Supercritical Compressor and Turbine Cascades with a Design Method for Transonic Flows, ASME paper 79-GT-30. Also J. of Engg. for Power, Trans. ASME (in press).

Scholz, N. and Klein, A. (1977) Aerodynamics of Cascades. AGARDograph no. 220, Chapter 9, :435-513.

Sells, C.C.L. (1968) Plane Subcritical Flow past a Lifting Airfoil. Proc. Royal Society, Part A, 308 :377-401.

Senoo, Y. and Nakase, Y. (1972) An Analysis of Flow Through a Mixed Flow Impeller. J. of Engg. for Power, Trans. ASME, 94, 1:43-50.

Shen, S.F. and Habashi, W.G. (1976) Local Linearization of the Finite Element Method and its Applications to Compressible

Flows. Int. J. for Numerical Methods in Engg., 10 :565-577.

Shen, S.F. (1977) Finite-Element Methods in Fluid Mechanics. Annual Reviews of Fluid Mechanics, 9 :421-445.

Singhal, A.K. and Spalding, D.B. (1978) A 2D Partially-Parabolic Procedure for Axial-Flow Turbomachinery Cascades. ARC R&M 3807.

Smith, D.J.L. (1970) Computer Solutions of Wu's Equations for the Compressible Flow through Turbomachinery. Symposium on the Fluid Mechanics and Design of Turbomachinery, Pennsylvania State University.

Smith, D.J.L. and Frost, D.H. (1969) Calculation of the Flow past Turbomachine Blades. Proc. of Inst. of Mech. Engrs., 184, paper 27.

Smith, L.H., Jr. (1966) The Radial-Equilibrium Equation of Turbomachinery. J. of Engg. for Power, Trans. ASME, 88 :1-12.

Stephens, H.E. (1978) Application of Supercritical Airfoil Technology to Compressor Cascades: Comparison of Theoretical and Experimental Results. AIAA paper 78-1138.

Thompson, D.S. (1973) Finite Element Analysis of the Flow through a Cascade of Aerofoils. CUED/A Turbo/TR 45.

Van den Braembussche, R.A. (1973) Calculation of Compressible Subsonic Flow in Cascades with Varying Blade Height. J. of Engg. for Power, Trans. ASME, 95 :345-351.

Veuillot, J.P. (1977) Calculation of the Quasi Three-Dimensional Flow in a Turbomachine Blade Row. J. of Engg. for Power, Trans. ASME, 99, 1:53-62.

Von Karman Institute for Fluid Dynamics Lecture Series 59 (1973) Transonic Flows in Turbomachinery, 3 and 4. Compressor and Turbine Tests.

Walitt, L., Liu, C.Y. and Harp, J.L. (1978) An Alternating Direction Explicit Method for Computing Three-Dimensional Viscous Flow Fields in Turbomachines. SAE Aerospace Meeting. San Diego, Paper 781001.

Walz, A. (1969) Boundary Layers of Flow and Temperature. MIT Press, Cambridge.

Wilkinson, D.H. (1968) A Numerical Solution of the Analysis and Design Problems for the Flow Past One or More Aerofoils in Cascade. ARC R&M No. 3545.

Wilkinson, D.H. (1970) Calculation of Blade-to-Blade Flow in a Turbomachine by Streamline Curvature. ARC R&M No. 3704.

Wu, C.H. (1952) A General Theory of Three Dimensional Flow in Subsonic and Supersonic Turbomachines of Axial, Radial and Mixed Flow Type. NACA TN 2604.

CHAPTER 9

LAMINAR AND TURBULENT BOUNDARY LAYER FLOW

A. J. Baker
University of Tennessee, Knoxville, TN USA

INTRODUCTION

For many flows of practical interest, the effects of viscosity and turbulence are confined to relatively small regions immediately adjacent to solid surfaces. Such flows are called boundary layer flows, and their analysis has demanded the attention of numerous researchers since the pioneering work of Prandtl (1904). With emergence of the high speed digital computer, numerical solution of the system of partial differential equations governing boundary layer flows has become quite commonplace, see Blottner (1970, 1975). The intent of this Chapter is to assess accuracy and convergence of the finite element-based implicit numerical solution algorithm, for laminar and turbulent boundary layer flows, and to compare performance of the linear element algorithm to the popular Crank-Nicolson finite difference algorithm.

Several very interesting and useful results are developed and discussed. In particular, in terms of ultimate accuracy of the equation system solution, a range of practically useful finite element discretizations for boundary layer flow prediction has been observed. Solutions employing coarse grid linear finite element discretizations generally display accuracy superior to those predicted by strict adherence to the convergence curve. Conversely, those solutions obtained with quadratic finite elements typically display coarse grid inaccuracy. On the other hand, on progressively refined discretizations, sources of error other than that associated with discretization. serve to obliterate the refined solution accuracy theoretically obtainable. The absolute error associated with solutions obtained employing quadratic elements is, however, uniformly smaller than that associated with linear element solutions on sufficiently refined grids. The accuracy obtainable using a non-uniform grid within the finite element algorithm is uniformly superior to that obtained on uniform grids. This is not uniformly the case for the Crank-Nicolson finite difference algorithm. Furthermore, while a non-uniform discretization solution exhibits better absolute error for use of both linear and quadratic finite elements, the presence of an optimum-accuracy grid was detected for linear element solution only. All turbulent flow analyses are performed on non-uniform discretizations, and accuracy and convergence character parallels the laminar flow results for the cases reported.

TWO-DIMENSIONAL LAMINAR BOUNDARY LAYER EQUATIONS

Prandtl (1904) established the concept of boundary layer flow and the corresponding boundary layer simplification of the parent Navier-Stokes equations. The character of the flow-field is illustrated in Figure 1. It is required to establish the two-dimensional velocity and pressure distributions, $\vec{u}(x,y)$ and $p(x,y)$, where

$$\vec{u}(x,y) \equiv u(x,y)\hat{i} + v(x,y)\hat{j} \tag{1}$$

Using the familiar laminar boundary layer order of magnitude analysis, cf., Schlichting (1968), the parent Navier-Stokes equations expressing conservation of mass and linear momentum are simplified in non-dimensional form to

$$L(\rho) = \frac{\partial u}{\partial x} + \frac{\partial v}{\partial y} = 0 \tag{2}$$

$$L(u) = u\frac{\partial u}{\partial x} + v\frac{\partial u}{\partial y} + \frac{1}{\rho}\frac{\partial p}{\partial x} - \frac{\partial}{\partial y}\left(\nu\frac{\partial u}{\partial y}\right) = 0 \tag{3}$$

$$L(v) = -\frac{1}{\rho}\frac{\partial p}{\partial y} = 0 \tag{4}$$

where ρ is the (constant) density and $\nu \equiv \frac{\mu}{\rho}$ is the kinematic viscosity. The x axis is assumed aligned with the direction of predominant flow, and y is the coordinate traversing the thickness of the boundary layer. Under the boundary layer assumption, the transverse momentum equation (4) is identically satisfied by a pressure distribution impressed uniformly across the boundary layer thickness, i.e., $p(x,y) = p_I(x)$. The potential flow solution exterior to $\delta(x)$, the boundary layer thickness, provides the data; hence, only $\vec{u}(x,y)$ is required established from solution of equations (2)-(3).

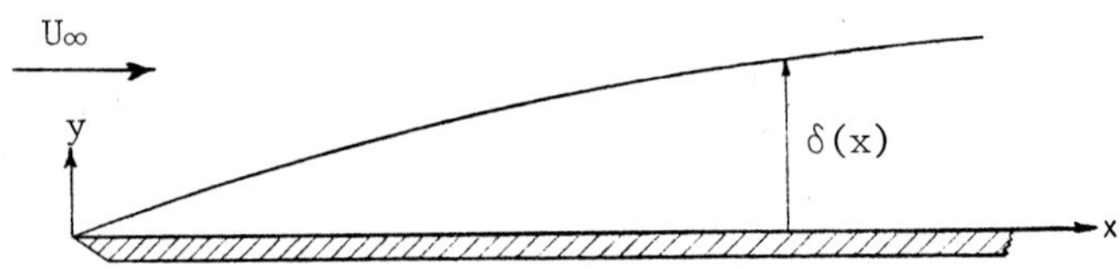

Figure 1. The Boundary Layer Flow Solution Domain

Closure of equations (2)-(3) requires the constitutive relationship for viscosity ν, and establishment of appropriate boundary and initial conditions for a well-posed problem. For laminar flow, kinematic viscosity is a property of the fluid constituting the boundary layer; hence, $\nu = \nu(T)$, where T is the static temperature. Equation (3) is an initial-valued, parabolic equation, while equation (2) is a purely initial-value problem on v. The appropriate boundary specifications at the solid surface are

$$u(x,0) = 0$$
$$v(x,0) = V_w(x) \tag{5}$$

which includes a mass flux. At the freestream, $y > \delta(x)$, appropriate conditions are either

$$u(x,\ y > \delta) = u_I(x)$$

or

$$\frac{\partial u(x,\ y > \delta)}{\partial y} = 0 \tag{6}$$

FINITE ELEMENT SOLUTION ALGORITHM

The finite element solution algorithm for equations (2)-(6) is established by assuming the dependent variables and parameters are interpolated on subdomains Ω_e of the global domain Ω, see Figure 2, and that $\Omega_e \equiv R_e \times \chi$ where χ is a generalized initial-value coordinate.

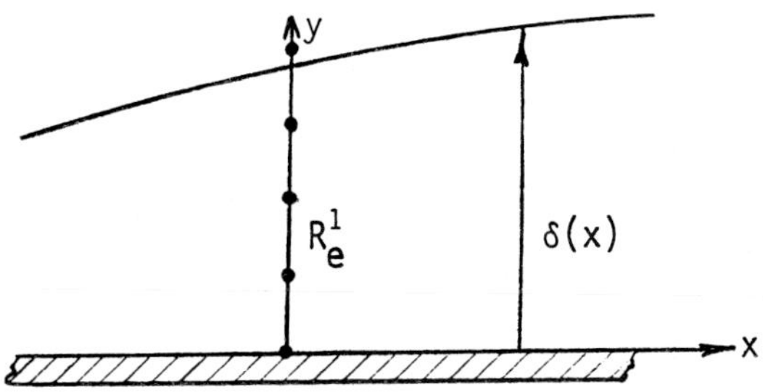

Figure 2. Finite Element Discretization of the Boundary Layer Region

Define the finite element approximation to the true solution q as q*, i.e.,

$$q^*(y,x) \equiv \sum_{e=1}^{M} q_e(y,x) \tag{7}$$

where M is the total number of finite element domains R^1_e spanning R^1. On each finite element domain Ω_e, then

$$q_e(y,x) = q_e(x_1,\chi) \equiv \{N_k(x_1)\}^T\{Q(\chi)\}_e \tag{8}$$

and the union of Ω_e forms the solution domain $\Omega \equiv R^1 \times \chi$. The elements of $\{N_k\}$ are interpolation polynomials complete to degree k, while the elements of $\{Q\}_e$ are the to-be-determined expansion coefficients. Using the familiar Galerkin-Weighted Residuals formulation, cf., Baker and Soliman (1977), and for the dependent variable q identified with u(x,y), the finite element solution algorithm for equation (3) becomes

$$S_e\Bigg[\{U\}_e^T\int_{R_e^1}\{N_k\}\{N_k\}\{N_k\}^T dx_1 \{U\}_e' +$$

$$\{XNU\}_e^T\int_{R_e^1}\{N_k\}\frac{d}{dx_1}\{N_k\}\frac{d}{dx_1}\{N_k\}^T dx_1\{U\}_e$$

$$+\int_{R_e^1}\{N_k\}\{N_k\}^T\{V\}_e\frac{d}{dx_1}\{N_k\}^T dx_1\{U\}_e + \frac{1}{\rho}\frac{dp_I}{dx}\int_{R_e^1}\{N_k\}dx_1\Bigg] \equiv \{0\} \tag{9}$$

Here, S_e is the finite element assembly operator. The integrals specified in equation (9) are readily evaluated; the efficient hypermatrix form is, see Baker and Soliman (1979),

$$S_e\Big[\Delta_e\{U\}_e^T[A3000]\{U\}_e' + \Delta_e\Big(\{XNU\}_e^T[A3011] + \{V\}_e^T[A3001]\Big)\{U\}_e + \Delta_e p_I'\{A10\}\Big] \equiv \{0\} \tag{10}$$

In equation (10), the matrix elements of $\{XNU\}_e$ are the nodal values of the viscosity ν. The solution specification is completed by the initial condition

$$u(x_0,y) \Longrightarrow \{U(\chi_0)\} \tag{11}$$

which corresponds to a mapping of the initial condition for u onto the nodal coordinates of $\sum_e u_e$. The rank of the global matrix system equation (10) is at least one less than its order, to account for the no-slip boundary condition u(x,0) = 0. Hence, application of the finite element algorithm to equation (2) has yielded a system of ordinary differential equations (10), with initial condition equation (11), for solution of the node point distribution of the discretized velocity u, i.e., $\{U\} \equiv \sum_e \{U\}_e$.

Solution of equation (2) deterines the discretized transverse velocity distribution {V}, hence $\{V\}_e$. The x dependence in v(x,y) is parameterized by u(x,y), resulting in the ordinary differential equation specification

$$\frac{\partial v(x,y)}{\partial y} \Longrightarrow \frac{dV(y)}{dy}\Bigg|_x \equiv V(x) = f(x,y) \Longrightarrow - \{U(\chi)\}' \tag{12}$$

Equation (12) is an ordinary differential equation; applying the trapezoidal rule integration algorithm yields

$$V_{n+1}(x) = V_n(x) + \frac{\Delta y}{2}\left[f_{n+1}(x) + f_n(x)\right]$$

In this equation, $\Delta y \equiv (y_{n+1} - y_n)$ is the integration step-size which corresponds exactly to the finite element measure (distribution) Δ_e.

An efficient implicit integration algorithm is required for equation (10), which is compactly written as, letting $\{U\}_e \Longrightarrow \{Q\}_e$

$$S_e\left[[C]_e\{Q\}_e' + [U]_e\{Q\}_e + \{f\}_e\right] = \{0\} \tag{13}$$

Here, $[C]_e$ is the matrix equivalent of the initial-value behavior and explicitly non-linear, $[U]_e$ contains both convection and diffusion contributions, and $\{f\}_e$ is the non-homogeneous pressure gradient term. The family of one-step, implicit finite difference integration algorithms is

$$\{Q\}_{j+1} = \{Q\}_j + h\left[\theta\{Q\}_{j+1}' + (1-\theta)\{Q\}_j'\right] \tag{14}$$

where j is the χ step index, h is the integration step-size, and $\theta = \frac{1}{2}$ yields the trapezoidal rule. Substitution of equation (13) into (14) eliminates the derivative matrix $\{Q\}_e'$ and produces the non-linear algebraic equation

$$S_e\{F(Q)\}_e = \{F(Q)_{j+1}\} = \{0\} \tag{15}$$

The Newton iteration algorithm for equation (15) is

$$\{Q\}_{j+1}^{p+1} = \{Q\}_{j+1}^{p} - \frac{\left\{F\left(\{Q\}_{j+1}^{p}\right)\right\}}{[J]_{j+1}} \tag{16}$$

which can be cast in the matrix solution form

$$\left[J\left(\{Q\}_{j-\ell}\right)\right]\{\delta Q\}_{j+1}^{p+1} = -\left\{F\left(\{Q\}_{j+1}^{p}\right)\right\} \tag{17}$$

where the dependent variable is the iteration vector,

$$\{\delta Q\}_{j+1}^{p+1} \equiv \{Q\}_{j+1}^{p+1} - \{Q\}_{j+1}^{p} \tag{18}$$

Here, p is the iteration index, and $\ell > 0$ retards evaluation of the Jacobian as an economy measure. The right side of equation (17) is equation (15) evaluated with the pth iterate,

$$\{F\}_{j+1}^{p} = S_e\left[[C]_e\left(\{Q\}_{j+1}^{p} - \{Q\}_j\right) + h\left(\theta\{g_e\}_{j+1}^{p} + (1-\theta)\{g_e\}_j\right)\right] \tag{19}$$

where

$$\{g_e\}^p \equiv [U]_e\{Q\}_e^p + \{f\}_e^p \tag{20}$$

Equations (19)-(20) involve matrix inner products over elements only, with the assembly operator yielding the equivalent global expression. The vanishing of $\{F\}$ to within definition of a computed zero yields equation (17) homogeneous, hence convergence of the iteration process for any evaluation of the Jacobian.

THEORETICAL ANALYSIS

The boundary layer equation system is explicitly non-linear in the initial-valued operator. Furthermore, for turbulent flow, the diffusion operator will also become strongly non-linear. An analysis of the linearized equation system can be employed to predict performance, and controlled numerical experiments can then be utilized to indicate validity of linear theory extension to the problem class.

An elementary estimate is facilitated using a truncation error analysis. For example, using linear ($k = 1$) interpolation polynomials in equation (8), it is a simple exercise to confirm that the diffusion matrix equivalent $\Delta_e\{XNU\}^T[A3011]\{U\}_e$ yields a second-order accurate, centered difference approximation on a non-uniform grid. Specifically, assembling this term in equation (10) over two adjacent elements, the finite difference recursion relation that results for the common node j is

$$\frac{1}{2\Delta^-}\left(\nu_{j-1} + \nu_j\right)\left(-u_j + u_{j-1}\right) + \frac{1}{2\Delta^+}\left(\nu_{j+1} + \nu_j\right)\left(+u_{j+1} - u_j\right)$$

where Δ^- and Δ^+ are the nodal separation distances on each side of y_j. For uniform discretization and viscosity, this becomes

$$\frac{1}{\Delta}\left[u_{j-1} - 2u_j + u_{j+1}\right]$$

which is Δ times the second-order accurate central difference operator. Hence, on this basis alone, one anticipates the linear element formulation to yield a spatially second-order accurate algorithm. The trapezoidal rule integration algorithm is also second-order accurate; hence, the results generated by the linear element embodiment should be comparable on an exact basis to the familiar Crank-Nicolson finite difference algorithm.

On the same truncation error analysis basis, for a uniform discretization and constant viscosity, the quadratic element-generated diffusion term under assembly yields a 4th order accurate central difference recursion relation. However, the non-linear convection influence, and the extension to turbulent flow prediction on non-uniform grids, involves the entire aspect of solution of systems of non-linear partial differential equations, which the elementary truncation error analysis studiously ignores. Therefore, the more appropriate approach may be to employ interpolation theory, and the concept of the energy norm which is intrinsic to analysis of linear elliptic partial differential equations.

The basic requirement is to determine how close the (finite element, or any other) solution q*,

$$q^* \equiv \sum_{e=1}^{M} q_e = \sum_{e=1}^{M} \{N_k(x_1)\}^T\{Q(\chi)\}_e \tag{21}$$

is to the true solution $q(x_1,\chi) = q(y,x)$. The fundamental theorem for elliptic partial differential equations, cf., Strang and Fix (1973), is that the finite element solution lies as close as possible to the exact solution, in the sense that the energy in the error ε, defined as

$$\varepsilon \equiv q - q^* \tag{22}$$

is minimized. For this case, the functional for minimization comparable to the extremization equivalent of the elliptic partial differential equation is

$$I(q,q) = \int_{R^n} \left[\frac{1}{2} \sum_{i=1}^{n} \left(\frac{\partial q}{\partial x_i} \right)^2 - gq \right] d\tau \tag{23}$$

where g is a non-homogeneity. The energy inner product for equation (23) is defined as the sum of quadratic terms appearing in I(q,q); hence, for a one-dimensional boundary value problem

$$E(q,q) = \frac{1}{2} \int_{R^1} \left(\frac{\partial q}{\partial x_1} \right)^2 dx_1 \tag{24}$$

The proof of the fundamental theorem, cf., Strang and Fix (1973), pp. 39-41, assures that the error in the energy is minimized by the finite element solution, and in particular, that the energy in the error satisfies the inequality

$$E(\varepsilon,\varepsilon) \equiv ||q-q_e,q-q_e|| \leq C_1 \Delta_e^{2(k+1-m)} ||q||_\alpha^2 \tag{25}$$

where C_1 is a constant independent of Δ_e, the measure of the largest finite element on R^1, the exponent of Δ_e is a function of the largest complete polynomial degree k in equation (8), $2m = 2$ is the order of the elliptic operator, equation (23), and $\alpha \equiv k + 1$ which is related to the required smoothness of the solution q.

Equation (25) states that the energy in the error of the finite element solution, obtained using linear interpolation polynomials for example, is bounded by a term of order Δ_e^2 (or, in finite difference terminology, h^2, i.e., the method is second-order accurate). By the same token, for $k = 2$, the error is bounded by Δ_e^4, i.e., fourth order accuracy. Hence, the form of equation (25) appears most appropriate. As a final note, $||q||_\alpha^2$ in equation (25) can be replaced by $||F||$, the L_2 norm of the data of the problem specification, i.e.,

$$||F|| = \left[\int g^2 dx\right]^{1/2} \tag{26}$$

Then, equation (25) states that the energy in the error of the linear finite element approximation to the true solution q decreases in proportion to the square of the measure of the largest finite element, and that the error depends continuously on the data of the problem specification. Hence, as a consequence of the fundamental theorem, the finite element solution converges in the energy norm, i.e.,

$$E(\varepsilon,\varepsilon) \rightarrow 0 \quad \text{as} \quad \Delta_e \rightarrow 0 \tag{27}$$

An extension in concept is required for the initial-valued parabolic boundary layer equations; i.e., the equation of interest is of the form

$$L(q) = \frac{\partial}{\partial \chi} q(x_1,\chi) + \nu \frac{\partial^{2m}}{\partial x_1^{2m}} q(x_1,\chi) + g(x_1,\chi) = 0 \tag{28}$$

Oden and Reddy (1976) develop error estimates for the fully discrete Galerkin approximation of an equation of the form (28). The components of the approximate solution scheme are defined as: $q(x_1,\chi)$, the solution to the semidiscrete (continuous dependence on "time" χ is still assumed) Galerkin approximation; and $q^*(x_1)$, the solution to the finite difference Galerkin approximation at "time" $\chi = n\Delta\chi$. The following definitions for error are employed:

$$\varepsilon(x_1,n\Delta\chi) = q(x_1,n\Delta\chi) - q^*(x_1) \quad = \text{approximation error}$$

$$\sigma(x_1,n\Delta\chi) = q(x_1,n\Delta\chi) - Q(x_1,n\Delta\chi) = \text{semidiscrete approximation error}$$

$$\tau(x_1,n\Delta\chi) = Q(x_1,n\Delta\chi) - q^*(x_1) \quad = \text{temporal approximation error} \tag{29}$$

For any choice of norm,

$$||\varepsilon(t)|| = ||\sigma+\tau|| \leq ||\sigma|| + ||\tau|| \tag{30}$$

For use of a forward difference integration algorithm, Oden and Reddy (1976) prove that if the components of the error are given by equation (29), then for equation (28) linear, the energy in the error satisfies the inequality

$$||\varepsilon(n\Delta\chi),\ \varepsilon(n\Delta\chi)|| \leq C_1 \Delta_e^{2(k+1-m)} ||q(n\Delta\chi)||_k^2 + C_5 \Delta\chi ||Q_0||_m^2 \tag{31}$$

Note the first term on the right side of equation (31) is identical to inequality (25). In the additional term, C_5 is a constant independent of Δ_e, $\Delta\chi$ is the "time" step, and $||Q^0||_m$ is the energy of the initial data.

Accuracy and convergence may also be measured in terms of the familiar boundary layer integral parameters. In particular, boundary layer displacement (δ^*) and momentum (θ) thickness are variables of primary interest in engineering evaluation of results, and are defined as

$$\delta^*(x) \equiv \int_0^\infty \left[1 - \frac{u(x,y)}{u_I(x)}\right] dy \tag{32}$$

$$\theta(x) \equiv \int_0^\infty \frac{u(x,y)}{u_I(x)} \left[1 - \frac{u(x,y)}{u_I(x)}\right] dy \tag{33}$$

A convenient error norm is shape factor H, defined as

$$H \equiv \delta^*/\theta \tag{34}$$

Skin friction is the engineering measure of drag, hence the distribution of viscosity induced shear stress. The skin friction coefficient C_f is also a useful error norm, and is defined as

$$C_f \equiv \frac{\tau_w}{\frac{1}{2}\rho_I u_I^2} \equiv \left(\frac{1}{2}\rho_I u_I^2\right)^{-1} \rho_w \mu_w \frac{\partial u}{\partial y}\bigg|_w \tag{35}$$

LAMINAR FLOW, ACCURACY AND CONVERGENCE

The first published attempt to formally quantize convergence, and compare finite difference and linear finite element relative solution accuracy for laminar flows, is by Popinski and Baker (1976). The test case corresponds to laminar boundary layer development on a flat plate in zero pressure gradient, for which the exact Blasius similarity solution exists for comparison, cf., Schlichting (1968). Accuracy and convergence was determined in a max norm

$$E_{max}(\chi) \equiv \max_{e=1}^{m+1} \left[\frac{u_e - u_\beta}{u_I}\right] \tag{36}$$

where $u_e(x_1,\chi)$ is the numerical solution, available at the nodes of the discretization, u_β is the corresponding (nodal) exact Blasius solution, and u_I is the (constant) freestream velocity. Equation (36) was evaluated at ten solution stations, and the results averaged to generate the results which are summarized in Figure 3. The various finite element notations correspond to different approximations to the finite element initial-value operator $[A3000]_e$, while FD-CN denotes the Crank-Nicolson finite difference algorithm.

These results fail to quantize solution differences, but do provide an overall estimation of acceptable performance. However, the predicted rate of convergence is nominally cubic in the E_{max} norm. It is well known that the Crank-Nicolson finite difference algorithm is second-order accurate; hence, the

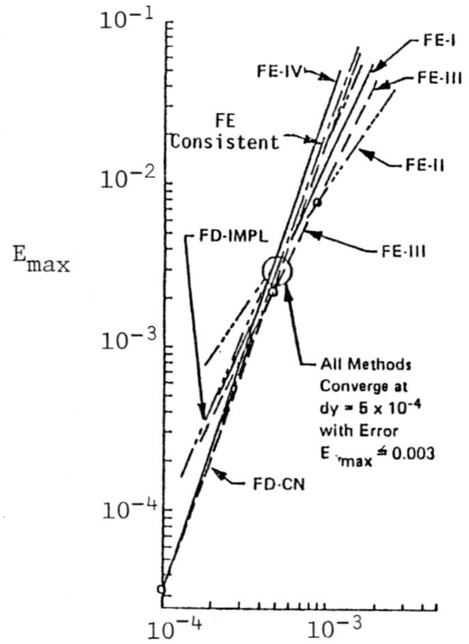

a) Error Map for FE and FD Methods

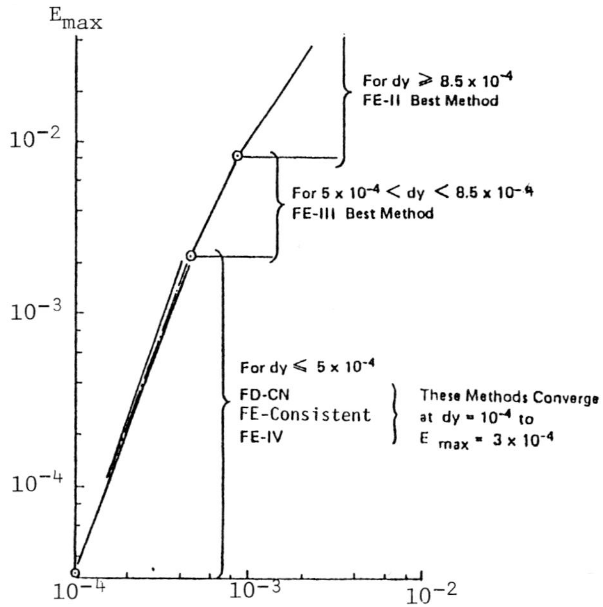

b) Best Methods as a Function of dy

Figure 3. Solution Error in Max. Norm for Boundary Layer Solutions convergence rate should be quadratic in any consistently defined norm. As discussed, the energy norm is the intrinsic error measure for a finite element algorithm, and results so obtained should be consistent with correctly defined alternative norms. Soliman (1978) presents an evaluation of the same (Blasius) test case, and concludes that, to quantify accuracy differences and convergence, numerical experiments must be conducted using initial data that are uniformly evaluable in the energy norm. Otherwise, the second term in the convergence statement (31) can swamp the significance of the discretization error.

Soliman and Baker (1979) present results of a series of controlled numerical experiments that has provided the data necessary to quantize accuracy and convergence. The initial data specification corresponds to a slug profile, for which the energy evaluation is a uniform constant for all (non-uniform) discretizations and for use of linear and quadratic polynomials in equation (8), see Soliman (1978). Baker and Manhardt (1977) confirmed the usefulness of a geometric progression for locating nodal coordinates on R^1; hence, coordinates were specified as

$$y_{e+1} = y_e + \frac{R^1}{\sum_{j=1}^{M}\left[\rho^{j-1}\right]} \rho^{(e-1)} \qquad 1 \leq e \leq M \tag{37}$$

where y_{e+1} is the extremum nodal coordinate of R_e^1, y_e is the coordinate of the first node of R_e^1, ρ is the geometric progression ratio, and M is the total number of finite elements spanning the solution domain R^1. The scope of the test series includes a range of applied axial pressure gradients normalized by the definition

$$\Delta p \equiv \frac{dp_I/dx}{\rho U_I^2/L} \tag{38}$$

where ρ_I, U_I and L are appropriately selected freestream constants. The zero pressure gradient case yields a 50% decrease in the level of the energy norm on the test interval. The smaller non-zero pressure gradient maintains an approximately constant energy on the interval, while the larger yields a 50% increase in the energy norm. Hence, a significant range for $||F||$ in equation (26) is evaluated.

The accuracy and convergence results reported by Soliman and Baker (1979) are summarized in Figure 4. Figure 4a presents the linear $(k=1)$ finite element solutions, while 4b contains the second order accurate Crank-Nicolson finite difference algorithm results. For both cases, the open symbols are for results obtained using uniform discretizations of R^1. The solid lines are drawn at an integer slope of two. Both solutions indeed converge quadratically in the energy norm, as predicted by the extension of linear theory as well as the truncation

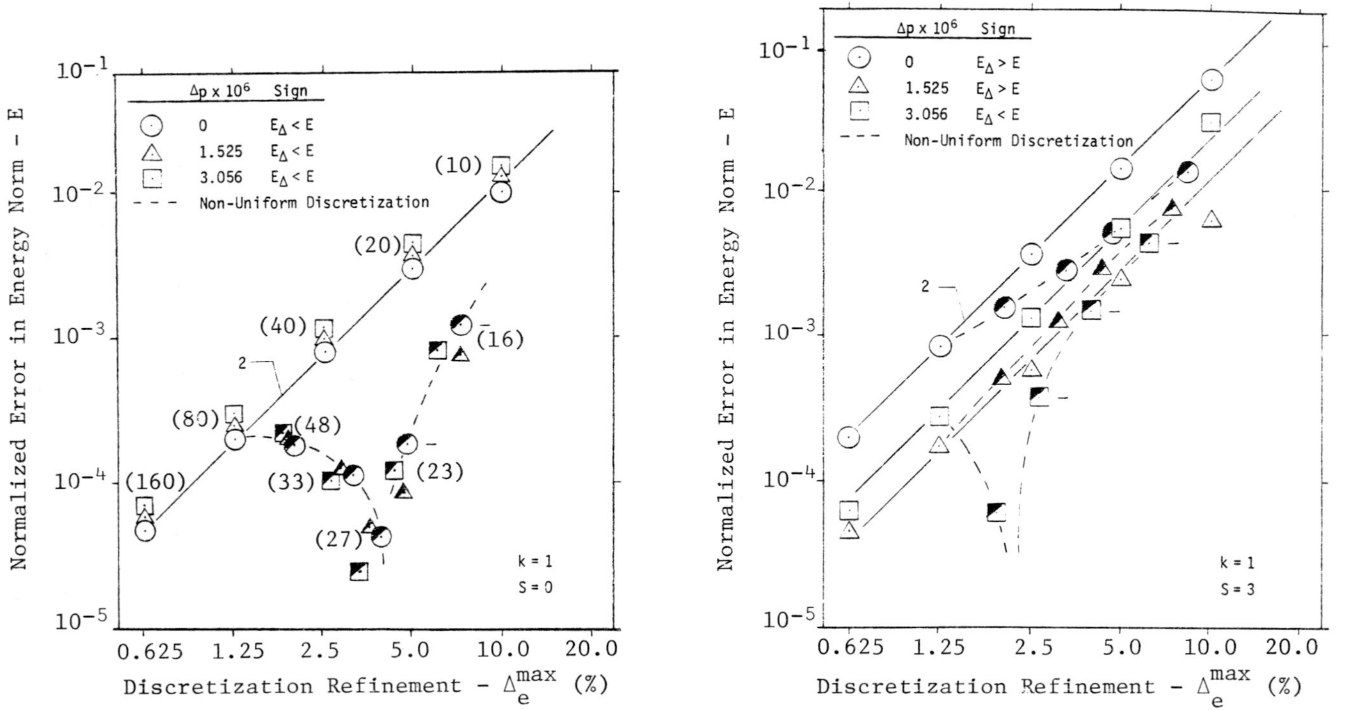

a) Linear Finite Element Algorithm

b) Crank-Nicolson Finite Difference Algorithm

Figure 4. Accuracy and Convergence with Discretization Refinement in Energy Norm, Laminar Boundary Layer with Pressure Gradient error analysis. It is interesting that the finite element results collapse onto a single line, for the range of pressure gradients evaluated, while the absolute accuracy of the Crank-Nicolson prediction is quite sensitive (in energy) to the level of pressure gradient. The linear finite element and Crank-Nicolson solutions are of comparable accuracy, for the cases of non-zero pressure gradient, while the finite element solution for zero pressure gradient is more accurate by about a factor of four. These illustrated differences are due solely to the nonlinear initial-value operator matrix structure. The form established by finite element theory is

$$[A3000]_0 \equiv \frac{1}{12}\begin{bmatrix} \begin{Bmatrix}3\\1\end{Bmatrix} & \begin{Bmatrix}1\\1\end{Bmatrix} \\ \begin{Bmatrix}1\\1\end{Bmatrix} & \begin{Bmatrix}1\\3\end{Bmatrix} \end{bmatrix} \tag{39}$$

while the Crank-Nicolson equivalent form is

$$[A3000]_3 \equiv \frac{1}{2}\begin{bmatrix} \begin{Bmatrix}1\\0\end{Bmatrix} & \begin{Bmatrix}0\\0\end{Bmatrix} \\ \begin{Bmatrix}0\\0\end{Bmatrix} & \begin{Bmatrix}0\\1\end{Bmatrix} \end{bmatrix} \tag{40}$$

It is an elementary operation to show that the assembly of $\{U\}_e^T[A3000]_3\{U\}_e'$, for a uniform discretization and reexpressed as a recursion relation, is $u_j u_j'$.

Significantly more pronounced differences in solution accuracy level, and convergence rate with discretization refinement, are illustrated for solutions obtained using non-uniform discretizations of R^1. These data correspond to the half-shaded symbols in Figure 4, which are interpolated in an approximate manner by the dashed curves. All results are obtained holding the span Δ_1 of the element nearest the wall equal to the $M = 80$ uniform grid value. All data for the linear finite element solutions again collapse onto a universal curve, and any non-uniform discretization produces a solution of accuracy superior to that of the corresponding uniform discretization result. (The number in parenthesis next to the symbols is the corresponding value of M.) A somewhat distressing observation is that the finite element solutions actually diverge from the correct solution for $27 \leq M \leq 80$. This is the direct consequence of the energy error actually passing through zero at $\Delta_e^{max} \approx 0.04$ (as denoted by the appended minus signs in Figure 4a). Conversely, no particular benefit accrues to the Crank-Nicolson algorithm with use of a geometric progression non-uniform discretization, see Figure 4b. Except for the largest applied pressure gradient at $M = 48$, convergence remains essentially quadratic with modest select differences in absolute accuracy level.

Several important conclusions can be drawn from these data. Foremost, convergence with discretization refinement as determined in an energy norm, is consistent with the elementary truncation error analysis. The numerical results tend to confirm that the finite element linear theory may be extensible to the more interesting problem classes in fluid mechanics. The finite element algorithm results appear to collapse onto a universal curve, over a substantial range of "data" in energy, The modest alteration required to reproduce the Crank-Nicolson algorithm destroys this universality even for uniform discretizations. No consistent benefit to the Crank-Nicolson accuracy in energy accrues for use of a geometrically non-uniform discretization, in clear distinction to the finite element results. Of ultimate impact, since the energy norm is explicitly sensitive to the fact that solution of a partial differential equation system is involved, it is a measure of total solution performance on the global basis. Hence, quantization of absolute differences in attainable solution accuracy in the intrinsic norm is achieved.

From the engineering standpoint, one is undoubtedly more interested in the accuracy of prediction of (say) boundary layer shape factor H and skin friction C_f. Figures 5 and 6 present the corresponding data, of Soliman and Baker (1979), comparing the linear finite element and Crank-Nicolson algorithm predictions on both bases. Again, the open symbols are interpolated at an integer slope and correspond to uniform discretization results. Essentially no differences exist in the absolute accuracy of the prediction of each algorithm in either H or C_f. The non-uniform discretization results are measurably different, but the overall consistency evidenced in the energy norm comparisons is lacking. The accuracy and convergence rate for the Crank-Nicolson algorithm is unaffected by a non-uniform discretization (half-shaded symbols) for vanishing pressure gradient. For the favorable pressure gradients evaluated, however, the Crank-Nicolson algorithm performance generally parallels the improved accuracy attainable by the linear element algorithm. If anything, the accuracy of the finite element results appears more uniform over the range of discretization non-uniformity evaluated. In all cases, the non-uniform discretization results are of superior accuracy, indicating that these grids should definitely be employed for practical engineering calculations. Furthermore, such discretizations employing as few as approximately 25 finite elements (or finite difference nodes) can yield solutions of superior accuracy with one-third the computer effort (CPU) of the uniform grid solutions.

TURBULENT BOUNDARY LAYER FLOW

The two-dimensional turbulent flow boundary layer equations are obtained by time-averaging the parent Navier-Stokes equations and then performing the boundary layer order of magnitude

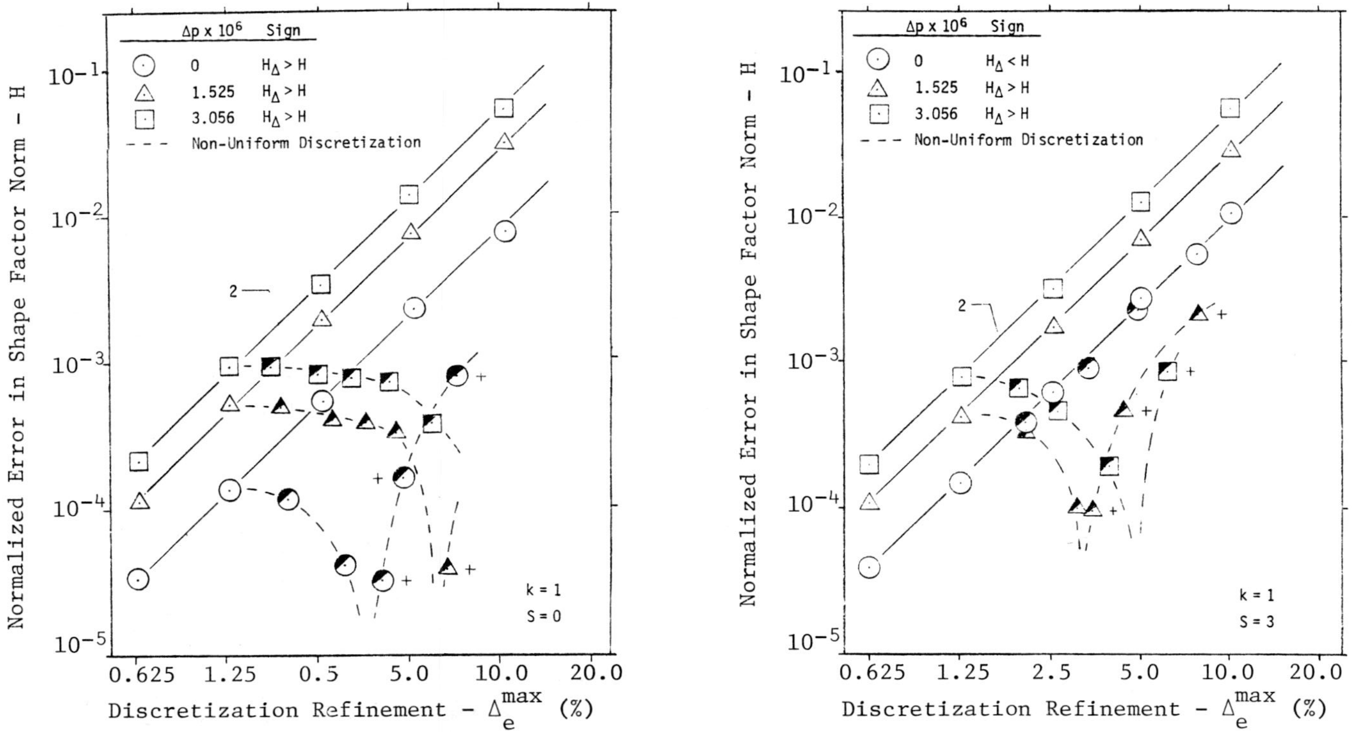

a) Linear Finite Element Algorithm

b) Crank-Nicolson Finite Difference Algorithm

Figure 5. Accuracy and Convergence with Discretization Refinement in Shape Factor Norm, Laminar Boundary Layer with Pressure Gradient

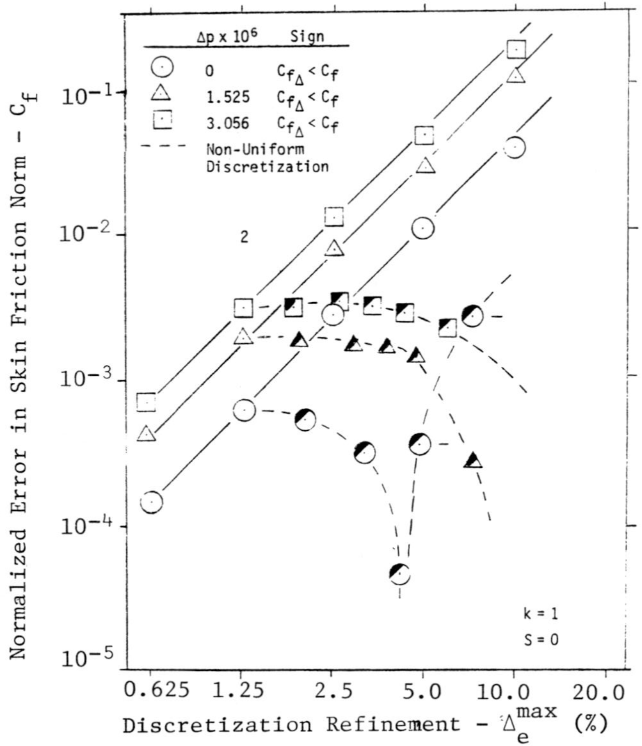

a) Linear Finite Element Algorithm

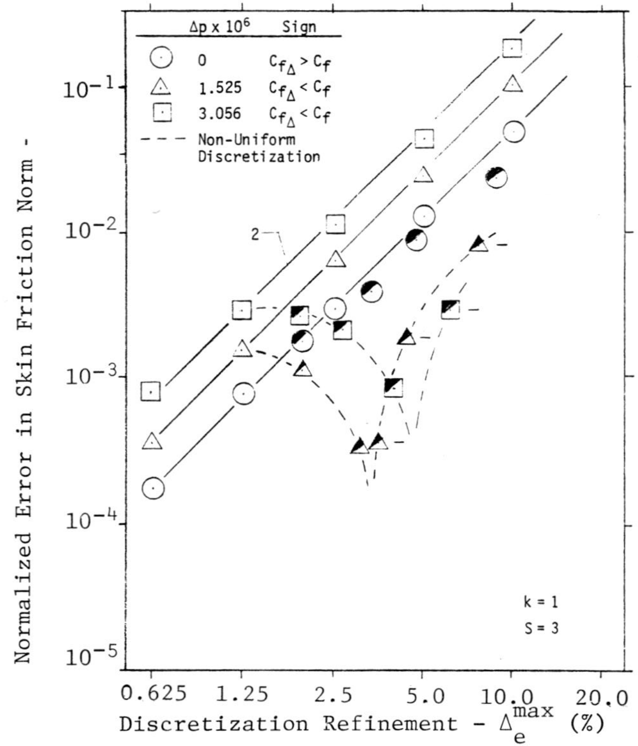

b) Crank-Nicolson Finite Difference Algorithm

Figure 6. Accuracy and Convergence with Discretization Refinement in Skin Friction Norm, Laminar Boundary Layer with Pressure Gradient analysis, cf., Cebeci and Smith (1974). The resultant boundary layer equations for incompressible turbulent flow, comparable to equations (2)-(4), are

$$L(\rho) = \frac{\partial \bar{u}}{\partial x} + \frac{\partial \bar{v}}{\partial y} = 0 \tag{41}$$

$$L(\bar{u}) = \bar{u}\frac{\partial \bar{u}}{\partial x} + \bar{v}\frac{\partial \bar{u}}{\partial y} + \frac{1}{\rho}\frac{d\bar{p}}{dx} - \frac{\partial}{\partial y}\left[\frac{\nu}{Re}\frac{\partial \bar{u}}{\partial y} - \overline{u'v'}\right] = 0 \tag{42}$$

$$L(\bar{v}) = \frac{\partial}{\partial y}\left[\bar{p} + \overline{v'v'}\right] = 0 \tag{43}$$

The overbar signifies time-averaged steady flow component, and $\overline{u'v'}$ is the time-average of the correlation of the fluctuating velocity components. Comparing equations (41)-(43) to the laminar form, shows the continuity equation unchanged for the mean-flow velocity, the addition of the Reynolds shear stress $\overline{u'v'}$ to the x-momentum equation, and a second-order effect of balancing of time-averaged pressure through the boundary layer by the Reynolds stress normal component $\overline{v'v'}$. Hence to first order, the inviscid freestream pressure gradient dp_I/dx is impressed unaltered upon the boundary layer. Therefore, the essential initial-boundary value character of the equation system is unaltered, and it remains to express the components of the Reynolds stress.

The elementary form for closure is to assume existence of an "effective" viscosity coefficient ν^e defined as

$$\nu^e \equiv \frac{\nu}{Re} + \nu^t \tag{44}$$

where ν^t is the effective kinematic viscosity used to correlate the Reynolds shear stress to the mean-flow velocity gradient, i.e.,

$$-\overline{u'v'} \equiv \nu^t \frac{\partial \bar{u}}{\partial y} \tag{45}$$

Hence, equation (42) takes the form

$$L(\bar{u}) = \bar{u}\frac{\partial \bar{u}}{\partial x} + \bar{v}\frac{\partial \bar{u}}{\partial y} - \frac{\partial}{\partial y}\left[\nu^e \frac{\partial \bar{u}}{\partial y}\right] + \frac{1}{\rho}\frac{dp_I}{dx} = 0 \tag{46}$$

Prandtl mixing length theory (MLT) can be invoked to establish the relation for ν^t, cf., Schlichting (1968), Cebeci and Smith (1974). According to a dimensional analysis, ν^t must involve the product of a scale velocity and a scale length. Using the mean velocity gradient for the former, and Prandtl's mixing length ℓ_p, the required relation is

$$\nu^t \equiv \omega^2 \ell_p^2 \left|\frac{\partial \bar{u}}{\partial y}\right| \tag{47}$$

The mixing length is defined as,

$$\ell_p \equiv \begin{cases} \kappa y & 0 \leq y \leq \lambda\delta/\kappa \\ \lambda\delta & y < \lambda\delta/\kappa \end{cases} \tag{48}$$

and the Van Driest function ω accounts for the damping influence of a wall. Following a rigorous analysis, Cebeci and Smith (1974), the damping function is of the form

$$\omega \equiv 1 - \exp(-y/A) \tag{49}$$

In equations (47)-(49), y is the coordinate normal to the surface, δ is the boundary layer thickness, λ and κ are constants (typically 0.09 and 0.435), and A is a function of factors influencing flow phenomena near the surface including axial pressure gradient and normal mass flow.

The turbulent boundary layer equation system can be established in a transformed coordinate system, such that stretching in the y direction can compensate for boundary layer growth. A useful transformation is

$$\eta \equiv \frac{y - f_1(x)}{\left[f_2(x) - f_1(x)\right]} \tag{50}$$

where η normalizes the span between piecewise continuous surfaces $f_1(x)$ and $f_2(x) > f_1$, that bound the solution domain Ω. Using the chain rule for differentiation, equations (41)-(45) become

$$L(\bar{v}) = \left[\frac{\partial}{\partial x} - (h_2 + \eta h_3)\frac{\partial}{\partial \eta}\right]\bar{u} + \frac{\partial \bar{v}}{\partial \eta} = 0 \tag{51}$$

$$L(\bar{u}) = \bar{u}\left[\frac{\partial}{\partial x} - (h_2 + \eta h_3)\frac{\partial}{\partial \eta}\right]\bar{u} + \bar{v}\frac{\partial \bar{u}}{\partial \eta}$$

$$- \frac{\partial}{\partial \eta}\left[\nu^e h_1 \frac{\partial \bar{u}}{\partial \eta}\right] + \frac{1}{\rho}\frac{dp_I}{dx} = 0 \tag{52}$$

The functions h_i are defined as

$$h_i \equiv \begin{Bmatrix} (f_2 - f_1)^{-1} \\ f_1' h_1 \\ (f_2' - f_1')h_1 \end{Bmatrix} \tag{53}$$

where the superscript prime denotes ordinary differentiation with respect to x.

The form of the finite element algorithm for equations (51)-(53) is only a modest extension on the laminar algorithm statement. Specifically, the x momentum equation (52) becomes, in the equivalent finite element hypermatrix form,

$$S_e \left[\Delta_e \{U\}_e^T [A3000] \{U\}'_e - \Delta_e \{U\}_e^T \left(h_2 [A3000] + \left(h_3 \{ETA\}_e^T - \{V\}_e^T \right) [A4001] \right) \{U\}_e + \Delta_e h_1 \{XNUEFF\}_e^T [A3001] \{U\}_e + \Delta_e p'_I \{A10\} \right] \equiv \{0\} \tag{54}$$

The additional terms introduced by the coordinate transformation are readily handled. The elements of the column matrix $\{ETA\}_e$ are the (fixed) nodal coordinates of the finite element discretization of Ω. The matrix elements of $\{XNUEFF\}_e$ are the element nodal values of the evaluation of equation (44).

TURBULENT FLOW, ACCURACY AND CONVERGENCE

The finite element algorithm is amply verified to produce numerical predictions in agreement with data, cf., Baker and Manhardt (1977). The fundamental requirement is to assess accuracy and convergence. Of particular impact for turbulent flow, since the "effective viscosity" now involves the shear of $\bar{u}$, the specific (non-linear) form of the energy norm E becomes, using equations (24) and (44),

$$E(\bar{u}_e, \bar{u}_e) \equiv \frac{1}{2} \sum_{e=1}^{M} \int_{R_e^1} \frac{\partial \bar{u}_e}{\partial y} \left(\frac{\nu}{Re} + \omega^2 \ell^2 \left| \frac{\partial \bar{u}_e}{\partial y} \right| \right) \frac{\partial \bar{u}_e}{\partial y} \, dy \tag{55}$$

Substituting the interpolation function (8) yields, equivalently

$$E(\bar{u}_e, \bar{u}_e) \equiv \frac{1}{2} \sum_{e=1}^{M} \{U\}_e^T \left(\{XNUEFF\}_e^T [A3011] \right) \{U\}_e \tag{56}$$

where the elements of $\{XNUEFF\}_e$ are ν^e evaluated at the nodal coordinates of the finite element discretization of R^1. The distribution of the eddy viscosity will exert a profound influence on the level of E, since $0 \leq \nu^t/\nu \leq 10^3$ for typical turbulent boundary layer flows. Hence, the robustness of the energy norm for quantizing accuracy and convergence should be considerable.

Soliman and Baker (1979b) report results of numerical experiments to quantize accuracy and convergence assessments for the turbulent equivalent of the laminar slug start in zero pressure gradient. Five different geometric progression non-uniform discretizations were used to study accuracy and convergence with discretization refinement. The number of elements M spanning the solution domain R^1, and the corresponding node coordinate progression ratios ρ are listed in Table 1 for the linear and quadratic finite element solutions. All computed solutions were identically initialized by employing the ramp-slug initial $\bar{u}$ velocity profile used for the laminar flow studies.

TABLE 1. Discretization Data - Turbulent Flat Plate Flow

k = 1	M	12	24	36	48	60
	ρ	1.627	1.222	1.125	1.083	1.061
k = 2	M	6	12	18	24	30
	ρ	2.814	1.510	1.271	1.176	1.110

The span of Δ_1 was constant for all discretizations. The number of elements between the plate and the knee of the velocity profile was always one-sixth the total number of elements spanning the solution domain. No turbulence transition model was employed; instead computational transition from laminar to turbulent flow was specified to occur when the boundary layer shape factor H achieved 90% of its fully developed laminar flow value.

Figure 7 summarizes computed solution error in the energy norm, as obtained with the linear element algorithm as a function of discretization refinement and initial-value matrix modification. Here, S = 0 defines finite element, S = 3 is the Crank-Nicolson finite difference algorithm, while S = 2 is the procedure employed by Baker and Manhardt (1977) for explicit integration. Convergence is essentially quadratic, with modest data scatter, for all three initial-value matrix structures. The finite element algorithm form S = 0 yields the smallest level of error for any M, and in all cases, convergence is from above. As evidenced in the laminar flow results, the S = 0 algorithm displays accuracy for a coarse grid discretization that is superior to strict adherence to the convergence curve. However, differences in the three solution procedures are certainly modest. Figure 8 presents error computed in the energy norm with discretization refinement for the linear and quadratic finite element algorithms, $1 \leq k \leq 2$. Convergence is from below and of fourth degree for the quadratic element solution, which predicts extension of the linear equation finite element theory to this very non-linear problem. However, the results for the finest discretization show a significantly larger absolute error than predicted by the convergence curve. This is interpreted by the authors as an indication of the accuracy limit of practically useful discretizations, i.e., on progressively finer discretizations, other error sources serve to obliterate the refined solution accuracy theoretically attainable from discretization refinement. Notwithstanding, note

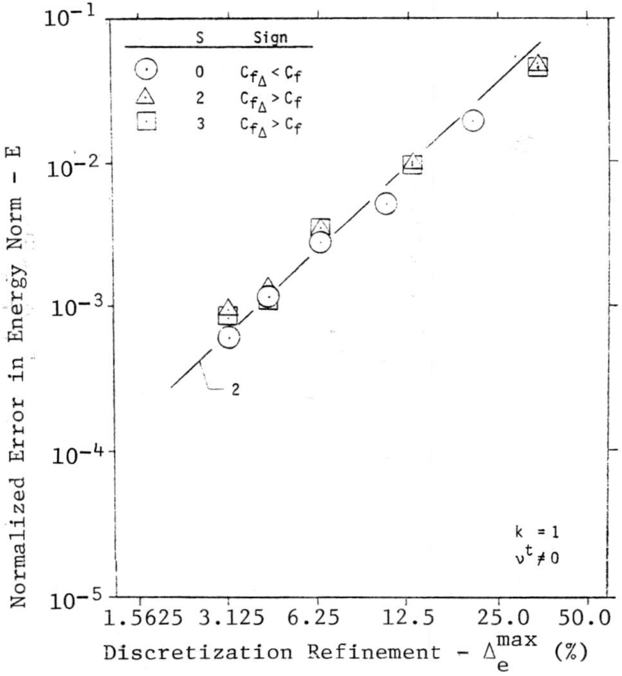

Figure 7. Accuracy and Convergence with Discretization Refinement in Energy Norm, Flat Plate Turbulent Flow, Linear Elements, MLT Closure

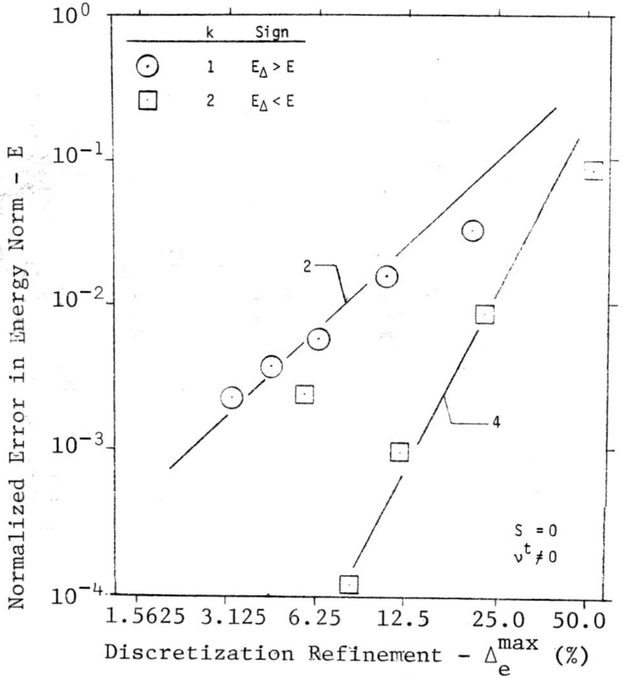

Figure 8. Accuracy and Convergence with Discretization Refinement in Energy Norm, Flat Plate Turbulent Flow, MLT Closure

that the accuracy of the quadratic element solutions can be up to a factor of 50 improvement over the corresponding linear element results.

SUMMARY

This chapter has presented the finite element solution algorithm applied to solution of the non-linear partial differential equation systems governing laminar and turbulent, two-dimensional boundary layer flows. Development of the solution procedures has provided ample illustration of the great versatility built into the finite element solution concept. The presented results have attested to the capability to solve complexly non-linear equations as well as document basic accuracy and convergence concepts.

Baker, A. J., and Manhardt, P. D. (1977) Finite Element Analysis of Low Speed Viscous and Inviscid Aerodynamic Flows. NASA CR-2908.

Baker, A. J., and Soliman, M. O. (1977) On the Accuracy and Convergence of Implicit Numerical Integration of Finite Element Generated Ordinary Differential Equations. Proceedings, International Symposium on Simulation, Software and Numerical Methods for Differential Equations, IMACS, pp. 1-8.

Baker, A. J., and Soliman, M. O. (1979) Accuracy, Convergence, and Utility of Finite Element Generated Solution Algorithms for Initial Value Problems. J. Comp. Phys., to appear.

Blottner, F. G. (1970) Finite Difference Methods of Solution of the Boundary Layer Equations. AIAA J., Vol. 8, No. 3, pp. 193-205.

Blottner, F. G. (1975) Computational Techniques for Boundary Layers. AGARD Lecture Notes No. 73.

Cebeci, T., and Smith, A. M. O. (1974) Analysis of Turbulent Boundary Layers, Academic Press, New York.

Oden, J. T., and Reddy, J. N. (1976) Introduction to Mathematical Theory of Finite Elements. John Wiley and Sons, New York.

Popinski, Z., and Baker, A. J. (1976) An Implicit Finite Element Algorithm for the Boundary Layer Equations. J. Comp. Phys., 21, No. 1, pp. 55-84.

Prandtl, L. (1904) Uber Flussigkeitsbewegung bei sehr kleiner Reibung. Proceedings 3rd. Int'l Math. Congr., Heidelberg.

Schlichting, H. (1968) Boundary Layer Theory, McGraw-Hill, New York.

Soliman, M. O. (1978) Accuracy and Convergence of a Finite Element Algorithm for Computational Fluid Dynamics. Ph.D. Dissertation, University of Tennessee, Knoxville.

Soliman, M. O., and Baker, A. J. (1979) Accuracy and Convergence of a Finite Element Algorithm for Laminar Boundary Layer Flow. J. Computers and Fluids, submitted.

Soliman, M. O., and Baker, A. J. (1979b) A High Order Accurate Numerical Solution Algorithm for Turbulent Boundary Layer Flow. AIAA Paper No. 79-0001.

Strang, G., and Fix, G. J. (1973) An Analysis of Finite Element Method. Prentice Hall, New Jersey.

CHAPTER 10

A FINITE ELEMENT MODEL OF ONE AND TWO EQUATION MODELS OF TURBULENT FLOW

C. TAYLOR[†], T.G. HUGHES[‡] and K. MORGAN[†]

† University of Wales, Swansea

‡ W.S. Atkins & Partners, Swansea

1. INTRODUCTION

The advent of digital computers has been accompanied by a considerable advancement in the development of computer based numerical models for analysing a large range of engineering problems. Over the past decade such models have been used to analyse turbulent flow where it has been possible to include sophisticated models of turbulence involving the utilisation of algebraic and differential transport equations to characterise the flow. The first models to use differential transport equations were suggested by Prandtl (1945) and Kolmogorov (1942). Finite difference techniques, well summarised in Launder and Spalding (1972) were then applied to such models and solutions obtained to a large range of flow problems. These are well documented and provided a means of investigating the effect of varying certain parameters on the prediction of a flow field bounded by irregular geometries. The main efforts relating to such investigations centred on the provision of numerical values for the coefficients inherent in the basic governing equations. The purpose of the present chapter is to present the finite element method as a technique which complements previous numerical methods. Indeed, where irregular geometries and the imposition of natural boundary conditions is required the method could prove to possess certain advantages over other techniques. This has already been demonstrated for potentialflow, De Vries and Norrie (1971) Taylor et al. (1973), laminar flow Taylor and Hood (1973, 1974) and turbulent flow K. Morgan et al. (1977). The solution of both the so-called one equation and two equation turbulence models by the finite element method has been shown to be possible, Barrow et al. (1976), Hutton (1976), Morgan (1977, 1978), Taylor et al. (1977, 1978), Schamber and Larock (1978), Baker (1978). The simultaneous solution of a time averaged form of the Navier Stokes equations and the transport equations for turbulence kinetic energy and mixing length required particular attention only when considering the generation and dissipation terms. Indeed it has been found that careful attention has to be given to the manner in which these terms are treated in the overall solution procedure, Hughes (1978). In this chapter, turbulent flow in pipes is analysed and the experimental results of Laufer (1954) are used as a basis for comparison.

2. MATHEMATICAL CONCEPTS

The time-averaged Navier-Stokes equation which may be used to describe steady state incompressible turbulent flow in a Cartesian coordinate system can be written as

$$\rho\, u_j \frac{\partial u_i}{\partial x_j} = \frac{\partial}{\partial x_j}\left(\tau_{ij} - \rho\, \overline{u_i' u_j'}\right) \qquad \left.\begin{matrix} i \\ j \end{matrix}\right\} 1,2,3 \tag{1}$$

Following the notation adopted in the current text, u_j are local time-averaged velocities in the x(1) and y(2) and z(3) cartesian coordinate directions, u_i' are the local fluctuating components of velocity, ρ the fluid density, τ_{ij} the mean viscous shear stress and $-\rho\, \overline{u_i' u_j'}$ the Reynolds shear stress.

The time averaged Navier-Stokes equation together with the continuity equation,

$$\frac{\partial u_i}{\partial x_i} = 0 \tag{2}$$

form the basis for the development of a number of techniques which have been used (e.g. Launder and Spalding (1972), Nee and Kouasznay (1968), Bradshaw and Ferris (1968) and Wolfshtein (1970)) to predict turbulent flow under various physical conditions. In equation (1) the mean shear stress is defined by,

$$\tau_{ij} = -\, p\, \delta_{ij} + \mu_e \left[\frac{\partial u_i}{\partial x_j} + \frac{\partial u_j}{\partial x_i}\right] \tag{3}$$

where μ_e is the fluid effective molecular viscosity and p the local time-averaged pressure. From this point in the development of the required equations the main difference in approach is the interpretation of the quantities $\rho\, \overline{u_i' u_j'}$.

In equation (3) this has been replaced by the expression,

$$-\rho \overline{u_i' u_j'} = \mu_T \left[\frac{\partial u_i}{\partial x_j} + \frac{\partial u_j}{\partial x_i}\right] \tag{4}$$

where μ_T is called the turbulent dynamic viscosity.

It is clearly apparent that once the turbulent viscosity model can be defined then equations (1) - (3) form a closed set which can be solved for u_i and p.

2.1 Mixing Length Model

By considering the turbulent transfer of momentum Prandtl (1925) suggested a turbulence viscosity model of the form

$$\mu_T = \rho \ell^2 |J| \tag{5}$$

where,

$$J^2 = 0.5 \left(\frac{\partial u_k}{\partial x_j} + \frac{\partial u_j}{\partial x_k}\right)^2 \tag{6}$$

where the local value of ℓ, the Prandtl mixing length, is flow dependent and has to be specified. This simple mixing length concept has been widely used and indeed forms the basis for many comprehensive studies of turbulent flow. However, in all applications some degree of empiricism has, of necessity, to be incorporated. This naturally places severe restrictions on the general applicability of an advocated empiricism and recourse has, usually, to be made to experimental results.

The above is considered to be the first step in trying to impart a rational measurable meaning to the turbulent viscosity. When such a model is introduced into equation (2) the solution becomes iterative since a knowledge of the flow field is necessary before the required velocity gradients can be evaluated to evaluate such quantities as

$$\mu_T = \rho \ell^2 (|J|) \tag{7}$$

A more detailed account of the iterative processes will be outlined subsequently.

2.2 One Equation Model

Although the mixing length concept as developed in section 2.1 can be used with measurable success, to predict turbulent flow behaviour, considerable defects in this simple assumption have been discussed, Launder and Spalding (1972). These were expounded by simply comparing the inherent assumptions in equation (4) and experimental observations. The fact that the random turbulent viscosity is assumed to be identically zero when the mean velocity gradient is zero has been disproved by observation. For instance for plane flow in a channel the value of μ_T has been shown to be of the same order of magnitude as its maximum value. This is of no great consequence as far as the shear stress is concerned but has a significant bearing on the turbulence exchange coefficient $\Gamma_{k,t}$ where,

$$\Gamma_{k,t} = \frac{\mu_T}{\rho\sigma_k}$$

in which σ_k is a constant. The significance of this value approaching zero where the velocity gradient approaches zero will become apparent when the so called one-equation model is developed.

The other main defect in the mixing length model is the purely local nature of the turbulent viscosity since its determination is influenced by local variations in the velocity field. This is a gross over-simplification which can be demonstrated by introducing a local disturbance into a steady state flow. This disturbance will subsequently affect the flow field at various points downstream. Indeed where separation exists 'downstream', in a general sense, could effect local velocities 'upstream'. The occurrence of such local disturbances or instantaneous fluctuations from the mean could, therefore, have a significant effect on the evaluation of a local turbulent viscosity. Prandtl (1945) and Kolmogorov (1942) were the first to suggest an improvement by relating the turbulent viscosity to a property of the turbulence, the local time averaged turbulence kinetic energy k, such that,

$$\mu_T = C_\mu \rho \ell k^{\frac{1}{2}} \tag{9}$$

where,

$$k = \frac{1}{2} (\overline{u_i'^2}) \qquad i = 1,2,3 \tag{10}$$

and C_μ is a constant.

This should obviate the deficiency that $\mu_T \neq 0$ when the local velocity gradient is zero, since the turbulence kinetic energy is not necessarily zero. The influence of neighbouring regions in a local turbulence kinetic energy is accommodated via a transport differential equation in which k is the dependent variable. This can be derived from the Navier Stokes equations, Wolfshtein (1970), leading to

$$\rho u_j \frac{\partial k}{\partial x_j} + \frac{\partial}{\partial x_j}\left[\frac{1}{2}\rho\,\overline{u'_j\,u'_i\,u'_i} - \overline{u_j\,p'}\right] + \rho\,\overline{u'_i\,u'_j}\,\frac{\partial u_i}{\partial x_j}$$

$$= \mu\left[\frac{\partial^2 k}{\partial x_j\,\partial x_j} - \overline{\frac{\partial u'_i}{\partial x_j}\,\frac{\partial u'_i}{\partial x_j}}\right] \tag{11}$$

Utilising (3), equation (11) can be re-written

$$\rho\, u_j \frac{\partial k}{\partial x_j} + \frac{\partial}{\partial x_j}\left[\frac{1}{2}\rho\,\overline{u'_j\,u'_i\,u'_i} - \overline{u'_j\,p'}\right] - \mu_T \frac{\partial u_i}{\partial x_j}$$

$$\left(\frac{\partial u_i}{\partial x_j} + \frac{\partial u_j}{\partial x_i}\right) = \mu\left[\frac{\partial^2 k}{\partial x_j\,\partial x_j} - \overline{\frac{\partial u'_i}{\partial x_j}}\,\overline{\frac{\partial u'_i}{\partial x_j}}\right] \tag{12}$$

where the term representing the transfer of energy from the mean to the fluctuating motion is now written in terms of the averaged mean values. An immediate concern in equation (11) is a successful definition of the terms containing time averaged instantaneous fluctuations. The first of these is the second term in the left hand side of equation (12). Representing

$$\frac{1}{2}\rho\,\overline{u'_j\,u'_i\,u'_i} - \overline{u'_j\,p'} = -\rho\,\Gamma_{k,t}\,\frac{\partial k}{\partial x_j} \tag{13}$$

Equation (13) represents a diffusion type process and is attributed to the transfer of turbulence energy and pressure fluctuations due to random turbulent eddies. A modified form of equation (12) now becomes,

$$\rho\, u_j \frac{\partial k}{\partial x_j} - \frac{\partial}{\partial x_j}\left[\frac{\mu_T}{\sigma_k}\frac{\partial k}{\partial x_j}\right] - \mu_T \frac{\partial u_i}{\partial x_j}\left(\frac{\partial u_i}{\partial x_j} + \frac{\partial u_j}{\partial x_i}\right)$$

$$= \mu\left[\frac{\partial^2 k}{\partial x_j\,\partial x_j} - \overline{\frac{\partial u'_i}{\partial x_j}}\,\overline{\frac{\partial u'_i}{\partial x_j}}\right] \tag{14}$$

The remaining term containing instantaneous fluctuations is replaced by assuming that the dissipation of turbulent energy into heat is proportional to the level of turbulence and usually proportional to the scale of turbulence such that,

$$\overline{\frac{\partial u'_i}{\partial x_j}}\,\overline{\frac{\partial u'_i}{\partial x_j}} = C_D\,\frac{k^{3/2}}{\ell} \tag{15}$$

where C_D is dimensionless and assumed to be constant.

Equation (14) can be re-written,

$$\rho\, u_j \frac{\partial k}{\partial x_j} = \frac{\partial}{\partial x_j}\left[\left(\frac{\mu_T}{\sigma_k} + \mu\right)\frac{\partial k}{\partial x_j}\right] + \mu_T \frac{\partial u_i}{\partial x_j}\left(\frac{\partial u_i}{\partial x_j} + \frac{\partial u_j}{\partial x_i}\right)$$

CONVECTION DIFFUSION PRODUCTION

$$- C_D \frac{k^{3/2}}{\ell}$$

DISSIPATION (16)

representing the transport equation for turbulence kinetic energy. If the basic momentum and continuity equations are used in association with (16) then the mixing length remains a local parameter which must be attributed a numerical value in order that the computation may proceed. Before developing further on the specification of the mixing length some observations on equation (4) are necessary.

The interpretation of the left hand side of equation (16) is straightforward and simply represents a convective rate of change of kinetic energy along a streamline. The first term on the right hand side represents a diffusion of kinetic energy. Clearly in a free stream, where μ_T/σ_k and μ are constant, then no contribution to the total energy is made. On a fixed boundary where $\frac{\mu_T}{\sigma_k} = 0$ and μ constant then again no contribution is made. Therefore the term represents a means of re-distributing energy in space without making a contribution to the overall energy level. The second term on the right hand side has been interpreted Launder and Spalding (1972) as representing the rate at which mean kinetic energy is converted into turbulence energy by the work done by the turbulent stress $\rho\, \overline{u_i' u_j'}$ against the mean velocity gradient. The last term has been explained previously and is assumed to represent the dissipation of turbulent energy into heat.

2.3 Two Equation Model

In the so-called two equation model both the turbulence energy, k, and the length scale, ℓ, are determined from corresponding transport equations. Some of the reasons why such an equation is necessary for the length scale, as opposed to utilising an algebraic relationship, have been described in detail elsewhere, Launder and Spalding (1972). However, a brief resumé of these reasons would be an advantage when accepting such an equation and the inherent difficulties when effecting a solution.

Consider a strongly accelerated flow situation, Figure 1, associated with a co-flowing stream in free shear flow. It has been demonstrated that an algebraic relationship for mixing length is perfectly adequate in developing flow within the nozzle or pipe. However such representation when the boundary layers have merged and when mixing is taking place downstream of the end of the pipe is highly unsatisfactory. For recirculating flows, Figure 2, convective transport is particularly important and is, indeed, the mechanism by which recirculation is propagated and sustained.

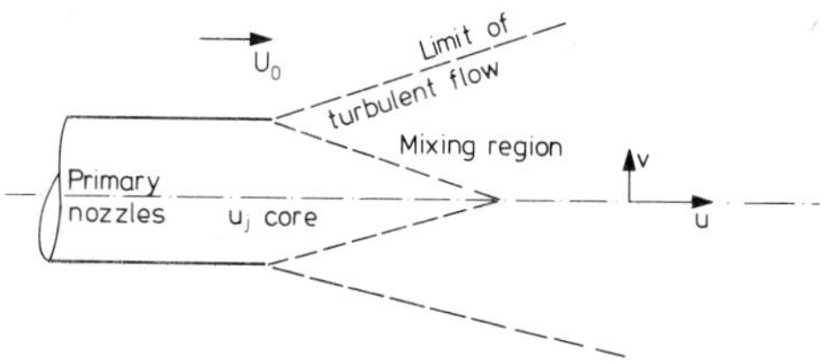

Figure 1 Turbulent jet in a co-flowing stream.

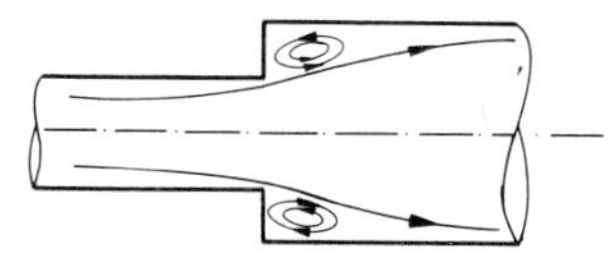

FIGURE 2 Flow through an expansion where significant recirculation is apparent.

Having, briefly, given reasons why such a transport equation is desirable the interested reader is referred to Ng and Spalding (1972) for further details. The resulting form of the transport equation employed to evaluate the length scale is,

$$u_j \frac{\partial}{\partial x_j}(k\ell) = \underbrace{\frac{\partial}{\partial x_j}\left(\frac{k^{3/2}\ell}{\sigma_1}\frac{\partial \ell}{\partial x_j}\right) + \frac{\partial}{\partial x_j}\left(\frac{k^{1/2}\ell^2}{\sigma_2}\frac{\partial k}{\partial x_j}\right)}_{\text{DIFFUSION}}$$

CONVECTION

$$+ \underbrace{C_B k^{1/2}\ell^2 \frac{\partial u_i}{\partial x_j}\left(\frac{\partial u_i}{\partial x_j} + \frac{\partial u_j}{\partial x_i}\right)}_{\text{PRODUCTION}} - \underbrace{C'_m k^{3/2}}_{\text{DISSIPATION}} \qquad (17)$$

where σ_1, σ_2, C_B are all assumed to be constant for fully developed turbulent flow, C'_m is a function of the normal distance from the wall given by $C'_m = C_M + C_w (\ell/x_2)^q$.

The use of the two equation model to predict turbulent flow requires therefore the simultaneous solution of the four equations, (1), (2), (16) and (17).

Summary

$$u_j \frac{\partial u_i}{\partial x_j} = -\frac{1}{\rho}\frac{\partial p}{\partial x_i}\delta_{ij} + \nu \frac{\partial}{\partial x_j}\left[\frac{\partial u_i}{\partial x_j} + \frac{\partial u_j}{\partial x_i}\right]$$

$$+ \frac{\partial}{\partial x_j}\left[\nu_T \left(\frac{\partial u_i}{\partial x_j} + \frac{\partial u_j}{\partial x_i}\right)\right] \quad , \quad \frac{\partial u_j}{\partial u_j} = 0$$

$$u_j \frac{\partial k}{\partial x_j} = \frac{\partial}{\partial x_j}\left[\left(\frac{\nu_T}{\sigma_k} + \nu\right)\frac{\partial k}{\partial x_j}\right] + \nu_T \frac{\partial u_i}{\partial x_j}\left(\frac{\partial u_i}{\partial x_j} + \frac{\partial u_j}{\partial x_i}\right)$$

$$- C_D \frac{k^{3/2}}{\ell}$$

$$u_j \frac{\partial}{\partial x_j}(k\ell) = \frac{\partial}{\partial x_j}\left(\frac{k^{3/2}\ell}{\sigma_1} \cdot \frac{\partial \ell}{\partial x_j}\right) + \frac{\partial}{\partial x_j}\left(\frac{k^{1/2}\ell^2}{\sigma_2} \cdot \frac{\partial k}{\partial x_j}\right)$$

$$+ C_p k^{1/2} \ell^2 \frac{\partial u_i}{\partial x_j}\left(\frac{\partial u_i}{\partial x_j} + \frac{\partial u_j}{\partial x_i}\right) - (C_m + C_w (\ell/x_2)^q k^{3/2}$$

In the above summary the momentum equation (1) has been modified by the introduction of (3) and $\nu_T = \mu_T/\rho$.

There now remain the development of a suitable solution procedure. Conceptually this is particularly straightforward and the only complexity over and above that required to solve for a single dependent variable in the coupled nature of the equations. This necessitates the specification of boundary conditions and initial conditions in order to start iterative procedure. A suitable choice of initial values requires some degree of skill when depicting ℓ and k. An unsuitable choice usually results in an excessive number of iterations to achieve convergence or, in some instances, a trivial solution could be obtained. An efficient re-solution process utilised for the two equation model in the solution of the governing equations can be summarised,

i) Specify boundary conditions for u, p, v, k and ℓ

ii) Estimate and impose initial values on velocity, pressure, turbulence energy and length scale

iii) Use the current values of k and ℓ to determine the pointwise distribution of turbulent viscosity

iv) Evaluate the distribution of u, p, v, k and ℓ

v) Determine whether the variable values calculated in (iv) satisfy imposed convergence criteria. If not repeat the process from (iii) until such criteria are satisfied.

There now remains the problem of solving the equations to provide the third step in the resolution process. As indicated in the introduction the generalised technique adopted is based on transforming the differential equations, utilising the finite element method, into an equivalent matrix form and then solving at discrete points within the regime of flow.

3. FINITE ELEMENT MODEL

There are numerous texts, using the finite element method, which develop solution techniques for laminar flow. However, an extension to solve problems where turbulence is a major contributing factor are not quite as numerous although, recently, these have begun to emerge. The additional difficulties of direct concern are

i) The inclusion of further non-linear terms due to the spatial variation in the effective viscosity

ii) The evaluation of pertinent flow parameters necessary in the determination of the turbulent viscosity

iii) For confined flow the evaluation of necessary parameters peculiar to the near wall region

iv) A judicious choice of the relative location of terms when formulating the matrix equations, and

v) Utilisation of non-linear boundary techniques.

The Galerkin weighted residual process is used as the first step in formulating a finite element computer based model of the governing equations (1), (2), (16) and (17). The continuity equation, as a specific example, would be written,

$$\iiint_{\Omega} W^k \left(\frac{\partial u_i}{\partial x_i}\right) d\ell = 0 \qquad k = 1,\ldots,nn \qquad (18)$$

where W^k is a weighting function and nn is the total number of discrete points where W^k is defined within the flow domain Ω. For each element or subdomain the variation in the variables across an element are defined in accordance with,

$$\{\lambda\} = N^J \lambda^J$$

and $$p = M^J p^J \tag{19}$$

where

$$\{\lambda\} = \begin{Bmatrix} u_i \\ k \\ k\ell \\ \nu_T \end{Bmatrix} \tag{20}$$

where N and M are the element interpolation functions and repetition of the indices J is understood. The choice of N and M is governed by the mixed interpolation concept, Taylor and Hood (1974), where the order of M is one lower than that of N. Using Green's theorem to reduce the second order terms a discrete form of the governing equations can be obtained and each element contribution summated over the whole domain. For simplicity the following is limited to one orthogonal direction only and two dimensions,

MOMENTUM

$$\sum_{e=1}^{Ne} \Bigg[\int_\Omega \Bigg[N^K \Big(N^J u_1^J \frac{\partial N^L}{\partial x_1} u_1^L + N^J u_2^J \frac{\partial N^L}{\partial x_2} u_1^L + \frac{1}{\rho} \frac{\partial M^J}{\partial x_i} p^J$$

$$+ \frac{\partial N^J}{\partial x_1} \nu_T^J \frac{\partial N^L}{\partial x_1} u_1^L + \frac{\partial N^J}{\partial x_2} \nu_T^J \frac{\partial N^L}{\partial x_1} v^L$$

$$+ N^J \nu_e^J \left(\frac{\partial N^K}{\partial x_1} . \frac{\partial N^L}{\partial x_1} . u_1^L + \frac{\partial N^K}{\partial x_2} . \frac{\partial N^L}{\partial x_2} u_1^L \right) d\Omega_e$$

$$- \int_{S_e} N^K N^J \nu_e^J \left(\frac{\partial u_1}{\partial x_1} . \ell_1 + \frac{\partial u_1}{\partial x_2} . \ell_2 \right) dS_e \Bigg] = 0 \tag{21}$$

where S_e describes the element boundary, ℓ_n denotes the direction cosine of the outward normal relating to the corresponding suffix and Ne the total number of elements. A similar equation can be written for the second suffix. Following the same procedure the equation of continuity becomes,

CONTINUITY

$$\sum_{e=1}^{Ne} \int_{\Omega_e} M^K \left(\frac{\partial N^L}{\partial x_i} u_i^L\right) d\Omega_e \qquad i = 1,2 \tag{22}$$

A convenient form of the turbulence energy transport equation (16) is,

TURBULENCE ENERGY

$$\sum_{e=1}^{Ne} \left[\int_{\Omega} N^K \left(CONVK + C_D \frac{(N^L k^L)^{3/2}}{N^J \ell^J} - GENK \right) \right.$$

$$+ \left(\frac{N^J \nu_T^J}{\sigma_k} + \nu \right) \left(\frac{\partial N^K}{\partial x_i} \cdot \frac{\partial N^L}{\partial x_i} k^L \right) d\Omega_e$$

$$\left. - \int_{S_e} N^K \left(\frac{N^J \nu_T^J}{\sigma_k} + \nu \right) \left(\frac{\partial k}{\partial x_i} \ell_i \right) dS_e \right] = 0 \tag{23}$$

and finally the equation for the length scale

LENGTH SCALE

$$\sum_{e=1}^{Ne} \left[\int_{\Omega} N^K \left(CONVKL + \left(C_M + C_w \left[\frac{N^I \ell^I}{N^I x_2^I} \right] \right) (N^L k^L)^{3/2} \right. \right.$$

$$\left. - GENKL \right) + \left[\frac{N^I \nu_T^I}{\sigma_1} + \nu \right] \left(\frac{\partial N^K}{\partial x_i} \cdot \frac{\partial N^L}{\partial x_i} (k\ell)^L \right)$$

$$\left. + \left[\frac{\sigma_1 - \sigma_2}{\sigma_1 \sigma_2} \right] \left[N^I \nu_T^T + \nu \right] N^I \ell^I \left(\frac{\partial N^K}{\partial x_i} \frac{\partial N^L}{\partial x_i} \right) (k)^L \right] d\Omega_e . \tag{24}$$

$$- \int_{S_e} N^K \left[\left(\frac{N^I \nu_T^I}{\sigma_1} + \nu \right) \left(\frac{\partial (k\ell)}{\partial x_i} \ell_i \right) \right.$$

$$+ \left(\frac{\sigma_1 - \sigma_2}{\sigma_1 \sigma_2} \right) \left(N^I \nu_T^I + \nu \right) N^I \ell^I \left(\frac{\partial (k)}{\partial x_i} \ell_i \right) dS_e$$

In (23) and (24)

$$CONV = (N^J u_i^J \frac{N^L}{x_i} k^L)$$

$$GENK = N^J \nu_T^I \left| 2 \left(\frac{\partial N^I}{\partial x_i} u_i^I \frac{\partial N^L}{\partial x_i} u_i^L \right) + \left(\frac{\partial N^L}{\partial x_j} u_i^L \right) \left(\frac{\partial N^I}{\partial x_j} u_i^I \right) \right.$$

$$\text{CONVKL} = N^J \, u_i^J \, \frac{\partial N^L}{\partial x_i} \, (k\ell)^L$$

$$\text{GENKL} = C_B \, N^I \, \ell^I \, \times \text{GENK}$$

and $\ell^I = \dfrac{(k\ell)^I}{k^I}$, $\quad \nu_T^I = C_\mu \, (k\ell)^I/(k^I)^{\frac{1}{2}}$

A matrix form of the above equations and the relevant choice of variables is apparent when the matrix coefficients are defined.

Consider the overall matrix

$$[H(B)^{n-1}] \ \{B^n\} = \{M^n\}$$

where n denotes the current iteration and

$$\{B_K\} = \begin{bmatrix} u_1^K \\ p^K \\ u_2^K \\ k^K \\ k\ell^K \end{bmatrix} \tag{25}$$

for two dimensional flow.

The matrix [H] comprises submatrices of the form

$$h_{KL} = \sum_{e=1}^{Ne} \int_\Omega \begin{bmatrix} h_1 & h_2 & h_3 & h_4 & h_5 \\ h_6 & h_7 & h_8 & h_9 & h_{10} \\ h_{11} & h_{12} & h_{13} & h_{14} & h_{15} \\ h_{16} & h_{17} & h_{18} & h_{19} & h_{20} \\ h_{21} & h_{22} & h_{23} & h_{24} & h_{25} \end{bmatrix} d\Omega_e \tag{26}$$

where the submatrix h_{KL} of [H] consists of terms associated with the Kth node which act on the Lth subvector of nodal variables B_L.

One method in which the discrete equations (21) - (24) can be represented is by defining the coefficients of [H] by,

$$h_1 = \alpha_1 + N^I \tilde{\nu}_e^I \alpha_2 - N^K \frac{\partial N^I}{\partial x_1} \tilde{\nu}_T^I \frac{\partial N^L}{\partial x_1}$$

where

$$\alpha_1 = N^K \left(N^I \tilde{u}_i^I \frac{\partial N^L}{\partial x_i} \right)$$

$$\alpha_2 = \frac{\partial N^K}{\partial x_i} \cdot \frac{\partial N^L}{\partial x_i}$$

$$h_2 = \frac{N^K}{\rho} \frac{\partial M^L}{\partial x_1}, \quad h_3 = - N^K \frac{\partial N^I}{\partial x_2} \tilde{\nu}_T^I \frac{\partial N^L}{\partial x_1}$$

$$h_q = 0 \text{ for } q = 4, 5, 7, 9, 10, 14, 15, 17, 20, 22, 24$$

$$h_6 = M^K \frac{\partial N^L}{\partial x_1}, \quad h_8 = M^K \frac{\partial N^L}{\partial x_2}, \quad h_{11} = - N^K \frac{\partial N^I}{\partial x_1} \cdot \tilde{\nu}_T^I \frac{\partial N^L}{\partial x_2}$$

$$h_{12} = \frac{N^K}{\rho} \cdot \frac{\partial M^L}{\partial x_2}, \quad h_{13} = \alpha_1 + N^I \nu_e^I \alpha_2 - N^K \frac{\partial N^I}{\partial x_2} \cdot \tilde{\nu}_T^I \frac{\partial N^L}{\partial x_2}$$

$$h_{16} = -N^K N^I \tilde{\nu}_T^I \left[2 \frac{\partial N^I}{\partial x_1} \tilde{u}_1^I \frac{\partial N^L}{\partial x_1} - \frac{\partial N^L}{\partial x_2} \left(\frac{\partial N^I}{\partial x_2} \tilde{u}_1^I + \frac{\partial N^I}{\partial x_1} \tilde{u}_2^I \right) \right]$$

$$h_{18} = -N^K N^I \tilde{\nu}_T^I \left[2 \frac{\partial N^I}{\partial x_2} \tilde{u}_2^I \frac{\partial N^L}{\partial x_2} - \frac{\partial N^L}{\partial x_1} \left(\frac{\partial N^I}{\partial x_1} \tilde{u}_2^I + \frac{\partial N^I}{\partial x_2} \tilde{u}_1^I \right) \right]$$

$$h_{19} = \alpha_1 + \alpha_2 \left(\frac{N_I \tilde{\nu}_T^I}{\sigma_k} + \nu \right) + C_D \frac{N^K (N^I \tilde{k}^I)^{\frac{1}{2}}}{N^I \tilde{\ell}^I} N^L$$

and $h_{21} = h_{16} \times C_B \, N^I \, \tilde{\ell}^I$

$$h_{23} = h_{17} \times C_B \, N^I \, \tilde{\ell}^I$$

$$h_{24} = \left[\frac{\sigma_1 - \sigma_2}{\sigma_2 \, \sigma_1}\right] N^I \, \tilde{\ell}^I \, (N^I \, \tilde{\nu}_T^I + \nu) \, \alpha_2$$

$$h_{25} = \alpha_1 + \alpha^2 \left[\frac{N^I \, \tilde{\nu}_T^I}{1} + \nu\right]$$

$$+ \left[C_M + C_w \left[\frac{N^I \quad I}{N^I \quad x_2^I}\right]^q\right] \frac{N^K \, N^L}{N^J \, (\tilde{k\ell})^J} \, (N^I \, k^I)^{3/2} \qquad (27)$$

in which $\sim$ denotes the previous iteration values.

The right hand side vector {M} contains only prescribed values or prescribed gradients of velocity, pressure, energy and length scales. The Mth subvector is given by

$$\{M_K\} = \sum_{e=1}^{Ne} \int_{S_e} \begin{bmatrix} b_1 \\ b_2 \\ b_3 \\ b_4 \\ b_5 \end{bmatrix} dS_e$$

where, $b_1 = N^K \left(\frac{\partial u_1^{(P)}}{\partial x_1} \ell_1 + \frac{\partial u_1^{(P)}}{\partial x_2} \ell_2\right) \left(N^I \, \tilde{\nu}_T^I + \nu\right)$

$$b_2 = 0$$

$$b_3 = N^K \left(\frac{\partial u_2^{(P)}}{\partial x_1} \ell_1 + \frac{\partial u_2^{(P)}}{\partial x_2} \ell_2\right) \left(N^I \, \tilde{\nu}_T^I + \nu\right) \qquad (28)$$

$$b_4 = N^K \left(\frac{\partial k^{(P)}}{\partial x_1} \ell_1 + \frac{\partial k^{(P)}}{\partial x_2} \ell_2\right) \left(\frac{N^I \, \tilde{\nu}_T^I}{\sigma_k} + \nu\right)$$

and $b_5 = N^K \left\{\frac{\partial}{\partial x_1} (k\ell)^{(P)} \ell_1 + \frac{\partial}{\partial x_2} (k\ell)^P \ell_2\right\} \left[\frac{N^I \tilde{\nu}_T^I}{\sigma_1} + \nu\right]$

$$+ N \left(\frac{\partial}{\partial x_1} (k)^{(P)} \ell_1 + \frac{\partial}{\partial x_2} (k)^P \ell_2\right) \left[\left(\frac{\sigma_1 - \sigma_2}{\sigma_1 \, \sigma_2}\right) \left(N^I \, \tilde{\nu}_T^I + \nu\right) N^I \tilde{\ell}^I\right]$$

where (P) denotes a prescribed value.

Although the above formulation need not be adhered to in all applications the general form of the matrix equations will not vary significantly. Further reference to the detailed formulation will be left to the interested reader, Hughes (1978).

4. ILLUSTRATIVE EXAMPLES

Both the one equation and two equation models are used in solving particular illustrative examples. The first is, necessarily, limited to cases where an adequate algebraic representation of the length scale can be defined.

4.1 One Equation Model

Amongst the simplest flow problems is the case of fully developed flow in pipes where both the mean velocity and level of turbulence can be assumed to remain constant and equal at successive sections along the pipe. The required length scale is calculated according to the relationship,

$$\ell = C_D^{1/4} \ell_N \tag{29}$$

where C_D is generally taken as 0.0916, Launder and Spalding (1972) and ℓ_N is Nikuradse's mixing length profile for fully developed pipe flow. The expression for ℓ was modified, as suggested by Wolfshtein (1969), both in the laminar sublayer and transition region which was necessitated by the proximity of the wall.

The constants C_μ and C_D were adjusted such that in the near wall region the length scale was equal to the normal distance from the wall,

$$\ell = x_2 \tag{30}$$

As the laminar sublayer is approached, however, the length scale distribution is represented by,

$$\ell_\mu = x_2 \left[1 - \exp^{(-A_\mu R_T)}\right]$$

and

$$\ell_D = x_2 \left[1 - \exp^{(-A_D R_T)}\right] \tag{31}$$

where A_μ and A_D are damping constants and R_T is the turbulent Reynolds number,

$$R_T = \frac{k^{\frac{1}{2}} x_2}{\nu} \tag{32}$$

The terms ℓ_μ and ℓ_D were introduced by Wolfshtein to specify the length scale at low Reynolds numbers in this algebraic expression for μ_T, ℓ_μ, and in the denominator of the dissipation term of the energy transport equation, ℓ_D.

Using values of ℓ_D = 0.416, A_μ = 0.016, A_D = 0.263, C_μ = 0.22 and σ_k = 1.53 the eddy viscosity is calculated from equation (9) and used to evaluate the distribution of velocity and pressure for fully developed flow in a pipe. The element distribution and boundary conditions are as shown on Figure 3 and the variable distribution is evaluated for a Reynolds number of 5 x 10^5. These are compared with the experimental results obtained by Laufer (1954). A comparison of the velocity distribution, Figure 4 (a) and (b) and turbulence kinetic energy, Figure 5, show that, within the limitations imposed by the turbulence model, a reasonable comparison has been obtained.

In Figure 4 the terms which have not been defined are,

$$u^* = \sqrt{\frac{\tau_w}{\rho}}$$ The friction velocity where τ_w is the wall shear stress

$$u_1^+ = \frac{u_1}{u^*}$$

and

$$y^+ = \frac{x_2 u^*}{\nu}$$

Further plots, Figures 6, 7, 8, illustrate the comparison of the near wall kinetic energy and the relative magnitude of the dissipation and generation terms observed numerically and those from experiment.

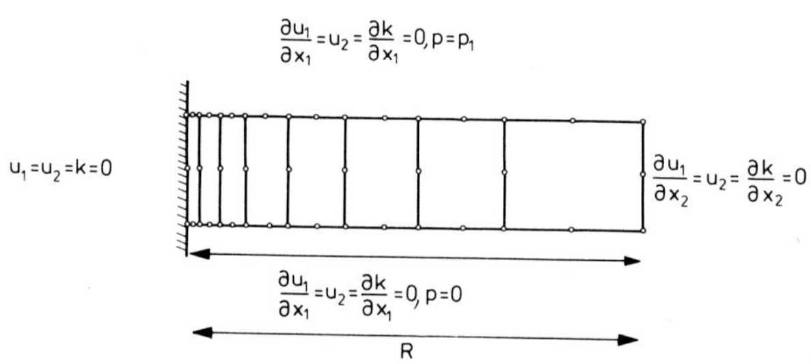

Figure 3 Region analysed and boundary conditions for fully developed pipe flow (elements not to scale)

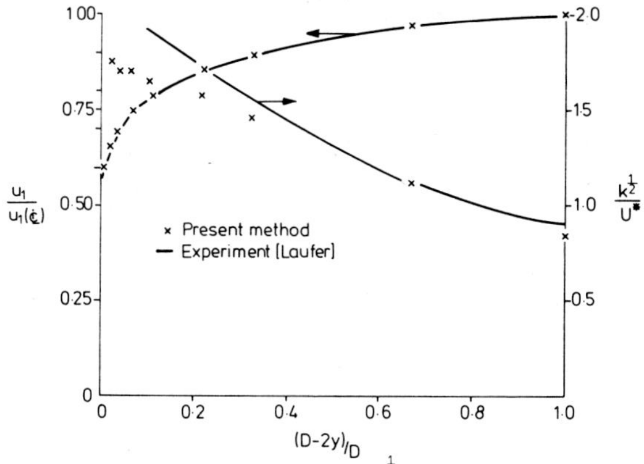

(a) Plot of $\frac{u_1}{u_1(\mathrm{℄})}$ and $\frac{k^{\frac{1}{2}}}{U^*}$

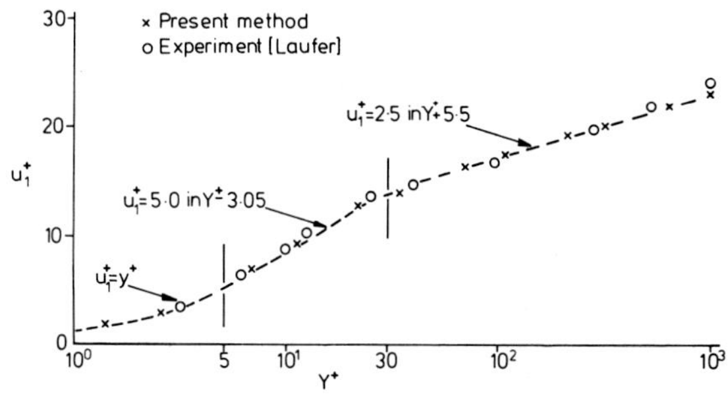

(b) Non dimensional near wall velocity plot.

Figure 4 Fully developed pipe flow $R_N = 5 \times 10^5$

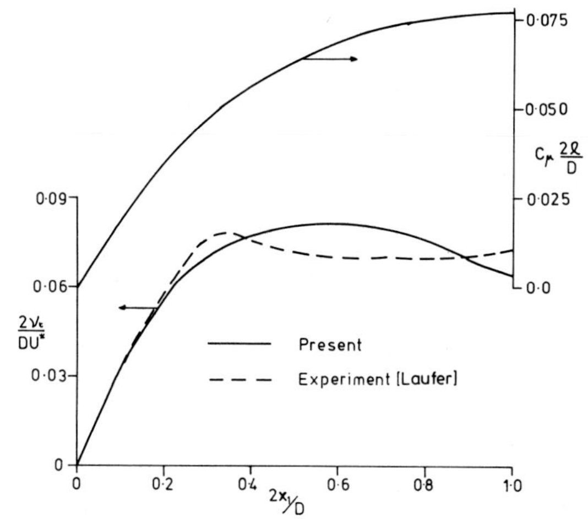

FIGURE 5 Specified lengh scale and predicted turbulent viscosity distribution.

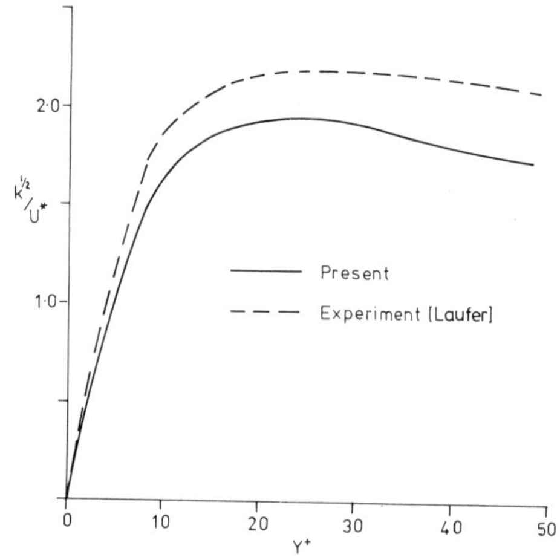

FIGURE 6 Near wall turbulent kinetic energy distribution.

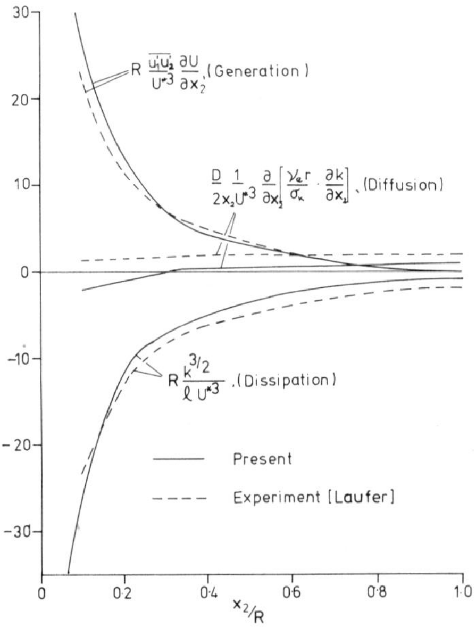

FIGURE 7 Kinetic energy balance in fully developed pipe flow, $R_N = 5 \times 10^5$

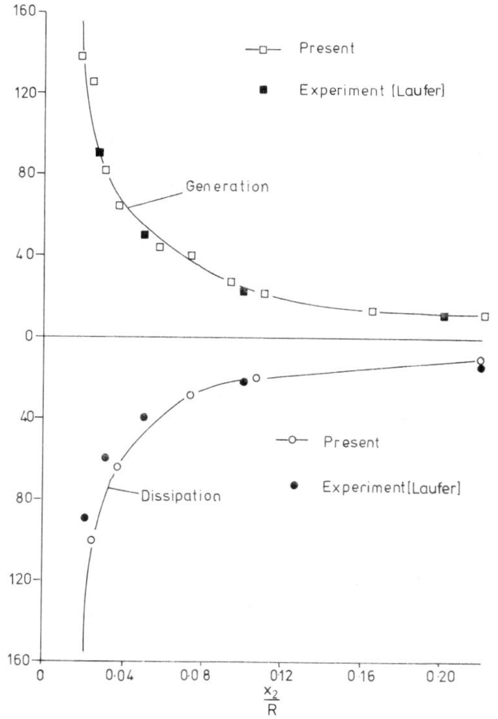

FIGURE 8 Energy balance in the near wall zone , $R_N = 5 \times 10^5$

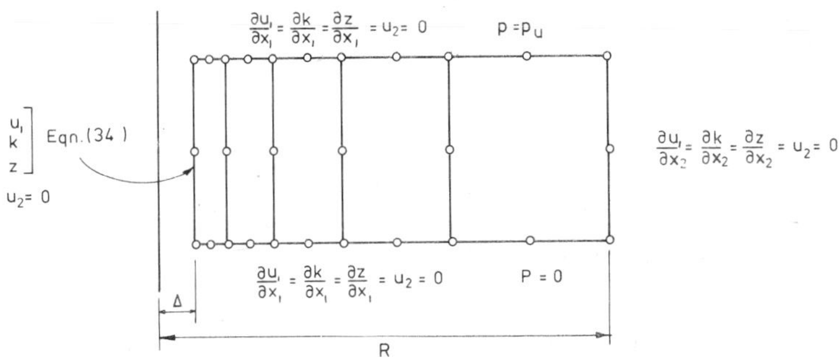

FIGURE 9 Finite element mesh and imposed boundary conditions for fully developed flow in pipes or channels.

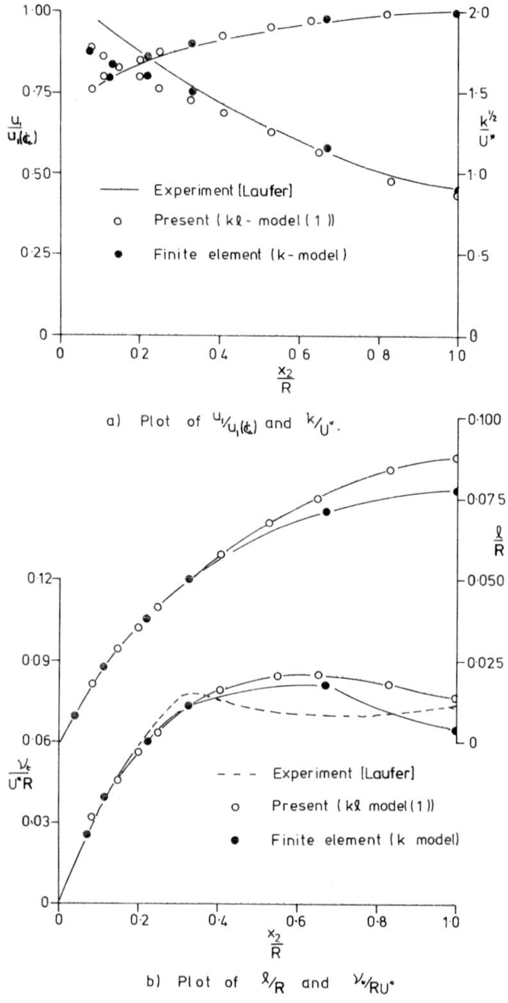

FIGURE 10 Fully developed pipe flow , $R_N = 5 \times 10^5$.

4.2 Two Equation Model

Again fully developed flow in a pipe is analysed. In this case the variation in length scale is depicted by equation (17) and all five equations are solved simultaneously.

In this case variations in the 'constants' of equations (16) and (17) are assumed to be significant in the near wall region and the solution domain is terminated some distance, Δ, away from the pipe wall, Figure 9. Ng and Spalding (1972) assumed the following relationships in the near wall region.

$$u_1 = \left(\frac{\tau_w}{\rho}\right)^{0.5}\left[2.5\ \ell_n \frac{x_2}{\rho\nu}\left(\frac{\tau_w}{\rho}\right)^{0.5} + 5.5\right]$$

$$k = \tau_w / C_D\ 0.5$$

$$\ell = 0.4\ C_D^{\ 0.25}\ x_2 \qquad (33)$$

where τ_w is the wall shear stress and is directly related to the applied pressure gradient.

The domain is discretised and the required boundary conditions are as shown in Figure 9. In the main, two models for the length scale equation can be used; Rodi and Spalding (1970), Ng and Spalding (1972) and Hanjalic and Launder (1972). The $k\ell$ model and the energy dissipation $\frac{k^{3/2}}{\ell}$ (Z) model. A full account of the relevance and comparison of these models has been discussed, Hughes (1978). However, the present study is limited to the $k\ell$ model in which the required constants are

$$\sigma_k = 2.0,\ \sigma_1 = 2.0,\ \sigma_2 = 1.2,\ C_D = 0.1\ ,\ C_M = 0.055$$

$$C_W = 22\ ,\ C_B = 0.84,\ q = 4,\ C_\mu = 1.0$$

Again, the results of Laufer (1954) are used for comparison. Figure 10(a) shows the variation in velocity and turbulent energy for a Reynolds number of 5×10^5. The turbulent viscosity is also compared with that obtained using the one equation model, Figure 10(b). A final plot, Figure 11 shows a comparison between the predicted make-up of energy balance in the main pipe region and the experimental results of Laufer (1954).

As in the case of the one equation model the comparisons are again quite favourable. An additional point is the slight increase in accuracy of the two equation model, as illustrated in Figure 11. The improvement was not expected to be significant owing to the invariant nature, section by section, of the characteristic lengths and constants for this type of flow problem.

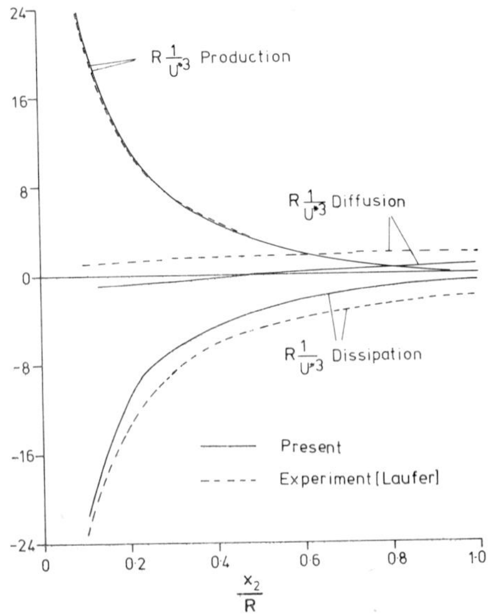

FIGURE 11 Turbulent energy balance in fully developed pipe flow.

REFERENCES

Baker,A.J. (1978) 'Finite element analysis of turbulent flows' Proc. Conf. Num. Meth. in Laminar and Turb. Flow, Swansea, : 203.

Barrow,H., Hornby,R.D. and Mistry,J. (1976) 'The prediction of fully developed turbulent flow in ducts by the finite element method' Maths. of F.E. and Applications, Academic Press, : 349.

Bradshaw,P. and Ferris,D.H. (1968) 'Derivation of a shear stress transport equation from the turbulent energy equation' Proc. A.F.O.S.R./I.F.P. Conf. on Comp. of Turb. Boundary Layers, 1, Stanford Univ.

De Vries,G. and Norrie,D.H. (1973) The Finite Element Method, Academic Press.

Hanjalic,K. and Launder,R.E. (1972) 'Fully developed asymmetric flow in a plane channel', J.Fluid Mech. 51 : 301.

Hill,V.W. and Kovasznay,L.S.G. (1968) 'The calculation of incompressible turbulent boundary layers by a simple theory' Proc. A.F.O.S.R./I.F.P. Conf. on Comp. of Turb. Boundary Layers, Stanford Univ. : 300.

Hughes,T.G. (1978) 'An investigation of turbulent flow problems using F.E. methods', Ph.D. Thesis, Univ. of Wales.

Hutton,A.G. (1976) 'Finite element analysis of turbulent, incompressible flow bounded by smooth walls' ICCAD, 2nd Int. Symp. on F.E. in Flow Problems, Santa Margharita, Italy [illegible].

Kolmagorov,A.N. (1942) Izv. Akad. Nauk SSSR Ser. Phys. No. 1-2, 6 : 56.

Laufer,J. (1954) 'The structure of turbulence in fully developed pipe flow', N.A.C.A. Report 1174.

Launder,B.E. and Spalding,D.B. (1972) Lectures in Mathematical Models of Turbulence, Academic Press, London.

Morgan,K., Hughes,T.G. and Taylor,C. (1977) 'An investigation of a mixing length and two-equation turbulence model utilising the finite element method', App. Math. Modelling, 1 : p.395.

Morgan,K., Hughes,T.G. and Taylor,C. (1978) 'A numerical model of turbulent shear flow behind a prolate spheroid', App. Math. Modelling, 2 : 271.

Ng,K.H. and Spalding,D.B. (1972) 'Turbulence model for boundary layers near walls', Phys. of Fluids, 15, No.1 : 20.

Prandtl,L. (1925) Bericht über untersuchungen zur ausgebildeten turbulenz, Z.A.M.M. 5, 136.

Prandtl,L. (1945) Über ein neues Formelsystem für die ausgebildeten turbulenz, Nachrichten van der Akad. der Wissenschaft in Göttingen.

Rodi,W. and Spalding,D.B. (1970) 'A two parameter model of turbulence and its application to free jets', Wärme und Stoffabertragung, Springer-Verlag, Berlin,3 : 85.

Schamber,D.R. and Larock,B.E. (1978) 'Computational aspects of modelling turbulent flow', Proc. Conf. Num. Meth. in Laminar and Turb. Flow, Swansea : 245.

Taylor,C., Zienkiewicz,O.C. and France,P.W. (1973) 'Some free surface transient flow problems of seepage and irrotational flow', Maths. of F.E. and Applications', Acade mic Press : 313.

Taylor,C. and Hood,P. (1972) 'A numerical solution of the Navier Stokes equations using the finite element technique' Comp. and Fluids, 1 : 73.

Taylor,C. and Hood,P. (1974) 'Navier Stokes equations using mixed interpolation' Proc. Conf. F.E. in Flow Problems, Swansea : 121.

Taylor,C., Hughes,T.G. and Morgan,K. (1977) 'A numerical analysis of turbulent flow in pipes', Comp. and Fluids, 5 : 191.

Taylor,C., Hughes,T.G. and Morgan,K. (1978) 'Finite element solution of one-equation models of turbulent flow', Comp. Phys. Academic Press, 29, 2 : 163.

CHAPTER 11

FULLY-DEVELOPED FLOW IN PASSAGES OF ARBITRARY CROSS-SECTION

A.D. GOSMAN, Mechanical Engineering Department, Imperial College

C.W. RAPLEY, Mechanical Engineering Department, Sunderland Polytechnic.

1. INTRODUCTION

The general field of 'duct' or 'passage' or 'confined' flows as they are variously called is a very large one, embracing as it does an infinite variety of geometrical configurations, fluids and flow rates, whose various combinations give rise to virtually every kind of flow structure imaginable. Thus, at the lower end of the scale of complexity there are the familiar steady, laminar, fully-developed single-phase flows in straight circular tubes, whose accessibility to straightforward analysis has made them a standard feature of introductory textbooks and courses on fluid mechanics. Towards the other end of the scale, where lies many of the duct flows encountered in practical circumstances, the fluid motion is almost invariably unsteady, due to the onset of turbulence and other phenomena and sometimes consists of more than one phase, while the enclosing duct is often curved and exhibits changes of shape and size of cross-section, which may be so abrupt as to provoke separation: by contrast with the earlier example, problems of this kind lie at the frontiers of current research and are only beginning to be accessible to predictive analysis.

The obstacles to prediction of all classes of duct flows are of two kinds: firstly, it is necessary to be able to translate them into a closed mathematical problem, consisting of the governing equations of motion and the associated boundary conditions. If the motion is laminar, this presents no especial difficulties, for the equations are well known: however it is equally known that the practical necessity to describe turbulent motion in terms of its time-averaged behaviour gives rise to formidable problems of closure (see, e.g. Bradshaw, 1978).

The second kind of obstacle relates to the solution of the equations, which must invariably be done numerically for cases of practical interest, notwithstanding the ingenuity of the developers of analytical techniques. Here again laminar flows present the least problems, although the inherent non-linearities and other complexities of the Navier-Stokes equations are continuing to occupy the attentions of numerous researchers as are the problems associated with the sometimes complex shapes of the passages. The inclusion of the additional equations of turbulence closures compounds the difficulties, for these introduce further non-linearity and simultaneity, as well as additional costs of solution.

In view of the above it is not surprising that greatest attention has been given to the simpler classes of duct flows. Indeed prior to the past decade, attention was almost exclusively focussed in the turbulent flow area on steady single-phase flows in straight round tubes, for these are at most two-dimensional in character, when incompletely developed and even then admit the boundary-layer simplifications, thus removing many of the complexities of the equations. This, together with the fact that the simple Prandtl mixing length type of turbulence closure is often found to be adequate for such flows allows straightforward analytical or numerical solution methods to be applied.

The present chapter is concerned with a class of problems one step up in generality from those just described, namely steady, fully-developed flows in straight ducts of non-circular cross-section, such as the general case illustrated in Figure 1. Here too the flow is always two-dimensional in character, with all the dependent variables apart from pressure functions of the cross-stream coordinates ξ_1 and ξ_2 only: while the pressure varies only with axial coordinate ξ_3, in a manner which is either prescribed or deducible from overall continuity considerations. However the equations for the non-circular case are inherently more complex, being elliptic in respect of both ξ_1 and ξ_2 coordinates: moreover the effects of turbulence are more far-reaching and require far more sophistication in mathematical modelling. Nevertheless, the past decade has seen the emergence of computer-based methods of prediction for this class whose accuracy is rapidly becoming comparable to the circular-duct procedures. It is these methods which are the focus of the present chapter.

Examples and practical relevance

Fully-developed flows in non-circular ducts are not merely of academic interest: indeed they occur in a wide variety of practical circumstances, examples involving laminar flow being the lubricant supply passages for bearings, the extrusion ducts of polymer processing plant and the compact heat exchangers of the food processing industries (it is of interest to note that in the latter two cases the fluid often exhibits non-Newtonian behaviour, which brings in additional physical and computational problems). Compact heat exchangers often operate in turbulent flow, as do the regenerators employed for waste heat recovery in the glass-making and other industries. Figures 2(a)-(d) illustrate typical passage cross-sections encountered in such equipment. A further important area of application involves regular arrays of parallel tubes, as sketched in Figures 2(e) and (f), as are found in boilers, condensers and nuclear reactors: here, although the passages, which are bounded by the surfaces of adjacent tubes and planes containing their axes, are intercommunicating, the flow in each will nonetheless become fully-developed when directed parallel to the tube axes.

In all of the fore-mentioned examples, designers are primarily interested in the overall pressure drop and heat transfer characteristics of the flows, as influenced by the parameters characterising the passage geometry (e.g. the spacing of the tube arrays). In many instances details of the internal structure are of interest as well: for example the level of turbulence at the interface between adjoining tube-array passages is influential in determining the level of turbulent mixing which occurs between them.

Experimental evidence

A large number of experimental studies have been performed of the present class of flows. Since however the present concern is primarily with prediction, rather than measurement, only those features which are salient to the former will be discussed here. Extracts from the available data will be shown later.

Firstly, in connection with laminar flows, measurements have indicated that in the absence of an external driving mechanism such as a transversely-sliding wall, the cross-stream velocities are everywhere identically zero. This behaviour, it can be shown, is a direct consequence of the fact that, in circumstances where the stress field is directly proportional to the rate of strain to some power, as it is in all Newtonian and many non-Newtonian fluids, as well as many eddy-viscosity representations of turbulent fluids, the stresses required to drive the cross-stream velocities are themselves absent when the latter are zero, i.e. there is no internal mechanism to sustain such motions. As a consequence, the mathematical description of the laminar problem is especially simple, and so consequently is the task of solution since the Navier Stokes equations reduce to a degenerate form of the axial momentum equation.

By contrast with the above, there is clear evidence (see, e.g. Brundrett and Baines, 1964 and Aly et al., 1978) that cross-stream motions always exist in turbulent flows in non-circular ducts, even in the absence of an external driving mechanism. The general character of the motions, as deduced from experiment for various shapes of section, is indicated in Figure 3: the tendency is for the core fluid to flow in towards narrow regions such as the corners of the triangular duct, thence parallel to the walls and finally back into the core in the wider regions. The magnitude of the secondary velocities is quite small, scarcely exceeding 1-2% of the mean axial velocity: nonetheless they exert a significant influence on all aspects of the flow behaviour including the level and distribution of wall shear (Brundrett and Baines, 1964) and heat transfer (Redman et al., 1966). Detailed investigations of the source of these motions reveal them to be, not surprisingly, due to a particular feature of the turbulent stress field, whereby the cross-stream stresses are connected to the axial strain rates (Gessner, 1973).

The existence of the secondary motions gives rise, it will be seen, to a much more formidable mathematical problem than the laminar case: for not only is it necessary to retain and solve the equations for all three velocity components, these must be supplemented by additional equations for the turbulent stresses, which have unfavourable properties from the computational point of view.

Contents of chapter

Now that the background has been established, the remainder of the chapter will be devoted to a more detailed analysis of the problem, commencing in the next section with a statement of the equations of motion applying to the laminar and turbulent cases and including, in the latter instance, an example of turbulence model equations for the stress components. Following this a brief survey will be given of the various numerical finite-difference/volume procedures which have been devised for solving the equations including the particular procedure which the authors have developed. Finally, an indication of the level of accuracy advisable will be conveyed via sample comparisons of predictions with experiment, and this will lead naturally into an assessment of the current status and directions where further research is needed.

2. MATHEMATICAL DESCRIPTION

Equations of mean motion

The equations of motion for steady, fully-developed laminar or turbulent duct flows will now be extracted from the Navier-Stokes equations. For this purpose, the instantaneous velocities

u_i appearing in the latter will be decomposed in the usual way into a time-mean value U_i and fluctuation u_i' about that value, i.e. $u_i \equiv U_i + u_i'$: then, according to Hinze (1975) and others, the momentum conservation equations for the time-averaged flow run, in Cartesian tensor notation:

$$\partial(\rho U_i U_j)/\partial x_i = - \partial P/\partial x_j - \partial(T_{ij})/\partial x_i \tag{1}$$

and the continuity equation is:

$$\partial(\rho U_i)/\partial x_i = 0 \tag{2}$$

Here the total stress tensor T_{ij} is the sum of the viscous components τ_{ij} and the turbulent (Reynolds) stress contribution γ_{ij}: the former is linked, for a Newtonian incompressible fluid, to the mean strain rate by the usual linear relation:

$$\tau_{ij} = - \mu(\partial U_i/\partial x_j + \partial U_j/\partial x_i) \tag{3}$$

while the latter, which arise from the velocity decomposition and time-averaging process, are linked to the fluctuating velocity components through:

$$\gamma_{ij} = \rho\overline{u_i' u_j'} \tag{4}$$

and represent additional unknowns. Here the overbar denotes time-averaging. The u' and γ_{ij} are of course identically zero in the laminar case, by definition.

If x_3 is now identified as the axial coordinate, and the fully-developed condition $\partial/\partial x_3 = 0$ is invoked on all variables apart from pressure, equations (1) reduce to:

$$\partial(\rho U_1^2)/\partial x_1 + \partial(\rho U_2 U_1)/\partial x_2 = -\partial P/\partial x_1 - \partial T_{11}/\partial x_1 - \partial T_{12}/\partial x_2 \tag{5}$$

$$\partial(\rho U_1 U_2)/\partial x_1 + \partial(\rho U_2^2/\partial x_2 = -\partial P/\partial x_2 - \partial T_{12}/\partial x_1 - \partial T_{22}/\partial x_2 \tag{6}$$

$$\partial(\rho U_1 U_3)/\partial x_1 + \partial(\rho U_2 U_3)/\partial x_2 = -dP/dx_3 - \partial T_{13}/\partial x_1 - \partial T_{23}/\partial x_2 \tag{7}$$

while the continuity equation and total stress tensor components become:

$$\partial(\rho U_1)/\partial x_1 + \partial(\rho U_2)/\partial x_2 = 0 \tag{8}$$

$$T_{11} = \rho\overline{u_1'^2} - 2\mu\partial U_1/\partial x_1 \tag{9}$$

$$T_{22} = \rho\overline{u_2'^2} - 2\mu\partial U_2/\partial x_1 \tag{10}$$

$$T_{12} = T_{21} = \rho\overline{u_1'u_2'} - \mu(\partial U_1/\partial x_2 + \partial U_2/\partial x_1) \tag{11}$$

$$T_{13} = \rho\overline{u_1'u_3'} - \mu(\partial U_3/\partial x_1) \tag{12}$$

$$T_{23} = \rho\overline{u_2'u_3'} - \mu(\partial U_3/\partial x_2) \tag{13}$$

Several useful observations may be made about the fully-developed flow equations. Firstly, it will be noted that dP/dx_3 appears in place of $\partial P/\partial x_3$ in the axial momentum equation (7), i.e. the pressure surfaces in successive $x_1 \sim x_2$ planes are assumed to run parallel to each other: reflection will confirm that this is a necessary condition for the flow to become fully-developed. The level of dP/dx_3 may therefore be specified a priori, in which case the total mass flow rate $\dot{m}$ is determined as part of the solution. Alternatively equation (7) may be integrated over the duct cross-section A to give:

$$0 = -A\frac{dP}{dx_3} - \oint_S \tau_s \, dS \tag{14}$$

where τ_s is the axial shear stress at the surface S and the closed integral is over the duct periphery. This, together with the integral continuity requirement which simply states that

$$\int_A \rho u_3 \, dA = \text{constant} = \dot{m} \tag{15}$$

forms the basis for obtaining solutions at prescribed flow rates.

A second important point which emerges is confirmation of the earlier statement that, in laminar flow, the condition of zero velocity on the walls of the duct, together with the fact that the viscous stresses T_{11}, T_{22} and $T_{12/21}$ depend solely on the cross-planar strain rates $\partial U_1/\partial x_1$, $\partial U_1/\partial x_2$ etc., leads to the conclusion that U_1 and U_2 must be everywhere zero in the fully-developed state. Thus for the laminar problem equations (5) to (7) reduce to, respectively: $\partial P/\partial x_1 = 0$, $\partial P/\partial x_2 = 0$ and:

$$\frac{\partial}{\partial x_1}\left(\mu\,\frac{\partial U_3}{\partial x_1}\right) + \frac{\partial}{\partial x_2}\left(\mu\,\frac{\partial U_3}{\partial x_2}\right) = \frac{dP}{dx_3} \tag{16}$$

It follows therefore that prediction of laminar flow simply requires the solution of a single Poisson equation (16) for U_3. Finally, it is clear from the above arguments and the earlier observations about the invariable existence of finite cross-stream velocities in turbulent flows that these must be the consequence of non-zero gradients in the turbulent Reynolds stress components $\rho\overline{u_1'^2}$, $\rho\overline{u_2'^2}$ and $\rho\overline{u_1'u_2'}$. Nothing further can be deduced about these until the following section, where their connection with the mean strain rates and other properties of the flow will be outlined. It is already clear however that in this case the full equation set (5) - (13) must be solved, together with whatever additional equations are required to extract the Reynolds stresses.

Heat Transfer

Although the fluid-mechanical aspects are of primary interest here, heat transfer is as noted earlier of considerable practical importance, so for the sake of completeness the appropriate form of the thermal energy equation for a low-speed incompressible fully-developed duct flow will be quoted (Hinze, 1975). It runs:

$$\partial(\rho U_1 T)/\partial x_1 + \partial(\rho U_2 T)/\partial x_2 = \partial(\Gamma\partial T/\partial x_1)/\partial x_1$$
$$+ \partial(\Gamma\partial T/\partial x_2)/\partial x_2 - \rho\overline{u_1'T'} - \rho\overline{u_2'T'} - \rho U_3 dT/dx_3 \tag{17}$$

where T and T' represent the time-mean and fluctuating temperatures respectively, Γ is the molecular thermal diffusivity ($\equiv \mu/P_R$, where P_R is the molecular Prandtl number) and the quantities $\rho\overline{u_1'T'}$ and $\rho\overline{u_2'T'}$ are the turbulent thermal-diffusion counterparts of the Reynolds stresses, which like the latter, represent additional unknowns. The existence of thermally-developed flow imposes certain constraints on the boundary conditions, the condition of constant heat flux (for which dT/dx_3 is then a known constant) being one which satisfies these. For a more extensive discussion of this matter see Shah and London (1978).

Equations for Turbulent Stresses

Background

The process of relating the turbulent Reynolds stresses $\rho\overline{u_i'u_j'}$ and heat fluxes to known or calculable quantities in such a way as to yield a closed set of equations for the fluid motion has come to be known as 'turbulence modelling' and the additional equations for the stresses as 'turbulence models'.

Although the existence of the turbulence-driven secondary motions and their origins in the Reynolds stress field have been known for some time, the absence of detailed models for the stresses led early workers such as Diessler and Taylor (1958), Rapier and Redman (1964), Bender et al. (1967) and others to ignore the explicit appearance of secondary velocities in the equations of motion (thereby reducing them to the single axial momentum equation (7) with $u_2=u_3=0$) and obtain the Reynolds stresses from an empirical relation based on the Newtonian equation (3), thus:

$$-\rho\overline{u_i'u_j'} = \mu_t(\partial U_i/\partial x_j + \partial U_j/\partial x_i) - \delta_{ij}\rho k \qquad (18)$$

where μ_t is a scalar 'eddy' or 'turbulent' viscosity and the 'turbulent pressure' term $\delta_{ij}\rho k$ has been added (but is usually later combined with the static pressure) to account for the possibility of the turbulent normal stresses being non-zero in the absence of mean strain; here $k \equiv \overline{u_i'u_i'}/2$ is (one-half) the sum of these stresses and is commonly referred to as 'the kinetic energy of turbulence' or 'turbulence energy'. Typically the turbulent viscosity was obtained from a mixing length or similar formulation. Unfortunately, predictions based on this concept were substantially in error, exhibiting for example, deviations from measured axial velocity distributions of up to 100%. Attempts to improve matters led to 'second-generation' methods such as those of Ibragimov et al. (1966, 1967), Meyder (1975) and Ramm and Johannsen (1975) in which attempts were made to introduce the secondary-flow effects indirectly, either via measured data such as wall shear stress distributions or by causing the eddy viscosity to be anisotropic. (It is of interest to note that direct calculation of the secondary motions would not in any case have been possible, for the use of equation (18), like its laminar counterpart, precludes their existence). Such methods gave improved agreement with experiment, but leaned so heavily on empirical input as to be more interpolative than predictive.

A similar criticism could be applied to third-generation procedures of the kind devised by Kokorev et al. (1971), Nijsing (1972) and Gerrard (1974), in which the terms in the equations of motion containing the secondary velocities were retained, but the velocities themselves were specified empirically from measurement rather than calculated. Here however the improved agreement which was obtained could at least be properly interpreted as being due to realistic representation of secondary flow effects; and it therefore became clear that any prediction method worthy of the name would have to be capable of calculating these via their fundamental origins in the turbulent stress field. It was this recognition that led to the next, contemporary, generation of methods, in which the ad-hoc stress-strain relation (18) was abandoned and replaced by direct calculation of the stresses from their fundamental transport equations. Some details will now be given of these 'stress transport models' as they are sometimes called, and the simplified versions which are employed in the present application.

Stress Transport Models

As already implied, this class of turbulence model has its origins in the basic equations of motion, which can be manipulated and time-averaged to yield equations for the Reynolds stresses themselves, of the general form (Hinze, 1975)

$$\rho U_k \partial(\overline{u_i' u_j'})/\partial x_k = -\rho(\overline{u_j' u_k'}\,\partial U_i/\partial x_k + \overline{u_i' u_k'}\,\partial U_j/\partial x_k)$$
$$- 2\mu\,\overline{(\partial u_i'/\partial x_k)(\partial u_j'/\partial x_k)}$$
$$- \partial(\overline{\rho u_i' u_j' u_k'})/\partial x_k + \text{other terms} \qquad (19)$$

Unfortunately, as the above equation reveals, this process does not by itself give closure, for it generates new unknowns, such as those contained in the second and third terms of the right-hand side. Further progress requires that these be replaced by expressions involving known or calculable quantities, an exercise which, involving as it does a mixture of rigour and inspired hypothesis, has resulted in a multiplicity of closures. It is beyond the scope of the present survey to review all of these, but the interested reader may refer to works by Gutfinger (1975), Reynolds (1976) and Bradshaw (1978) for this purpose. Of particular interest within the specific context of fully-developed duct flows are the stress closures derived by Hanjalic and Launder (1972), Launder and Ying (1973), Naot et al. (1974) and Reece (1976). Inasmuch as the bulk of the analyses and predictions which will be shown later are based on developments of the first-named authors' model, which is in any case representative of the other alternatives, some details will here be provided about it.

The full form of the Hanjalic and Launder stress transport equation is:

$$\rho U_k \partial(\overline{u_i' u_j'})/\partial x_k = -\rho(\overline{u_j' u_k'}\ \partial U_i/\partial x_k + \overline{u_i' u_k'}\ \partial U_j/\partial x_k)$$

I II

$$- 2\delta_{ij}\ \rho\varepsilon/3 - C_{\phi 1}\ \rho\varepsilon/k)\ (\overline{u_i' u_j'} - 2\delta_{ij}\ k/3)$$

III IV

$$+ \rho a^{mi}_{\ell j}\ \partial U_\ell/\partial x_m + \rho a^{mj}_{\ell j}\ \partial U_\ell/\partial x_m$$

IV IV

$$+ C_s\ (\rho k/\varepsilon)\ \partial\left[\overline{u_i' u_\ell'}\ \partial\overline{u_j' u_k'}/\partial x_\ell + \overline{u_j' u_\ell'}\ \partial\overline{u_k' u_i'}/\partial x_\ell\right.$$

V

$$\left. + \overline{u_k' u_\ell'}\ \partial\overline{u_i' u_j'}/\partial x_\ell\right]/\partial x_k \qquad (20)$$

V

where

$$a^{mi}_{\ell j} \equiv \alpha\overline{u_m' u_i'}\ \delta_{\ell j} + \beta(\overline{u_m' u_\ell'}\ \delta_{ij} + \overline{u_m' u_j'}\ \delta_{i} + \overline{u_i' u_j'}\ \delta_m$$

$$+ \overline{u_i' u_\ell'}\ \delta_{mj}) + \gamma[\delta_{mi}\ \delta_{\ell j} + \eta(\delta_{m\ell}\ \delta_{ij} + \delta_{mj}\ \delta_{i\ell})]k$$

$$+ C_{\phi 2}\ (\overline{u_m' u_i'}\ \overline{u_\ell' u_j'} - \overline{u_m' u_j'}\ \overline{u_i' u_\ell'} - \overline{u_m' u_\ell'}\ \overline{u_i' u_j'})/k \qquad (21)$$

Here k and ε are respectively the kinetic energy of turbulence (defined earlier) and its dissipation rate, defined by:

$$2\nu\overline{(\partial u_i'/\partial x_k)(\partial u_j'/\partial x_k)} \equiv 2\delta_{ij}\ \varepsilon/3 \qquad (22)$$

while the quantities α, β, γ and η are functions of the empirical coefficients $C_{\phi 1}$ and $C_{\phi 2}$, details of which are given in the Appendix to this chapter. Names indicating their physical origins are commonly attached to the various terms in equation (20) labelled with Roman numerals thus:
I - convection; II - generation; III - dissipation; IV - redistribution; V - diffusion.

Six equations may be generated from the above tensorial stress transport equation for the six Reynolds stress components, although it is common to replace one of these by an equation for the scalar turbulence energy k, which is simply the trace of equation (20), i.e. the result of setting i = j and performing the indicated summations, to yield

$$\partial(\rho U_i k)/\partial x_i = \partial[(\mu_t/\sigma_k)\partial k/\partial x_i] + P - \rho\varepsilon \tag{23}$$

where: $P \equiv \rho\overline{u_i'u_j'}\ \partial U_i/\partial x_j$ (24)

is the turbulence production rate and

$$\mu_t = C_\mu\ \rho k^2/\varepsilon \tag{25}$$

is a turbulent viscosity, while σ_k and C_μ are empirical coefficients.

This modelling process has however given rise to one further unknown, namely the turbulence dissipation rate ε, for which a further equation must therefore be found. Two alternative approaches have been employed for this: one is to note from dimensional analysis that ε is linked to the local length scale ℓ_ε of the dissipating eddies by

$$\varepsilon = C_D\ k^{3/2}/\ell_\varepsilon \tag{26}$$

where C_D is an empirical constant, and to then specify the ℓ_ε distribution empirically. The other, more general approach is to derive a different transport equation for ε by operating on the Navier-Stokes equations in an analogous fashion to that for the stress components, the particular version arrived at by Hanjalic and Launder (1972) in this way being:

$$\partial(\rho U_i\varepsilon)/\partial x_i = \partial[(\mu_t/\sigma_\varepsilon)\ \partial\varepsilon/\partial x_i] + \varepsilon(C_{\varepsilon 1}P - C_{\varepsilon 2}\ \rho\varepsilon)/k \tag{27}$$

in which σ_ε, $C_{\varepsilon 1}$ and $C_{\varepsilon 2}$ are further empirical coefficients, also defined in the Appendix.

Algebraic Stress Models (ASM)

In principle stress transport equations of the above kind can of course be applied as they stand to the present class of flows, as indeed was done by Naot et al. (1974) and Reece (1976), both of whom invoked only those simplifications allowed by the fully-developed flow condition (Reece in fact considered the developing region as well). However, as will be discussed later, the complexity and costs engendered by such calculations provide a powerful incentive to exploit whatever further simplifications the applications allow.

Launder and Ying (1973) were evidently the first to recognise that the smallness of the secondary velocities in fully-developed duct flows provides, in conjunction with the fully-developed condition, justification for ignoring the terms in the stress transport equations (20) representing transport by convection and diffusion (i.e. terms I and V). This is an important simplification, for it results in the removal of all differential coefficients of $\rho\overline{u_i u_j}$; i.e. the equations become algebraic in the latter: hence the 'ASM' appelation. When employed in conjunction with the differential transport equation (27) for k, and the algebraic equation (26) for ε, with ℓ_ε calculated from the Buleev (1963) length scale formula (ℓ_ε having been equated with this length scale)

$$\frac{1}{\ell_\varepsilon} = 0.5 \int_0^{2\pi} \frac{d\phi}{s} \tag{27}$$

where s the distance from the point in question to the duct wall in direction ϕ, this combination enabled Launder and Ying to obtain for the first time true predictions of the detailed flow structure in a square duct: moreover they agreed well with the experimental data in most respects, including the pattern and level of secondary flows. This particular ASM will hereafter be referred to as 'LY1'.

The LY1 model was later employed by Aly et al. (1978) to predict turbulent flow in an equilateral-triangular duct with similar success, albeit after some adjustments to the empirical coefficients in the equations. On the other hand Carajilescov and Todreas (1975) decided in applying this method to axial flow in a tube assembly that it was better to use the measured length scale distribution to procure adequate agreement, thus implying that the use of equations (26) and (27) might not be adequate in all cases.

The merits of replacing the algebraic calculation of ε in the original Launder and Ying approach by solution of its own differential transport equation to yield what will be termed the LY2 version were explored by Tatchell (1975) who obtained a similar level of accuracy of prediction of the square-duct case as with the algebraic version: however the application was confined to this geometry. Subsequently Benodeker and Date (1978) applied the LY2 version to tube-assembly geometries with only moderate success, although some doubts must be attached to their results due to the limited fineness of grid they were able to employ in their numerical calculations. The experiences of the present authors (Gosman and Rapley 1978, 1980) in applying this version to a wide variety of duct geometries suggest that it performs adequately in virtually all cases: some evidence to this effect will be given in a later section.

Thus far the equations of the Launder and Ying ASM have not been presented, an omission which will now be corrected. They will not however be quoted in the form they were originally proposed, for this entailed solution of a set of implicit simultaneous algebraic equations for the individual stress components; whereas it was subsequently shown by Gessner and Emery (1976) that further algebraic manipulation yields an explicit set for the stresses which is more computationally convenient and physically informative. The explicit set runs as follows:

$$\rho\overline{u_1'^2} = C_3\rho k - C_2\mu_t(k/\varepsilon)\ (\partial U_3/\partial x_1)^2 \tag{28}$$

$$\rho\overline{u_2'^2} = C_3\rho k - C_2\mu_t(k/\varepsilon)\ (\partial U_3/\partial x_2)^2 \tag{29}$$

$$\rho\overline{u_1'u_2'} = -\ C_2\mu_t(k/\varepsilon)\ (\partial U_3/\partial x_1)\ (\partial U_3/\partial x_2) \tag{30}$$

$$\rho\overline{u_1'u_3'} = -\ \mu_t(\partial U_3/\partial x_1) \tag{31}$$

$$\rho\overline{u_2'u_3'} = -\ \mu_t(\partial U_3/\partial x_2) \tag{32}$$

It is of interest to compare these with the corresponding laminar Newtonian contributions in the total-stress equations (8) through (13). This will reveal that the axial stress $\rho\overline{u_1'u_3'}$ and $\rho\overline{u_2'u_3'}$ do depend on the turbulent viscosity and axial strain rate in a Newtonian fashion, as was supposed in earlier studies, but the cross-planar stress equations are decidedly different from the Newtonian relations, depending as they do on <u>axial</u>, as opposed to cross-planar, strain rates: indeed it is this feature that explains the existence of the turbulence-driven secondary motions. A further point to note is that the 'viscosity' field is effectively anisotropic, for μ_t is multiplied by the local ratio k/ε in the cross-planar relations.

Other Stress Models

Although details cannot be provided here, mention should nonetheless be made of two further stress models which have been applied to fully-developed duct flows, one being the differential transport model of Launder, Reece and Rodi (1975) which was applied in this form to square-duct prediction by Reece (1976), and exhibited measurable, but not overwhelming, superiority over the earlier Hanjalic and Launder-based versions, albeit at the price of appreciable increase in complexity. Rodi (1975) has also demonstrated that an ASM variant of the Launder-Reece and Rodi equations can be derived but it would appear that it is not possible to cast it into the convenient explicit form of equations (28) to (32), nor has it been applied to the present class of problems. The other stress model of interest is a derivative of the Launder and Ying (1973) ASM devised by Gessner and Emery (1976) and applied to square duct flow analysis by Gessner and Po (1976). In the former work it was shown that by invoking local turbulence equilibrium assumptions the differential transport equation (23) for k can be replaced by an algebraic relation, thereby yielding a wholly algebraic turbulence model for the Reynolds stresses. The results of the second paper suggest however that this simplification may be counter-productive.

The Near-Wall Region

All of the turbulence models described thus far are only valid in circumstances where the viscous stresses are negligible in comparison with the Reynolds stresses, and therefore all cease to be applicable in near-wall regions, where the turbulent fluctuations are progressively damped as the wall is approached. Moreover, although Reynolds stress models applicable in viscous regions have emerged in recent times (see, e.g. Hanjalic and Launder, 1975) it is clear that their solution in the characteristically steeply-varying velocity and turbulence fields in the near-wall region imposes an unacceptable computational burden, at least for practical calculations. Accordingly it has been the universal practice in duct flow applications to use special simplified representations of this region and match them to the full turbulence-model representation at locations sufficiently far removed from the wall for the viscous effects to be negligible.

The simplified representations devised to date are almost invariably based on the notion of one-dimensional flow and local turbulence equilibrium, conditions which give rise to the familiar logarithmic 'law of the wall' velocity distribution and similar universal profiles for the other variables of interest: these typically have the following form (Launder and Spalding, 1972):

$$U_r = U^* \kappa^{-1} \ln (En^+) \tag{33}$$

$$\partial k/\partial n = 0 \tag{34}$$

$$\varepsilon = C_\mu^{3/4} k^{3/2}/\kappa n \tag{35}$$

where

$$U^* \equiv \tau_s/\rho\, C_\mu^{1/4} k^{1/2} \tag{36}$$

$$n^+ \equiv C_\mu^{1/4} k^{1/2} \rho n/\mu \tag{37}$$

and U_r is the tangential resultant velocity at a height n along a normal to the surface, at which τ_s is the resultant shear stress.

Heat Transfer

Although there exist counterparts of the Reynolds stress equations for the turbulent heat flux components $\rho\overline{u_i'T'}$ appearing in the energy conservation equation (17) and also algebraic versions of same (Bradshaw, 1978) they have not yet been extensively applied to duct flows, one of the few examples being the calculations of an asymmetrically-heated round duct performed by Samaraweera (1978). It would appear however that if the circumferential temperature variations are not too non-uniform, the heat fluxes may be adequately calculated via simple Fourier's Law expressions of the form:

$$-\rho\overline{u_1'T'} = (\mu_t/\sigma_h)\ \partial T/\partial x_1 \tag{38}$$

$$-\rho\overline{u_2'T'} = (\mu_t/\sigma_h)\ \partial T/\partial x_2 \tag{39}$$

in which σ_h is a turbulent Prandtl number of order unity and the quantity μ_t/σ_h is a scalar eddy conductivity. Consistent with this practice is the use of the following logarithmic temperature profile expression (Launder and Spalding, 1972) for the near-wall region:

$$T-T_s = (\dot{q}_s''/\rho U^* C_p)\ [\sigma_h(\kappa^{-1}\ln\ (En^+) + P^*] \tag{40}$$

where T_s and $\dot{q}_s''$ are the surface temperature and heat flux respectively and:

$$P^* \equiv 9.24\ [(P_R/\sigma_h)^{3/4}-1]\ [1+0.28\ \exp(-0.007\ P_R/\sigma_h)] \tag{41}$$

Some evidence of the performance of this model will be provided later.

Summary

The assembly of the fundamental equations governing laminar and turbulent fully-developed flow and heat transfer in straight ducts of arbitrary cross section has now been completed. For the laminar case, the single unique set consists of equations (16) and (17), with the former yielding the distribution of the only non-zero velocity component U_3 and the latter the temperature distribution, after setting U_1,U_2 and the turbulent heat fluxes to zero. In the case of the equations of the time-averaged turbulent flow there is no unique closed set, due to the inherent need for approximation brought about by the time-averaging process. Thus, in this case, in addition to the equations of motion (5) to (8) it is necessary to solve for the Reynolds stress components appearing in the total stress relations (9) through (13). Several workable models exist for these and it is suggested that the alternative currently offering the best compromise between accuracy and complexity is the ASM based on the Hanjalic and Launder (1973) stress model, in which the components are calculated from algebraic equations (29) to (32) and the quantities k and ε appearing in these are obtained from their own differential transport equations (23) and (27) respectively (i.e. the 'LY2' model). For heat transfer the recommendation is to use the turbulent flux relations (38) and (39) in conjunction with the energy equation (17). Finally, the use of the model equations should be restricted to the inner fully-turbulent core of the flow and their solution matched to the near-wall relations (33), (34), (35) and (40).

NUMERICAL ANALYSIS

Appreciation of task

In this section the various numerical procedures of the finite-difference and finite-volume variety which have been devised to solve the differential equations of fully-developed duct flows will be briefly reviewed. Before this is done however it will be useful to isolate certain features of the problem which have particularly occupied the attentions of the investigators.

(a) Irregular geometries and wall effects. The variety of duct shapes shown thus far are indicative of the need for numerical procedures which are capable of handling with ease and efficiency shapes of arbitrary complexity. This is very much a function of the flexibility of the computational grid network which is employed, the ideal arrangement being one where the grid may be distorted to fit the duct cross section and may be locally refined near the walls in order to adequately resolve the locally steep gradients in velocities and other variables which typically occur there (by contrast the core region often exhibits relatively small gradients and therefore only requires relatively coarse resolution).

(b) Complexity, non-linearity and simultaneity. These features are generally present in all fluid flow calculations, although the present laminar flow equations are an exception in this regard: for the equations for U_3 and T are both linear, and although the equation for the latter variable contains the former, the converse is not true. (Of course non-linearity and simultaneity may exist if the fluid has strongly temperature-dependent properties or is non-Newtonian, but such problems lie outside the scope and concern of the present chapter). Thus the laminar flow equations present little challenge to the numerical analyst in these respects and are primarily useful as a preliminary testing ground for the more complex turbulent case.

The turbulent flow equations are of course in general highly complex, non-linear and closely coupled, because these features are not only inherent in the basic equations of motion, but also in the turbulence model equations. The differential stress models are particularly challenging in this regard, being numerous (a full model comprises at least seven differential transport equations) and containing within themselves connections between the stresses and other turbulence properties, as well as the external links to the mean velocity field.

The ASM equations are fortunately much less formidable, with fewer differential equations, simpler stress interlinkages and, in the case of the Launder-Ying model, stress-strain relations which are much closer in character to the traditional eddy viscosity type of formulation: nonetheless they do offer some new challenges, which will now be outlined.

(c) Destabilising effects of stress equations. Arising from (b) above are two features of differential stress models which make numerical calculations much more prone to failure than those based on the traditional eddy-viscosity representation (18). One is the loss of the stabilising effect which the latter confers by virtue of the fact that even at large Re theimplicit stress terms like $(\mu_t+\mu)\partial U_i/\partial x_j$ remain significant, because μ_t increases with Re and is orders of magnitude larger than μ: this effect is of course lost when $\rho u_i' u_j'$ is inserted explicitly.

The other drawback is that, as noted by Naot et al. (1974) the dominant stress component in a particular transport equation may change from one location to another within the flow, in reflection of changes in the dominant transport mechanisms. Thus the conventional practice of extracting each component from its 'own' transport equation may break down. Taken together, these features imply that the popular sequential type of iterative solution algorithm, wherein each of the dependent variables is calculated in turn from its own set of discretised equations, may encounter difficulties. Fortunately, the first of these problems is only partially encountered in the ASM (in the cross-stream momentum equations, whose turbulent stresses are functions of U_3 only - see equations (28) to (32)), while the second is totally absent: thus the destabilising effects are less severe for this class of model.

Characteristic Features of Available Methods

A brief outline will now be given of the numerical methods which have been developed for solution of the equations of fully-developed duct flows, in which emphasis will be laid on the difficulties just referred to.

Firstly, on the matter of matching the computational mesh to the duct geometry, many workers have either confined attention to rectangular ducts, for which a simple Cartesian mesh is appropriate (see e.g. Launder and Ying (1973), Naot et al. (1974), Gessner and Po (1976) and Reece (1976) or employed Cartesian or cylindrical polar meshes for irregular passage shapes such as tube assemblies and accepted the penalties, which include inefficient distributions of mesh points and additional complications in applying the boundary conditions: examples of this approach include Carajilescov and Todreas (1976), Trupp and Aly (1978), Aly et al. (1978) and Benodeker and Date (1978). Others however recognised the possibility of using more flexible coordinate frames such as the general curvilinear-orthogonal system to produce a more optimum mesh arrangement according to the criteria mentioned earlier: in particular Meyder (1975) employed this approach in a second-generation anisotropic eddy viscosity-based calculation of tube-assembly flows; and the present authors (Gosman and Rapley, 1978, 1980) have used such a system, in conjunction with the numerical mesh generation procedure of Antonopoulos, Gosman and Issa (1978), in their ASM-based calculations of flows in a range of non-rectilinear duct configurations. Some examples of the meshes employed will be displayed later in this chapter.

Turning now to the solution procedures themselves and associated practices for overcoming the instability difficulties referred to earlier, a diversity of approaches have been employed, all of which have involved ad-hoc practices of one kind or another. Several groups have based their methods on the numerical procedure of Gosman et al. (1969) for solving the two-dimensional flow equations in stream function vorticity variables: this procedure is of the finite-volume variety, uses centred differencing for all terms apart from those describing convection which are upwind-differenced, (which is usually not in fact necessary, due to the smallness of the secondary velocities), and employs point iteration with under-relaxation to solve the discretised equations in a sequential manner, i.e. alternating in turn between the vorticity, stream function and other fields. This procedure was used by Launder and Ying (1973) in their LY1-based square duct calculations with apparently little difficulty, apart from the need to experiment with under-relaxation factors, although no doubt stability was enhanced by the fact that they obtained the ε field from the algebraic equation (26) rather than the differential version(27).

When however the same numerical procedure was independently employed in conjunction with the LY1 model by Carajilescov and Todreas (1976) and Trupp and Aly (1978) for a particular tube-assembly flow, they obtained different results; and although the latter authors abscribed this to incomplete convergence of the earlier calculations they themselves were forced by numerical instability problems to impose an artificial immutable sign constraint on the turbulence stress-induced vorticity sources, which must have compromised the generality of their method. The only application to date of this numerical procedure to the LY2 model equations appears to be that of Benodeker and Date (1978), although the doubts expressed earlier about the fineness of their grid should be borne in mind.

The remaining stress-model-based calculations have, with one exception, been performed with variants of the finite-volume solution procedure for three-dimensional developing duct flows of Patankar and Spalding (1972) or the recirculating-flow version thereof described by Caretto et al. (1972). These procedures work with the primitive variables (i.e. velocity and pressure), use staggered velocity and pressure grids and hybrid central-upwind differencing, and approach the solution in a stepwise or iterative fashion by obtaining estimates of the velocity field from the momentum equations and then inferring corrections to it and the pressure field via a continuity-based pressure correction calculation. The algorithm is essentially of the sequential kind and solution for the individual variables is by a form of ADI. Tatchell (1975) employed the developing-flow variant and LY2 model in his calculations of square duct flow and Reece (1976) used the same variant to solve the full differential stress equations of the Launder, Reece and Rodi (1976) model for the same application. In such calculations the size of the forward step employed in the marching solution plays the role of an under-relaxation parameter, with too large a value resulting in failure. It is indicative of the difficulty of achieving results by this and other sequential methods that the step size was often limited to a few percent of the duct height. The present authors used a two-dimensional version of the recirculating flow variant in their calculations (Gosman and Rapley, 1978, 1980) and also found it necessary to employ under-relaxation and other devices to procure solutions, as will be described elsewhere.

Finally, the remaining 'exception' referred to earlier is the finite-difference method devised by Naot et al. (1974) to predict square-duct flow using a modified version of the Launder, Reece and Rodi (1975) differential stress transport model. Certain facets of the method are conventional, notably the vorticity-stream function formulation of the two-dimensional flow equations, the upwind differencing of convection terms and central differencing elsewhere and the solution by iteration. It is unique however in employing a local block inversion method to solve the linearised flow and stress equations simultaneously, rather than sequentially as in other methods: the overall iteration is necessary because of the nonlinearities and the fact that the blocks are confined to small clusters of grid nodes in order to avoid the otherwise prohibitive computer storage and time requirements which a whole-field block solution would entail. The authors report that their method proved superior to one of the sequential kind, but was nonetheless not unconditionally stable.

Summary

Evidently there do now exist several numerical methods capable of solving the equations of fully-developed turbulent duct flows, notwithstanding the difficulties referred to earlier. Of the approaches devised for coping with the complex duct geometries, the use of a flexible curvilinear-orthogonal computing mesh seems most satisfactory, for it leads to a more optimum mesh distribution than the other alternatives. However none of the approaches devised for dealing with the problems of numerical stability engendered by the stress equations are entirely satisfactory, for all require ad-hoc measures such as under-relaxation. There is little to choose in this respect between the methods.

APPLICATIONS

Examples to be shown

Some comparisons will now be shown between predictions obtained with several of the methods just described and experimental data for rectangular, triangular and elliptical ducts and tube assembly passages. The bulk of the comparisons are drawn from the present authors' work because it covers a wider range of duct geometries than earlier studies: however predictions from the latter are also included where available.

The method employed by the present authors is that outlined earlier and described more extensively by Gosman and Rapley (1978). Grid refinement tests were performed and indicated that, for the bulk of the cases shown, a network comprising some 160 nodes yielded solutions to the turbulent flow equations of acceptable accuracy, each such solution typically requiring some 45s of CDC 6600 central processor time. The numerical stability of the calculations was found to be quite sensitive to such factors as the detailed treatment of the non-linear source/sink terms in the discretised equations, the starting practices, and the level of under-relaxation employed in the various equations. Thus, for example: the source terms in the k and ε equations were linearised in such a way as to increase the magnitude of the main diagonal in the coefficient matrix and also ensure these variables could never assume physically unrealistic negative values; the calculations were initiated with the secondary-flow driving stresses suppressed and these were then later released after the axial flowfield had 'settled'; and under-relaxation factors of 0.7 were typically employed for mean flow variables and 0.3 for the cross-flow velocities.

The necessity for these measures is indicative of the difficulties of solving the turbulent flow equations. A further important point to be made about these solutions is that they were all obtained with the single set of turbulence model coefficients quoted in the Appendix, i.e. no effort was made to adjust them to procure optimum agreement for each geometry, on the grounds that this would distort the assessment of the generality of the method.

Laminar Flow

As already remarked, the laminar-flow cases provide a useful preliminary testing-ground for numerical procedures. Examples of such tests of the Gosman and Rapley (1978) method will now be presented. The forms of the computer-generated computational grids employed in these calculations are shown in Figure 4 below.

In each case the flow exhibits symmetry about one or more axial planes and this has been exploited to minimise the computing requirements. However in the equilateral-triangular duct application a region embracing half the duct cross-section, and therefore three symmetry elements, was also calculated in order to ensure that the method would itself produce the expected symmetries: the fact that the computational grid was not symmetrical in the same sense made this an especially stringent test.

Figures 5 and 6 show some results for the elliptical and equilateral-triangular ducts respectively, in the form of axial velocity profiles. In the former case agreement with the analytical solution due to Knudsen and Katz (1958) is excellent and superior to the earlier finite-difference calculation of Cain et al. (1972), while in the latter the same high level of agreement is obtained with the Sparrow (1962) analytical results for both velocity and wall shear stress: moreover the expected symmetry in U_3 about the centre of the bisector is also obtained with this half-duct calculation, thus further confirming the low level of error produced by the method. Results for a narrow-angled triangular duct are shown in Figure 7, which displays the numerical analytical and measured axial velocity profiles: in this case the first two are indistinguishable, but the measurements fall below them in the core region due, it is believed, to the presence of turbulence there. Figure 8 contains numerically and analytically-derived axial velocity profiles in tube assemblies having various pitch-to-diameter ratios and again the agreement is excellent. Finally, Table 1 below contains numerically- and analytically-derived values for the various ducts of the product of the friction factor f and Reynolds number Re based on the mean velocity in, and hydraulic diameter of, the duct, this product being a constant for a given shape of cross section. This provides a compact indication of the very acceptable level of accuracy which the method provides.

Passage	Predicted Gosman & Rapley	Analytical or Predicted	Reference
Elliptical aspect ratio = 2	16.88	16.82	Tao (1961)
= 5	18.55	18.6	
Isoceles Triangular duct, apex angle = 60°	13.31	13.33	Sparrow (1962)
22.1°	12.7	12.85	
11.7°	12.3	12.54	
Triangular rod assembly pitch/dia. = 1.1	20.14	20.377	Sparrow & Loeffler (1959)
1.2	24.64	24.95	
1.5	30.51	30.04	

Table 1 Comparison of fRe parameters for laminar flows

Turbulent Flow

The amount of information contained in each turbulent-flow calculation is too great to allow a full display here for each geometry examined (in general more details can be found in the references cited). Accordingly extensive results will be shown for just one case - the square duct - and brief extracts will be given for the others. In the case of the square and equilateral-triangular ducts the results of further symmetry tests will also be shown.

Square Duct

In this case the smallest symmetry element is the triangular region formed by the duct wall and the corner and wall bisectors, as shown in the inset of Figure 9. The accompanying plot displays calculated axial velocity and turbulence energy profiles along the corner bisector derived for this element, and the larger one encompassing the quarter-section of the duct. The two results are in very close agreement, which is most encouraging considering that the smaller zone required a curvilinear grid while the larger element could be fitted with a simple rectilinear arrangement.

The plots of Figures 10 and 11 show respectively the secondary flow and axial velocity distributions in the duct cross-section, the former illustrating the characteristic motion into the corner and back towards the core via the wall and its bisector, and the latter the resultant bulging of the high-velocity contours into what would otherwise be a region of relatively low velocity in the corner. Also shown is the wall axial shear-stress distribution, which is made more uniform by the effects just described. These results serve to confirm earlier observations that the secondary flows are very small in magnitude, but nonetheless exerts significant influence on the main flow behaviour. Figures 11 and 12 also allow the performance of the several other prediction methods and turbulence models to be measured against each other and experiment, and suggests that, in this instance, the ASM-based methods perform as well as, if not better than, those based on the full stress equations, the general level of agreement being quite good.

Further details of the interior flow structure are provided in Figures 13, displaying profiles of one of the secondary velocity components, and 14, showing certain of the Reynolds stress components. Here the picture is somewhat inconsistent, with satisfactory prediction of the secondary velocities but substantial discrepancies in the stress comparisons: some of this error can be attributed to the ASM, which is known, for example, to underpredict the degree of normal stress anisotropy for the values of the empirical coefficients which yield optimum performance in other respects (Launder, 1976). However the remaining sources of error, be they in measurement or modelling, have yet to be identified.

The final plot for this case, Figure 15, contains measurements and predictions of the friction factor versus Reynolds number characteristics. Here the ASM-based predictions are in accord with each other and lie slightly below the data and the empirical correlation of Blasius (Schlichting, 1968) the maximum discrepancy being about 10% at the lower range of Re. This slight underprediction of f is a characteristic feature of both versions of the Launder-Ying ASM, as will be seen in the later examples (and is also present in circular duct calculations, which are not shown) but it is not known whether the differential stress models would perform better in this respect. The error could, it should be observed, be reduced somewhat by 'tuning' some of the coefficients of the model, but this can have other, sometimes unpredictable repercussions and was therefore not attempted.

Rectangular Ducts

Figures 16(a) and (b) show respectively the secondary and axial velocity distributions in a rectangular duct of aspect ratio 3, for which the only predictions available are those of the present authors. The pattern of secondary motion appears different at first sight from that in the square duct but examination will confirm that the tendency for flow into the corner along its bisector is repeated, as are the other characteristics and this is confirmed by the available data (not shown). The axial velocity distribution has its characteristic distortion into the corner which is reproduced reasonably well by the predictions. As for the friction factor characteristics, these are shown in Figure 17 for a range of aspect-ratios, in both measured and predicted form. The pattern of slight underprediction is repeated, and there is a suggestion that the calculations overestimate the effect of aspect ratio, but this is in any case of the same order as the experimental accuracy, so no firm conclusions can be drawn.

Triangular Ducts

The equilateral triangular duct provides another useful test case for symmetry checks on the numerical procedure, since the cross-section can be sub-divided into six identical elements. Figure 18 shows the predicted secondary flow field for the half-section, which was obtained using the type of asymmetrical grid distribution illustrated in Figure 4. Encouragingly, the flow is seen to form the expected identical vortices in each element, and detailed assessment reveals the values predicted at corresponding locations to lie within a few percent of each other. The axial velocity and turbulence energy distributions in a single element are shown in Figures 19 and 20, which also contain the measurements and predictions of Aly et al. (1978): in their calculations, it will be recalled the Buleev (1963) length scale formula (27) was used to calculate ε algebraically via equation (26) and the model coefficients were 'tuned' in an effort to optimise accuracy. Neither calculation reproduces the axial velocity contours closely (although it should be borne in mind that the variations are in any case small) and there is little to choose between them. Similar comments can be made about the turbulence energy distributions. Both plots show once again the tell-tale distortions produced by the secondary flows.

Figure 21 displays the friction factor behaviour for this duct. Both of the two ASM-based predictions lie best to the data, with that of Aly et al. (1978) agreeing as a consequence of the coefficient-tuning process.

It should be mentioned that similar successes have been achieved in predicting flows in narrow isoceles triangular ducts, but space does not allow these results to be presented here.

Elliptical Ducts

The predicted secondary flow pattern for an elliptical duct of aspect ratio 2 is displayed in Figure 22: here, although there is no 'corner' as such, the narrow side of the duct produces a similar effect and the motion is therefore towards the wall along the major axis and away from it along the minor one. The axial velocity profiles along these same axes are plotted in Figures 23(a) and (b), which contain the measurements and predictions of Cain et al. (1972) as well as the predictions of the present authors which, in this case, are also shown for the hypothetical situation of no secondary motions, these having been suppressed in order to further illustrate their influence. As would be expected, the secondary flows tend to augment the velocities as the wall is approached along the major axis and diminish them along the minor one, although here the effect is less marked. The full ASM calculations are clearly superior to the other two predictions and in fair agreement with the data, the large errors in the Cain et al. result being probably due to the fact their method is of the second-generation type which both ignores secondary flows and employs a mixing-length eddy viscosity formulation. Figure 24 shows the variation of friction factor with Reynolds number for ducts of this and other aspect ratios. Here there is a trend in the measurements for f to increase with aspect ratio which is reproduced by the predictions, but the effect is small. The general level is as usual underpredicted, in this case by 10-15% which is rather more than in the other examples: this is due, in part, it is suspected, to incomplete flow development in the experiments.

Tube Assemblies

For this, the final geometry of the series a single excerpt is shown from a group of calculations performed of fully-developed axial flow in tube assemblies of different arrangements and spacings. The example selected is an equilateral-triangular array of pitch to diameter ratio 1.123, for which experimental data and predictions from other sources exist. The secondary and axial flow patterns and turbulence energy distributions in a single symmetry element are displayed in Figures 25, 26 and 27 respectively. The predicted secondary motion is, as would be expected from the earlier examples, directed into the narrowestgap along the symmetry plane and then flows outwards along the tube wall and into the core, forming a single large eddy. The earlier prediction of Carajilescov and Todreas (1975) suggested a multiple eddy pattern, but that of Trupp and Aly (1978) also produced a single eddy, although, as mentioned earlier, their numerical method imposed an immutable sign on the sense of rotation and thus would have prevented the formation of multiple eddies. No direct measurements of secondary velocities are available to resolve this issue: however the fact that the predicted axial velocity contours are both qualitatively and quantitatively close to the data suggests that the single-eddy pattern may be correct. Similar comments apply to the turbulence energy distribution of Figure 27.

The performance of the other predictions of this flow is illustrated in the axial velocity profiles of Figure 28 across the narrow ($\theta=0^{\circ}$) and wide ($\theta=30^{\circ}$) gaps of the passage.

The calculations of Carajilescov and Todreas (1975) are seen to agree best, but this is to be expected since they are based on the measured length scale distribution and 'tuned' constants. Of the other two calculations that of the previous authors is marginally closer to the data, although both underpredict the velocities in the narrow gap: however the experimenters attribute this to incomplete flow development.

Friction factor data and predictions for a range of pitch to diameter ratios are collected together in Figure 29. The main bulk of the data show the effect of this parameter to be relatively small, the exception to this rule being the measurements of Trupp and Azad (1975) and Kjellstrom (1971): however it can be argued that these may be influenced by such effects as incomplete flow development or rod misalignment, since they lie above an upper limit identified by Rehme (1972) for large spacings. The present author's predictions tend to agree with the bulk data, whereas those of Carajilescov (1975) and Trupp and Aly (1978) lie some 10-20% higher.

Heat Transfer

Heat transfer measurements in non-circular duct flows are far less extensive than data on the flow itself, and are mainly limited to the local or circumferential-average heat transfer coefficients. Some examples will now be presented of comparisons between such measurements and predictions obtained by the present authors.

Elliptical duct Figures 30 and 31 relate to an elliptical duct of aspect ratio 2, the fluid being air. The first figure shows the distribution of the local Nusselt number NuL, normalised by the average value $\overline{Nu}$. Evidently the variation is small, with the maximum value occurring at the widest part of the duct, as might be expected. The prediction exhibits similar trends to the data and shows a significant effect of secondary flow in comparison to the result for the hypothetical no cross flow case. The variation of Nu with Reynolds number Re is contained in Figure 31 for ducts of aspect ratio 1.5 and 2.0: in this instance both data and predictions suggest insensitivity to the latter parameter within this range. Two predicted curves are shown, corresponding to values of the turbulent Prandtl number σ_h (appearing in the equations (38) and (39) for the turbulent heat fluxes) of 0.8 and 0.9, which span the range of commonly-employed values: evidently the best agreement is obtained with the lower value, but this is not a conclusive argument for embracing it since some degree of underprediction might have been expected on the basis of the earlier underprediction of friction factor (cf Figure 24).

Tube Assemblies

The local Nusselt number distributions for both air and water flows in triangular arrays of around 1.1 pitch to diameter ratio are shown plotted in Figure 32: as in the previous example the degree of variation is small and the maxima and minima correspond to the regions of largest and smallest gap, respectively. The predictions can be said to be in reasonable agreement with the data, especially when the scatter is taken into consideration.

An example of average heat transfer predictions for this geometry is shown in Figure 33, which contains data and predictions for water flowing through arrays of aspect ratios from 1.12 to 1.27. The agreement is quite satisfactory in all respects, including the trend for Nu to increase with P/D.

Triangular ducts For this, the final example of the series, only average heat transfer data are available. Figure 34 displays comparisons for air flowing through an equilateral triangular duct and the two values of turbulent Prandtl number mentioned earlier. As usual the agreement is not bad and is best at the lower σ_h value.

CLOSURE

Summary of Progress

The present survey has revealed that a considerable body of experimental and theoretical knowledge now exists about the hydrodynamic and heat-transfer properties of fully-developed flows in straight, non-circular ducts and passages. On the experimental side, the existence of the turbulence-driven secondary motions has been confirmed both directly and indirectly in a variety of cross sections and has served to explain the observed greater uniformity of axial velocity and wall shear stress and heat transfer coefficients and higher mean values of the latter two properties than would be expected to prevail in the absence of such motions.

On the theoretical side, the advances made in turbulent stress modelling have assisted the identification of the mechanisms producing the secondary motions, the main one being the generation of cross-stream Reynolds shear stress gradients by the axial strain field, which tends to be most steeply-varying in regions where the duct walls converge. The deduction of this mechanism, which does not exist in Newtonian or quasi-Newtonian flows, has been facilitated by the extraction of the algebraic stress equations from the much more complex differential transport versions, with the Launder and Ying (1973) ASM version of the Hanjalic and Launder (1972) differential model being particularly convenient in this regard. Equally importantly, the emergence of the

ASMs has considerably eased the task of solving the equations, as has the approach of using special simplified models of the near-wall region.

The remaining link in the chain leading to 'prediction' methods for these flows which are worthy of the name has been provided by the development of numerical finite-difference/volume procedures which can solve the equations of motion in conjunction with the stress models and which employ flexible computational meshes capable of fitting arbitrary shapes of duct. The numerical errors of such methods can be made negligibly small, as has been demonstrated by comparison with exact analytical solutions for laminar flows and by symmetry tests for turbulent cases. The overall accuracy of the combined numerical procedures and turbulence models can reasonably be claimed, on the basis of a relatively large number of comparisons with data, to be adequate for many purposes in respect of both magnitudes and trends, and the computing requirements are likewise not unreasonable.

Future Development

Notwithstanding the above remarks, it is clear that there remains considerable scope for improvement in the measurement and prediction of this class of flows. Thus, in respect of the measurements, there are many anomalies and gaps in the existing data, some of which can be attributed to limitations in techniques and others to imperfections in design or construction. More specifically, the smallness of the secondary velocities and stress components renders them difficult to measure and easily altered by incomplete flow development or small variations in the shape of the cross section. Heat transfer data is particularly sparse and incomplete, particularly for circumstances of substantial peripheral variations in wall temperature or heat flux.

The uncertainties and deficiencies in the area of turbulence modelling are far too numerous to list here, but are adequately discussed in the specialised literature, to which references have been made. Suffice it to note that all of the models described are known to be deficient in one respect or another in the present context. For example in the discussion of the Gessner and Po (1976) paper it is acknowledged by one of the originators of the Launder and Ying (1973) ASM that adjustment of the coefficients to give the correct anisotropy of the near-wall normal stresses leads to overprediction of the secondary flow velocities. This and other problems are the subject of current research.

The scope for computational improvements is equally wide as in the other areas just described. One of the more straightforward areas is grid flexibility, which could be enhanced by relaxing the orthogonality constraint. It is by no means certain, however, that the overall economy of the methods would be thereby improved, for the complexity of the equations, and thus the number of computer operations required per mesh point, are both considerably increased in moving from an orthogonal to non-orthogonal system.

Less straightforward are the problems of numerical stability, which, as already noted increase in severity as one moves from the eddy-viscosity class of turbulent stress models, through ASMs to, finally, full differential stress transport models. Almost certainly the key to improvement here lies in moving towards block simultaneous solution algorithms in the spirit, if not necessarily the detail, of the approach of Naot et al. (1974). Undoubtedly this is the area where the greatest impact may be made.

REFERENCES

Aly,A.M.M., Trupp,A.C. and Gerrard,A.D. (1978) Measurement and Prediction of Fully Developed Flow in an Equilateral Triangular Duct. J. Fluid Mech. 85: 57-83.

Antonopoulos,K., Gosman,A.D. and Issa,R. (1978) A Simple Method for Generating Curvilinear Orthogonal Grids. Imperial College, Mech. Eng. Fluids Section Report.

Barrow,H. and Roberts,A. (1970) Flow and Heat Transfer in Elliptical Ducts. Paper No. FC4.1, 4th Int. Heat Transfer Conf., 1970.

Bender,D.J., Switick,D.M. and Field,J.H. (1967) Turbulent Velocity Distribution in a Rod-Bundle. G.E.A.P. 5411.

Benodeker,R.W. and Date,A.W. (1978) Prediction of Fluid Flow and Heat Transfer in Nuclear Rod Clusters. Indian Inst. of Tech. Report.

Bradshaw,P. (Editor) (1978) Turbulence. Topics in Applied Physics 12, Springer-Verlag.

Brundrett,E. and Baines,W.D. (1964) The Production and Diffusion of Vorticity in Duct Flow. J. Fluid Mech. 19: 375-394.

Buleev,N.I. (1963) Theoretical Model of the Mechanism of Turbulent Exchange in Fluid Flows. AERE Trans. 957.

Cain,D. and Duffy,J. (1971) An Experimental Investigation of Turbulent Flow in Elliptical Ducts. Int. J. Mech. Science 13: 451-459.

Cain,D., Roberts,A. and Barrow,H. (1972) A Theoretical Study of Fully Developed Flow and Heat Transfer in Elliptical Ducts. Paper No. C120/72, Proc. I. Mech. E. Conf. on Compact High Duty Heat Exchangers.

Cain,D., Roberts,A. and Barrow,H. (1973) An Experimental Investigation of Turbulent Flow and Heat Transfer in Elliptical Ducts. Warme-und-Stoffubertragung, 2: 101-107.

Carajilescov,P. and Todreas,N.E. (1975) Experimental and Analytical Study of Axial Turbulent Flows in an Interior Sub-Channel of a Bare-rod Bundle. Trans. ASME 75-WA/HT-51.

Caretto,L.S. et al. (1972) Two Numerical Procedures for Three Dimensional Recirculating Flows. Proc. 2nd Int. Conf. Numerical Methods: 60-68.

Deissler,R.G. and Taylor,M.F. (1958) Analysis of Turbulent Flow and Heat Transfer in Non-circular Ducts. NACA Tech. Note 4384.

Dingee,D.A.et al. (1955) Heat Transfer from Parallel Rods in Axial Flow. Battelle Mem. Inst. Report BM1 1026.

Eckert,E.R.G., Irvine,T.F. and Eichhorn,R. (1954) Laminar Transitional and Turbulent Flow in Triangular Pasages. WADC Report 54-443.

Gerrard,R. (1974) Finite Element Solution for Flow in Non-circular Conduits. Proc. ASCE 100 (3): 425-441.

Gessner,F.B. (1973) The Origin of Secondary Flow in Turbulent Flow Along a Corner. J. Fluid Mech. 58 1-25.

Gessner,F.B. and Emery,A.F. (1976) A Reynolds Stress Model for Turbulent Corner Flows Part I, J. Fluid Eng. ASME 76-FE-C.

Gessner,F.B. and Po,J.K. (1976) A Reynolds Stress Model for Turbulent Corner Flows Part II. J. Fluid Eng. ASME 76-FE-D.

Gosman,A.D. et al. (1969) Heat and Mass Transfer in Recirculating Flows. Academic Press.

Gosman,A.D. and Rapley,C.W. (1978) A Prediction Method for Fully Developed Flow Through Non-Circular Passages. Paper No. FS/78, Proc. Int. Conf. Numerical Methods.

Gosman,A.D. and Rapley,C.W. (1980) The Analysis of Fully Developed Turbulent Flow in Non-Circular Passages. In preparation.

Gutfinger,C. (Editor) (1975) Topics in Transport Phenomena. J. Wiley.

Hanjalic,K. and Launder,B.E. (1972) A Reynolds Stress Model of Turbulence and its Application to Asymmetric Boundary Layers. J. Fluid Mech. 52: 609-638.

Hanjalic,K. and Launder,B.E. (1975) Contribution Towards a Reynolds Stress Closure for Low-Reynolds Number Turbulence. J. Fluid Mech. 74: 1-18.

Hartnett,J.P., Koh,J.C.Y. and McComas,S.T. (1962) A Comparison of Predicted and Measured Friction Factors for Turbulent Flow Through Rectangular Ducts. Trans. ASME 84: 82-88.

Hinze,J.O. (1975) Turbulence. McGraw Hill.

Ibragimov,M.Kh. et al. (1966) Calculation of the Tangential Stresses at the Wall of a Channel and the Velocity Distribution in the Turbulent Flow of a Fluid. Soviet Atomic Energy, 21: 731-739.

Kays,W.M. (1966) Convective Heat and Mass Transfer, McGraw-Hill.

Kjellstrom,B. (1971) Studies of Flow Parallel to a Rod Bundle of Triangular Array. AB Atomenergi Report AE/RV/196.

Knudsen,J.G. and Katz,D.L. (1958) Fluid Dynamics and Heat Transfer. McGraw-Hill.

Kokorev,L.S. et al. (1971) Effects of Secondary Flows on the Velocity Distribution and Hydraulic Drag in Turbulent Liquid Flow in Non-circular Channels. Heat Trans. Soviet Res. 3: 66-78.

Launder,B.E. (1976) Discussion to Paper, J. Fluid Eng. ASME 76-FE-D.

Launder,B.E., Reece,G.J. and Rodi,W. (1975) Progress in the Development of a Reynolds Stress Turbulence Closure. J. Fluid Mech. 68: 537-566.

Launder,B.E. and Spalding,D.B. (1972) Mathematical Models of Turbulence. Academic Press.

Launder,B.E. and Ying,W.M. (1972) Secondary Flows in Ducts of Square Cross-section. J. Fluid Mech., 54: 289-295.

Launder,B.E. and Ying,W.M. (1973) Prediction of Flow and Heat Transfer in Ducts of Square-Section. Proc. I.Mech.E. 187: 455-461.

Leutheusser,H.J. (1963) Turbulent Flow in Rectangular Ducts. Proc. ASCE 89: 1-19.

Lowdermilk,W.H., Weiland,W.F. and Livingood,J.N.B. (1954) Measurements of Heat Transfer and Friction Coefficients for Flow of Air in Non-circular Ducts at High Surface Temperatures. NACA Res. Mem. E53J07.

Meyder,R. (1975) Turbulent Velocity and Temperature Distribution in the Central Sub-channel of Rod Bundles. Nuclear Eng. Design, 35: 181-189.

Naot,D., Shavit,A. and Wolfshtein,M. (1974) Numerical Calculationof Reynolds Stresses in a Square Duct with Secondary Flow. Wärme-und-Stoffubertragung, 7: 151-161.

Nijsing,R. (1972) Heat Exchange and Heat Exchangers with Liquid Metals. AGARD-LS-57-72.

Patankar,S.V. and Spalding,D.B.(1972) A Calculation Procedure for Heat, Mass and Momentum Transfer in 3-D Parabolic Flows. Int. J. Heat Mass Trans., 15: 1787-1806.

Ramm,H. and Johannsen,K. (1975) A Phenomenological Turbulence Model and its Application to Heat Transport in Infinite Rod Arrays with Axial Turbulent Flow. J. Heat Trans. ASME-97,75-HT-MM: 231-237.

Rapier,A.C. and Redman,J.D. (1964) The Calculation of Velocity Distributions in Rod Clusters. UKAEA Report TRG 838.

Redman,J.D., McKee,G. and Rule,I.C. (1966) An Experimental Study of Turbulent Forced Convection Heat Transfer to Gases in Axial Flow Rod Cluster Passages. UKAEA Report TRG931.

Reece,G.J. (1976) A Generalized Reynolds Stress Model of Turbulence. Ph.D. Thesis, University of London.

Rehme,K. (1972) Pressure-drop Performance of Rod Bundles in Hexagonal Arrangements. Int. J. Heat Mass Trans. 15: 2499-2517.

Reynolds,W.C. (1976) Computation of Turbulent Flows. Ann. Rev. Fluid Mech. 8: 183-208.

Rodi,W. (1975) A Note on the Empirical Constants in the Kolmogorov Eddy Viscosity Expression. J. Fluid Eng., ASME: 386-389.

Samaraweera,D.S.A. (1978) Turbulent Heat Transport in Two and Three Dimensional Temperature Fields. Ph.D. Thesis, University of London.

Schlichting,H. (1968) Boundary Layer Theory. McGraw-Hill.

Shah,R.K. and London,A.L. (1978) Laminar Flow Forced Convection in Ducts. Academic Press.

Sparrow,E.M. (1962) Laminar Flow in Isosceles Triangular Ducts. A.I.Ch.E.J. 8(5): 599-604.

Sparrow,E.M. and Loeffler (1959) Longitudinal Laminar Flow between Cylinders Arranged in Regular Array. A.I.Ch.E.J. 5: 325-329.

Subbotin,V.I. et al. (1971) Velocity Field of Turbulent Fluid Flow in a Longitudinal Streamline of Clusters of Rods. AEC-tr-7189.

Tao,L.N. (1961) On Some Laminar Forced Convection Problems. J. Heat Trans., ASME 83: 466-472.

Tatchell,D.G. (1975) Convection Processes in Confined Three-dimensional Boundary Layers. Ph.D. Thesis, University of London.

Thomas,D.L. and Easter,P.G. (1972) Measurements of Wall Shear Stress in a Duct of Square Cross-Section. CEGB Report RD/B/N2477.

Trupp,A.C. and Aly,A.M.M. (1978) Predicted Turbulent Flow Characteristics in Triangular Rod Bundles. Mech. Eng. Report ER25.26, University of Manitoba.

Trupp,A.C. and Azad,R.S. (1975) The Structure of Turbulent Flow in Triangular Array Rod Bundles. Nuc. Eng. Design, 32: 47-84.

APPENDIX

The constants in the Hanjalic and Launder stress transport equations (20 and 21) are related to the coefficient $C_{\phi 2}$ as follows:

$$\alpha = (10-8C_{\phi 2})/11, \qquad \beta = -(2-6C_{\phi 2})/11$$
$$\gamma = -(4-12C_{\phi 2})/55, \qquad \eta = (6-18C_{\phi 2})/55 \tag{A1}$$

The constants in the Algebraic Stress Model (equations 28 to 32) are related to the coefficients $C_{\phi 1}$ and $C_{\phi 2}$ as follows:

$$C_2 = (12C_{\phi 2}-4)/11\,(C_{\phi 1}-2C_{\phi 2}) \tag{A2}$$

$$C_3 = (22C_{\phi 1}-36C_{\phi 2}-10)/33\,(C_{\phi 1}-2C_{\phi 2}) \tag{A3}$$

$$C_\mu = (44C_{\phi 1}-22C_{\phi 1}C_{\phi 2}-128C_{\phi 2}-36C_{\phi 2}^2+10)/165(C_{\phi 1}-2C_{\phi 2})^2 \tag{A4}$$

In the calculations of Gosman and Rapley, the empirical constants used were taken from previous studies. No attempt was made to optimise any of these constants, which are listed below, together with the sources used to select the values shown:

Constant	Value	SOURCE(S)
$C_{\phi 1}$	2.78	Launder & Ying (1973), Aly et al. (1978)
$C_{\phi 2}$	0.358	
$C_{\varepsilon 1}$	1.55	Launder & Spalding (1972), Tatchell (1975)
$C_{\varepsilon 2}$	2.0	
σ_k	1.0	Launder & Spalding (1972)
σ_ε	1.22	
E	9.025	Schlichtling (1968)
κ	0.4	

TABLE 2 Values of the constants employed in the calculations of Gosman and Rapley.

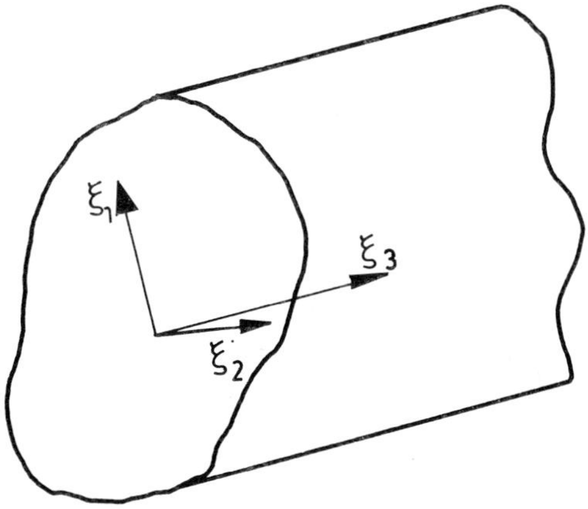

FIGURE 1 General co-ordinate frame in an arbitrary shaped passage

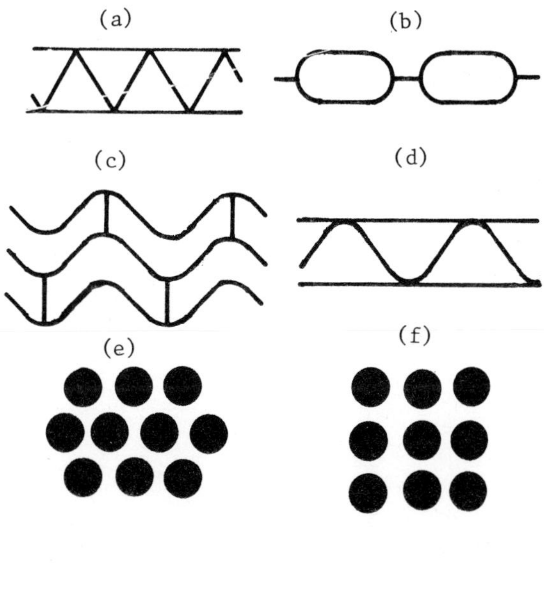

FIGURE 2 Typical non-circular passages encountered in practical equipment

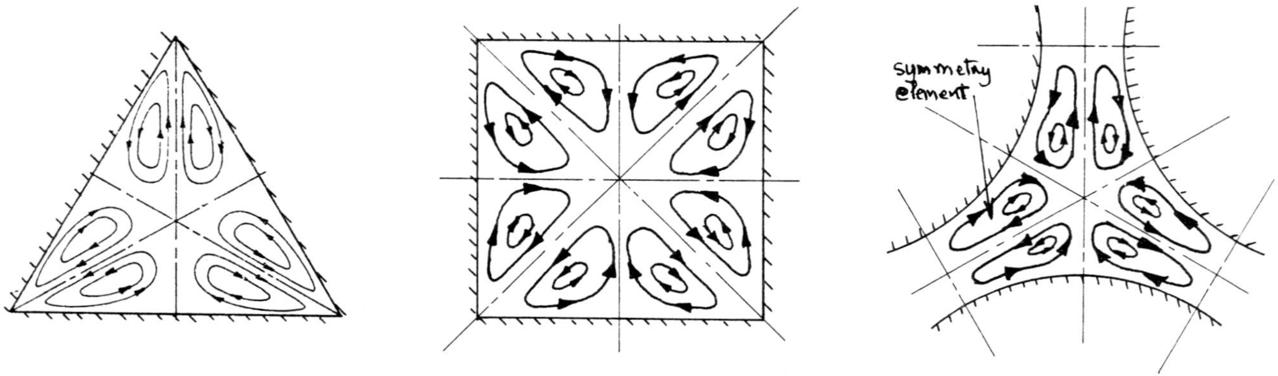

(a) Equilateral triangular duct

(b) square duct

(c) triangular tube assembly

FIGURE 3 Sketches of secondary flow patterns observed in turbulent flows in various geometries

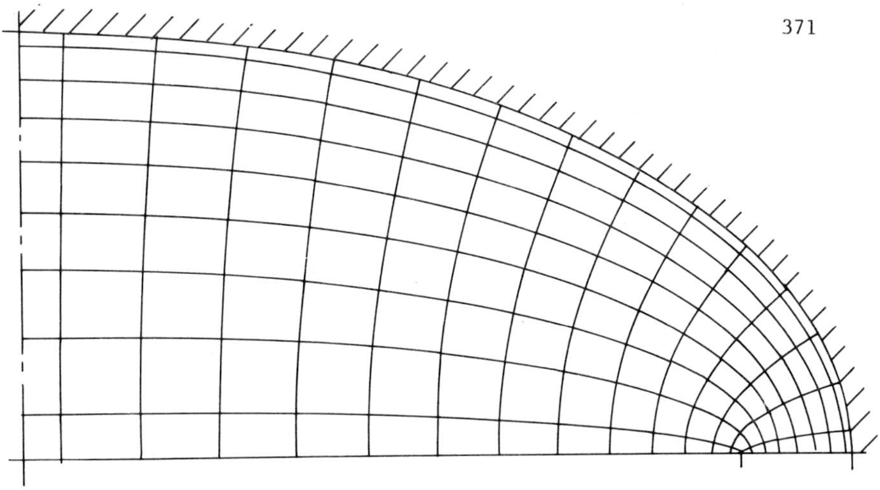

(a) Elliptical duct (quarter-duct)

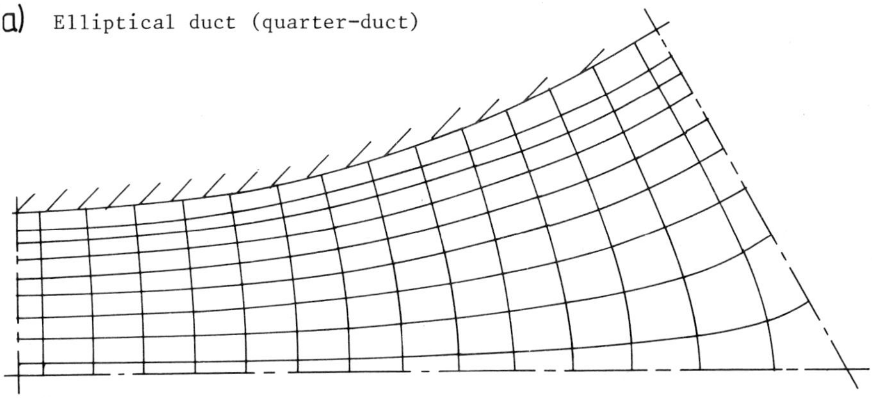

(b) Triangular tube-assembly (symmetry element - see Figure 3)

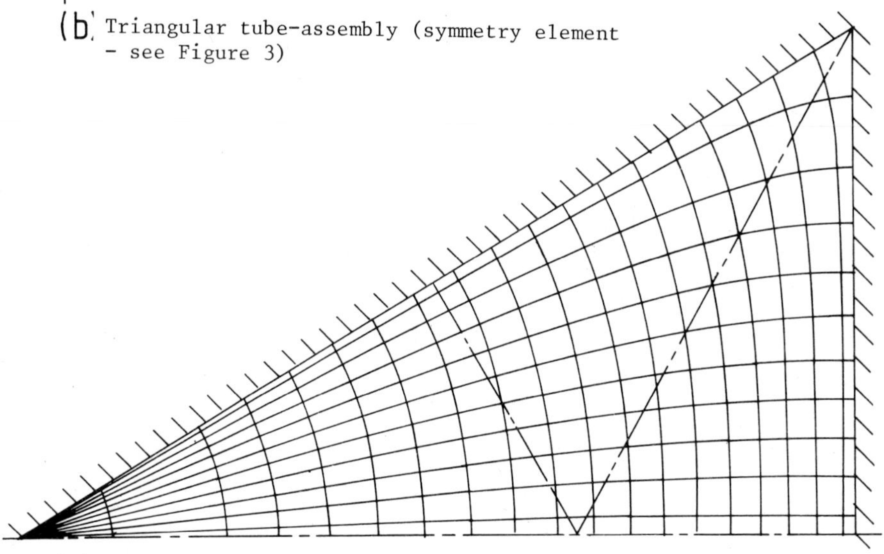

(c) Equilateral triangular duct (half-duct)

FIGURE 4 Illustrations of computational grids used in laminar flow calculations

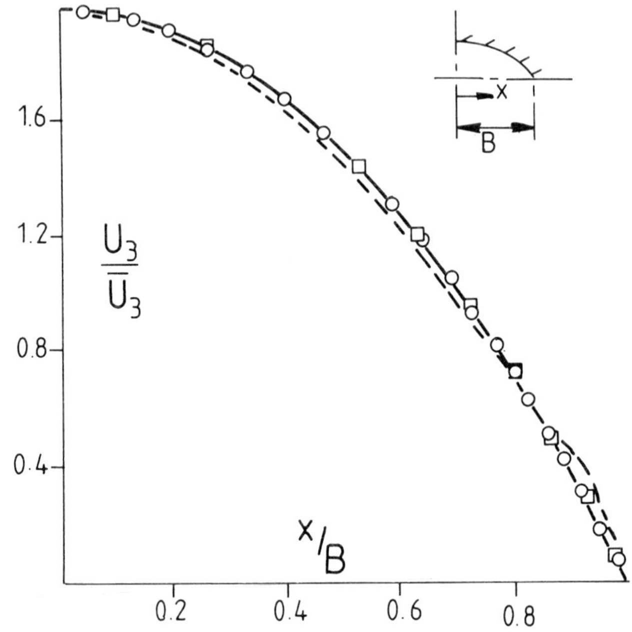

FIGURE 5 Axial velocity profiles in an elliptical duct of aspect ratio 2.0 with laminar flow

——— analytical, Knudsen & Katz (1958)

– – – predicted, major axis, Cain et al. (1972)

○ major axis } predicted, Gosman & Rapley
□ minor axis }

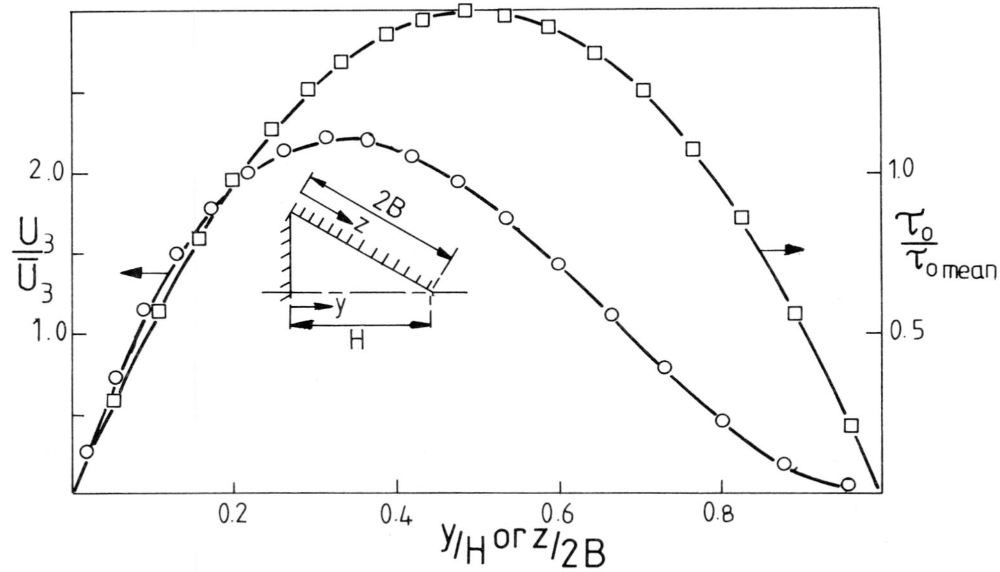

FIGURE 6 Axial velocity and wall shear stress profiles in an equilateral-triangular duct with laminar flow

——— analytical, Sparrow (1962)

○ □ } predicted, Gosman & Rapley

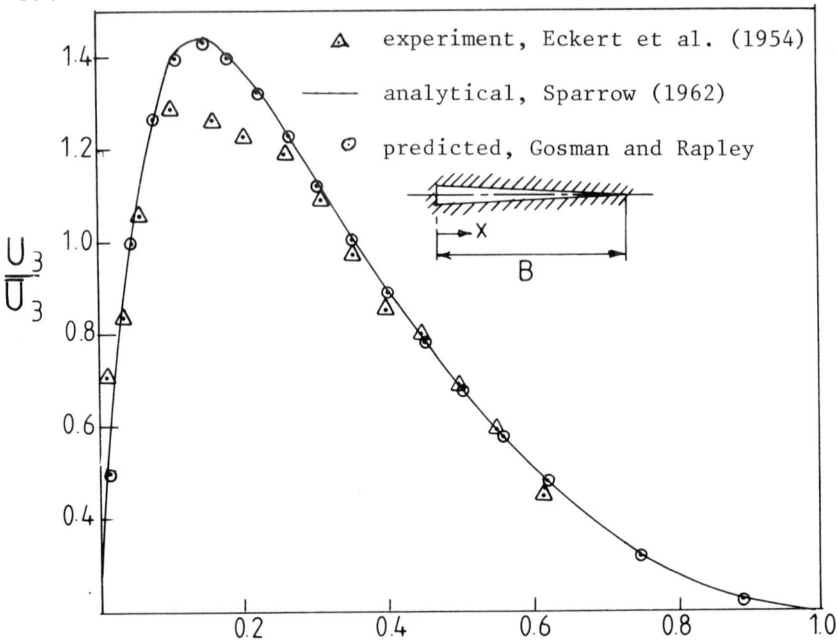

FIGURE 7 Centre-plane axial velocity profile in an isoceles-triangular duct with laminar flow

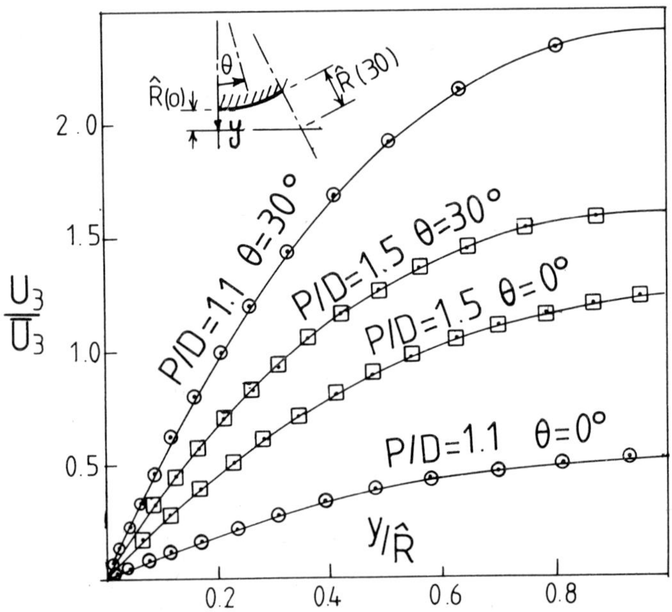

FIGURE 8 Axial velocity profiles in tube-assembly passages with laminar flow

——— analytical, Sparrow & Loeffler (1954)

⊙ ⊡ predicted, Gosman and Rapley.

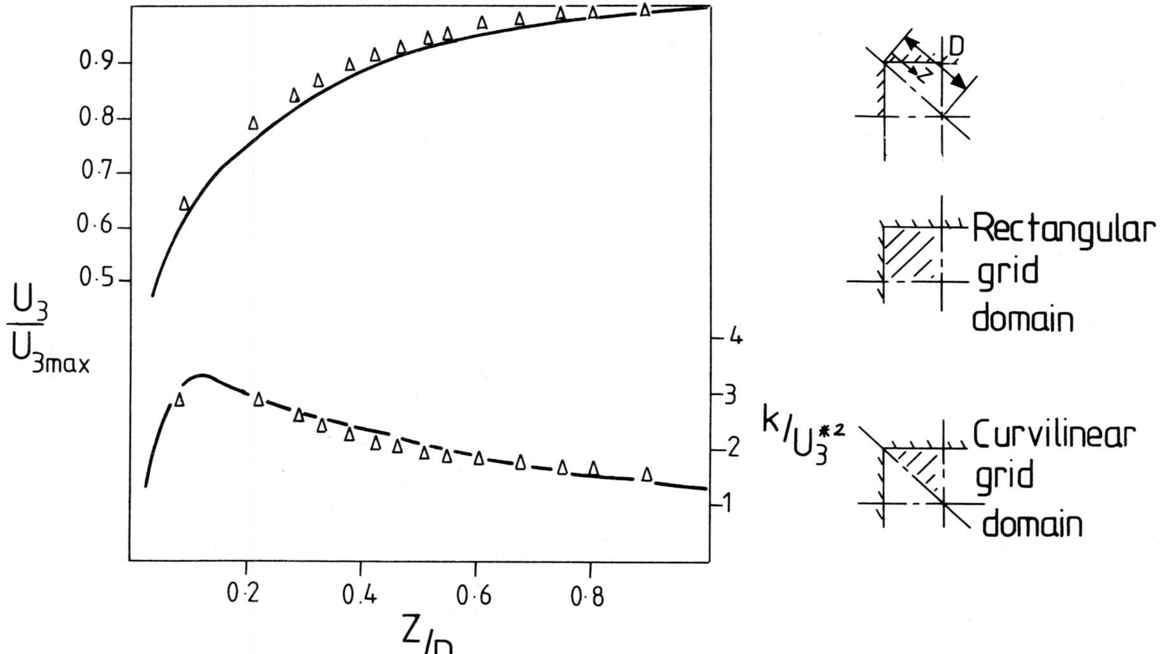

<u>FIGURE 9</u> Comparison of axial velocity and turbulence energy profiles calculated with two different grids in a square duct.

——— rectangular grid Δ curvilinear grid,

Re=3.4 x 10^4 Gosman and Rapley

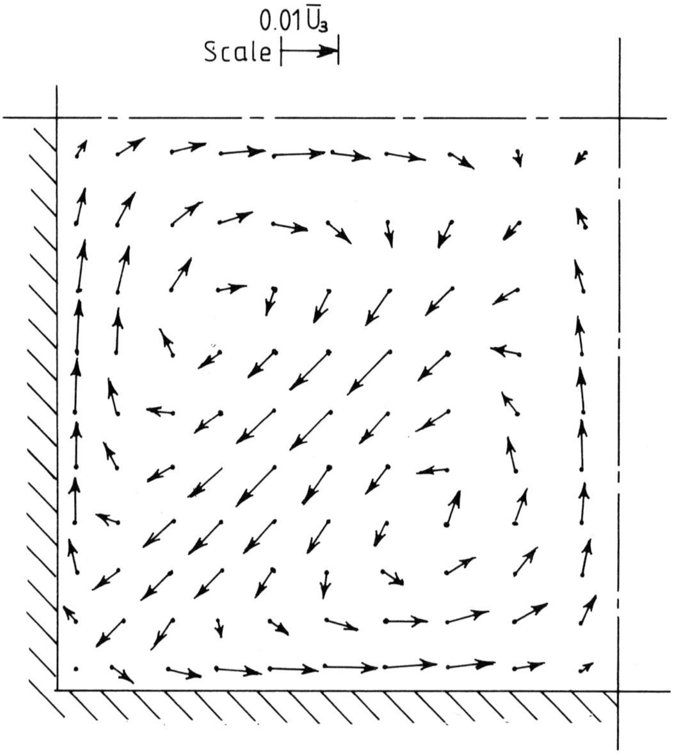

FIGURE 10 Secondary flow pattern in a square duct, Re = 2.15 x 10^5

Predicted, Gosman & Rapley

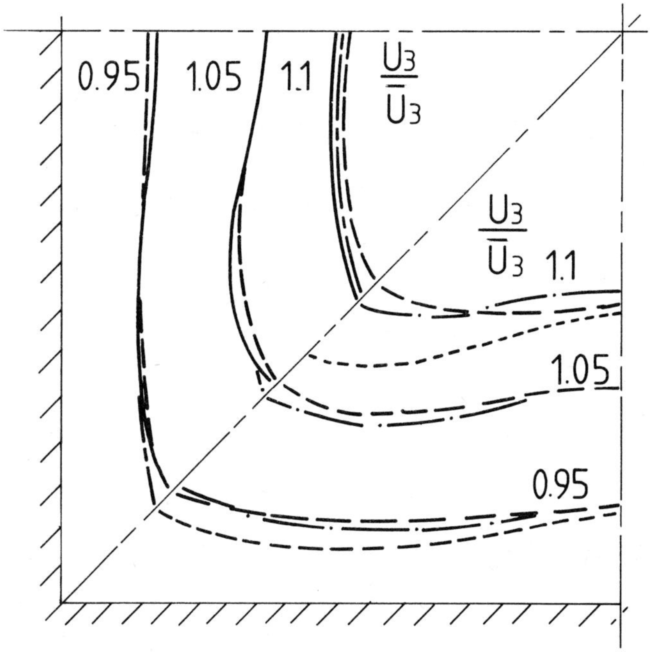

FIGURE 11 Axial velocity contours in a square duct with Re = 2.15 x 10^5

— — — — — experiment, Launder & Ying (1972)
—— . —— prediction, Launder & Ying (1973)
--------- prediction, Naot et al. (1974)
—— – —— prediction, Reece (1976)
————— prediction, Gosman & Rapley.

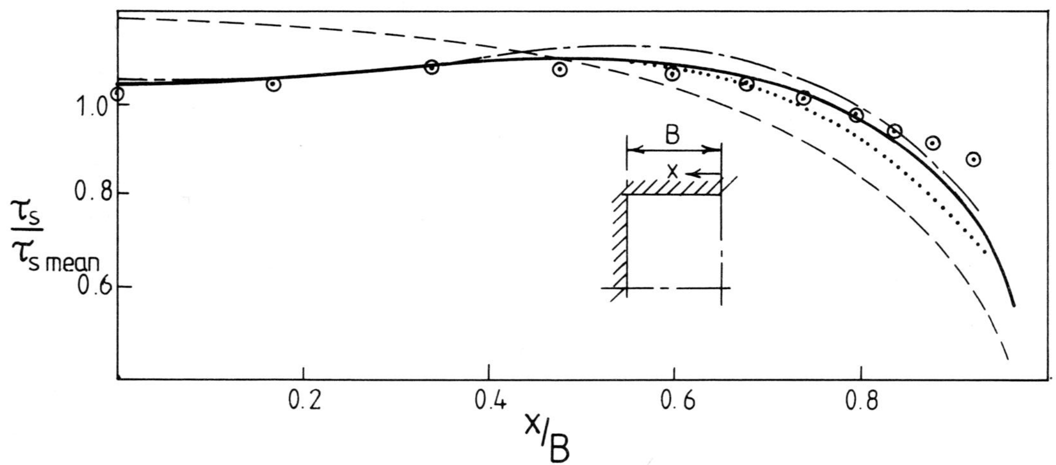

FIGURE 12 Wall shear stress profiles in a square duct with Re = 3.4 x 10^4

- ⊙ experiment, Leutheusser (1963)
- — – — predicted, Launder & Ying (1973)
- — · — predicted, Reece (1976)
- ——— predicted, Gosman & Rapley
- — – — predicted (no cross-flow), Gosman & Rapley

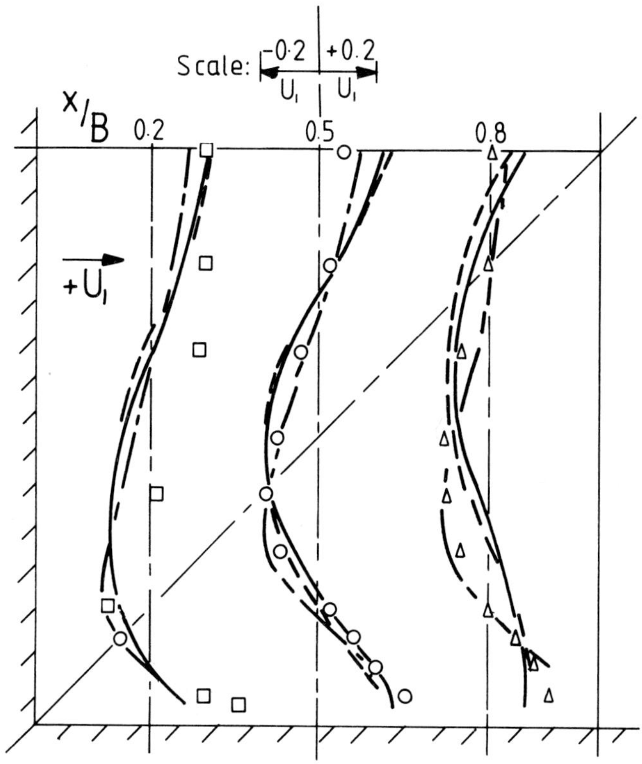

FIGURE 13 Secondary velocity profiles (U_1/U_3^*) in a square duct Re = 2.15 x 10^5

□ ○ △ Experiment, Launder & Ying (1972)

– – – – predicted, Launder and Ying (1973)

—— – —— predicted, Reece (1976)

———— predicted, Gosman & Rapley

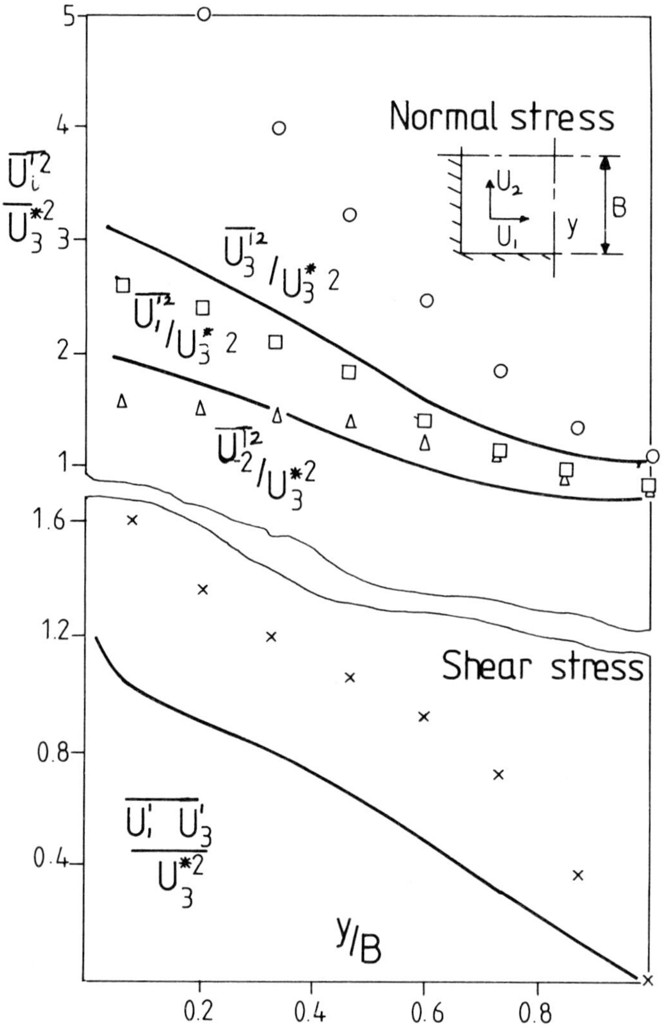

FIGURE 14 Reynolds stress profiles in a square duct Re = 8.3 x 10^4

○ $\overline{U_3'}$, □ $\overline{U_1'^2}$, △ $\overline{U_2'^2}$, × $\overline{U_1'U_3'}$ experiment

Brundrett and Baines (1964)

——— prediction, Gosman and Rapley

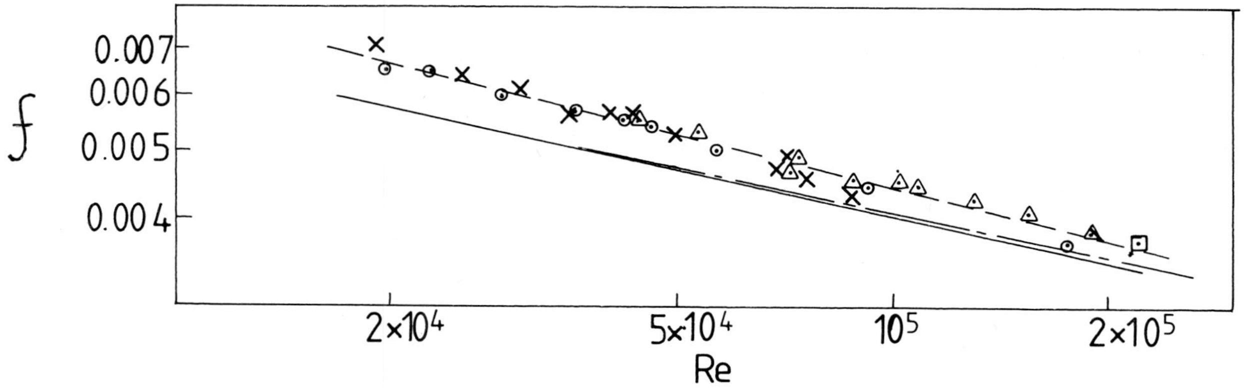

FIGURE 15 Friction factor characteristics for a square duct.

Experiment:

- ⊙ Hartnett et al. (1962)
- × Leutheusser (1963)
- △ Thomas & Easter (1972)
- ⊡ Launder & Ying (1973)
- – – – – Blasius equation

Prediction:

- —— – —— Launder & Ying (1973)
- ———— Gosman & Rapley

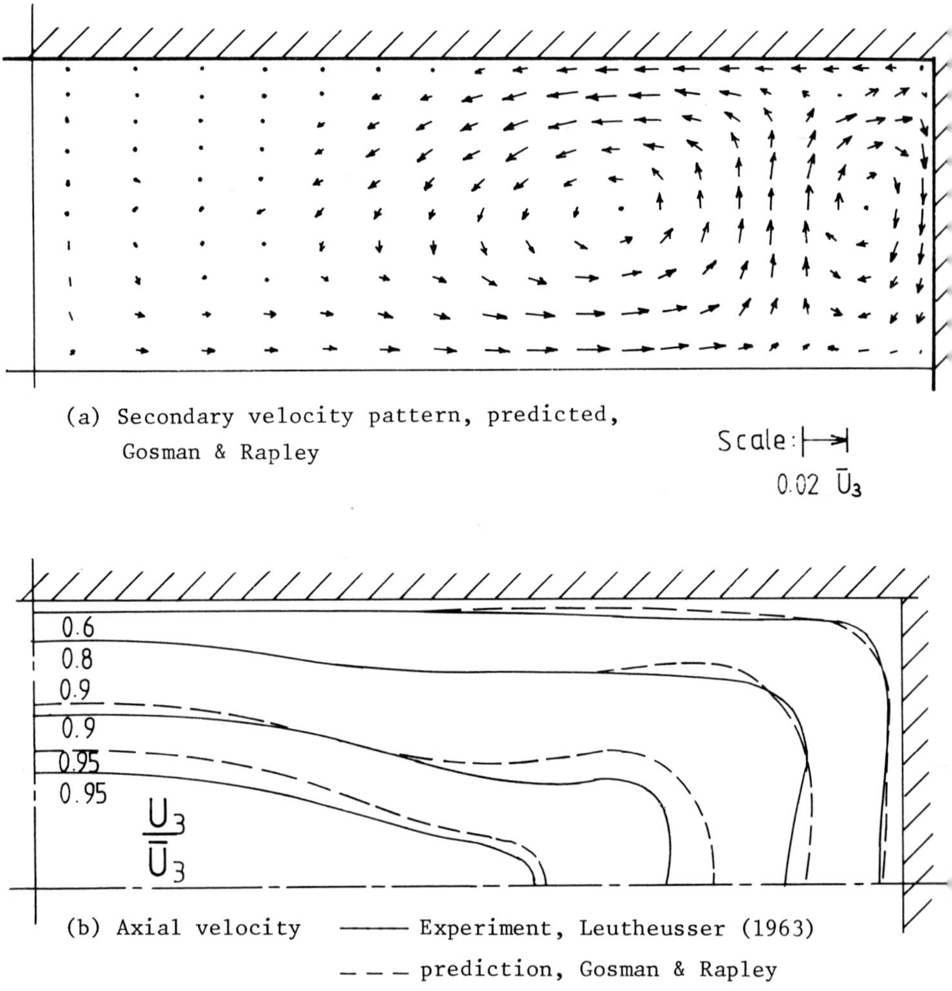

(a) Secondary velocity pattern, predicted, Gosman & Rapley

(b) Axial velocity

FIGURE 16 Velocity fields in a rectangular duct of aspect ratio 3

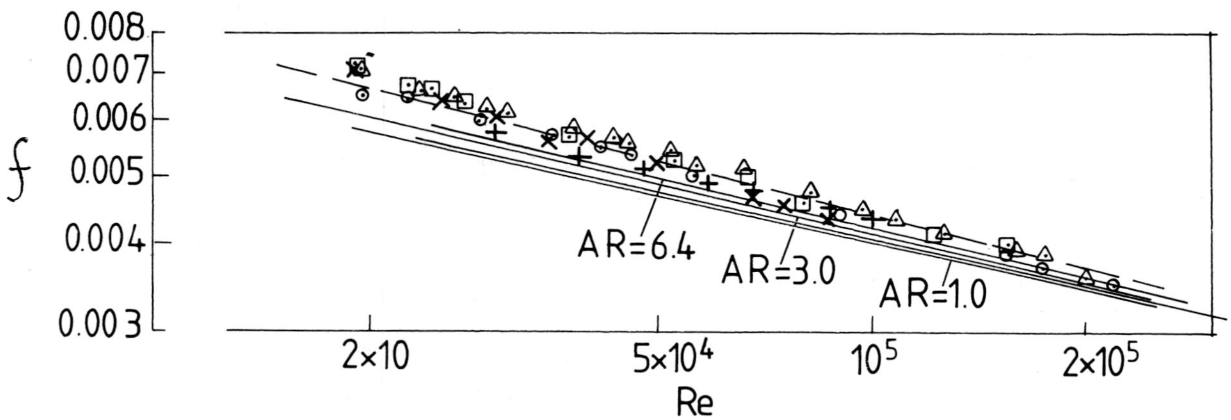

FIGURE 17 Friction factor characteristics for rectangular ducts

Experiment:

Aspect	Ref.
⊙ 1	Hartnett et al. (1962)
△ 5	
⊡ 10	
× 1	Leutheusser (1963)
+ 3	

– – – – Blasius equation

Prediction:

——— Gosman & Rapley

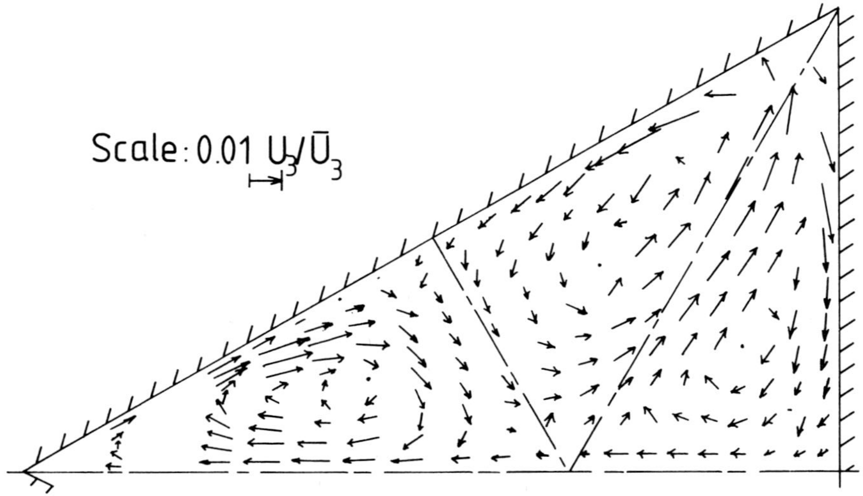

FIGURE 18 Secondary flow patterns in an equilateral triangular duct Re = 5.3×10^4

Predicted, Gosman & Rapley

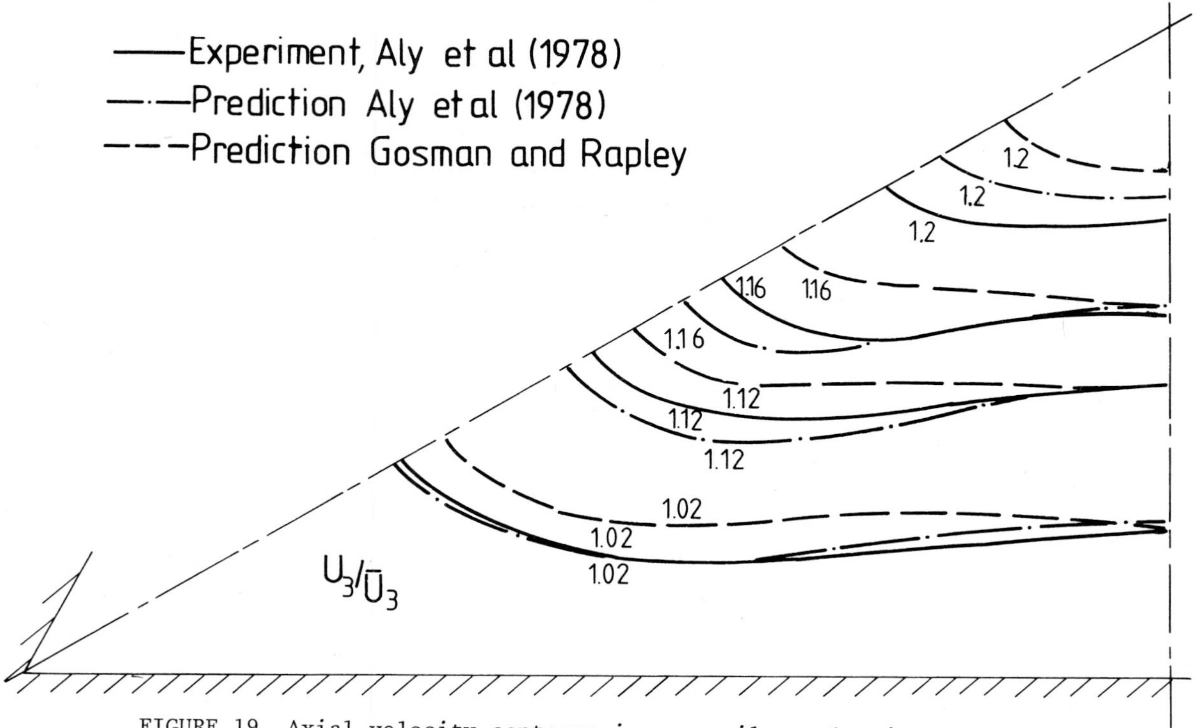

FIGURE 19 Axial velocity contours in an equilateral-triangular duct Re = 5.3 x 10^4

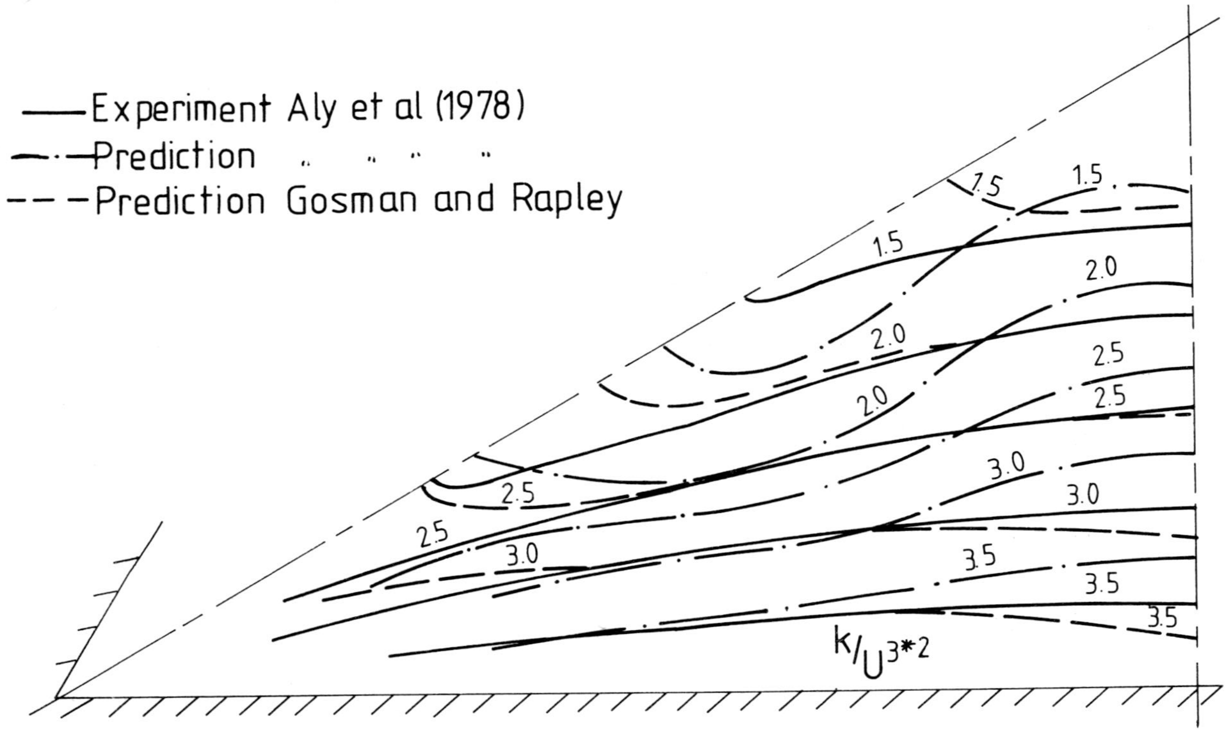

FIGURE 20 Turbulence energy contours in an equilateral-triangular duct Re = 5.3 x 10^4

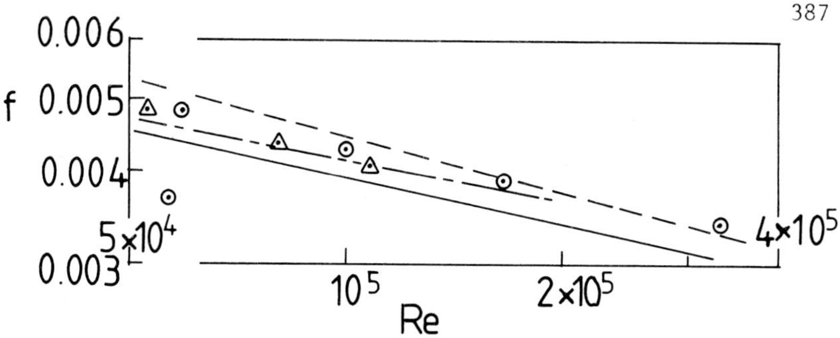

FIGURE 21 Friction factor characteristics for an equilateral triangular duct

⊙	experiment, Lowdermilk (1954)
△	experiment, Aly et al. (1978)
— — — —	Blasius equation
——·——	prediction, Aly et al. (1978)
————	prediction, Gosman & Rapley

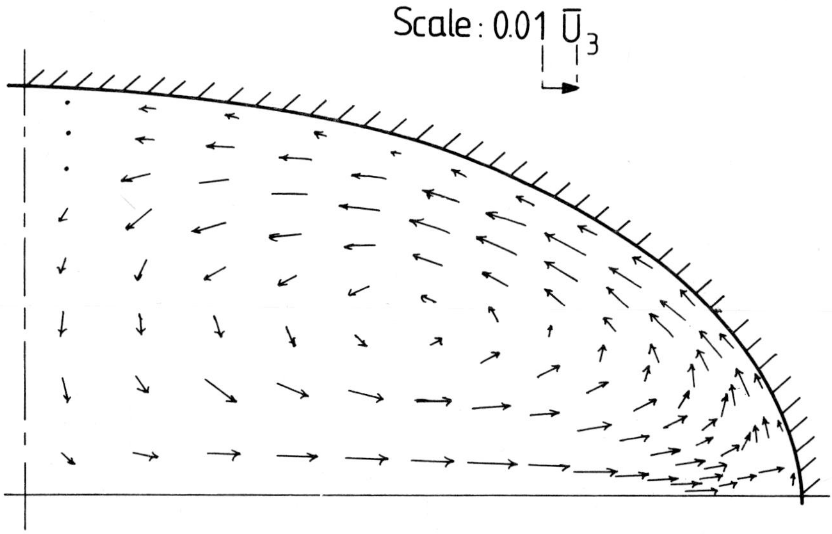

FIGURE 22 Secondary flow pattern in an elliptical duct of aspect ratio 2.0, Re = 2.15 x 10^5

Predicted, Gosman & Rapley

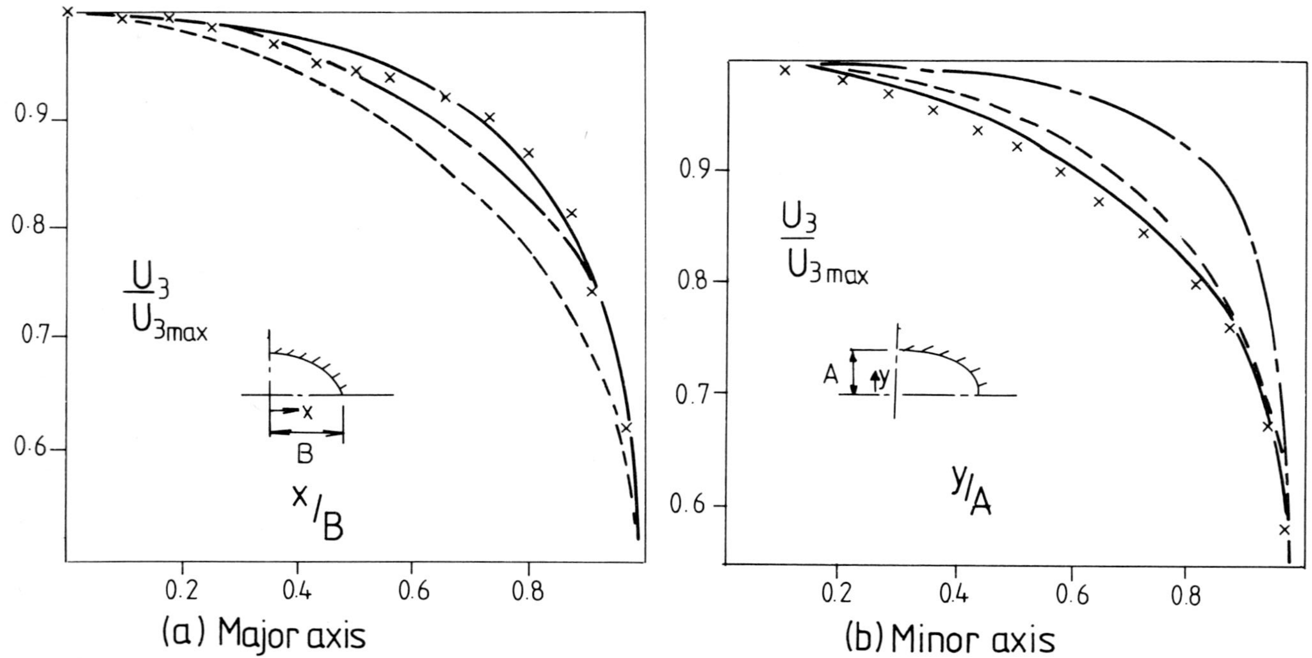

FIGURE 23 Axial velocity profiles in an elliptical duct of aspect ratio 2.0 Re = 2.15×10^5

x experiment, Cain & Duffy (1971); —— – —— prediction, Cain et al. (1972);

——— prediction, Gosman & Rapley; – – – – – prediction (no cross-flow), Gosman & Rapley

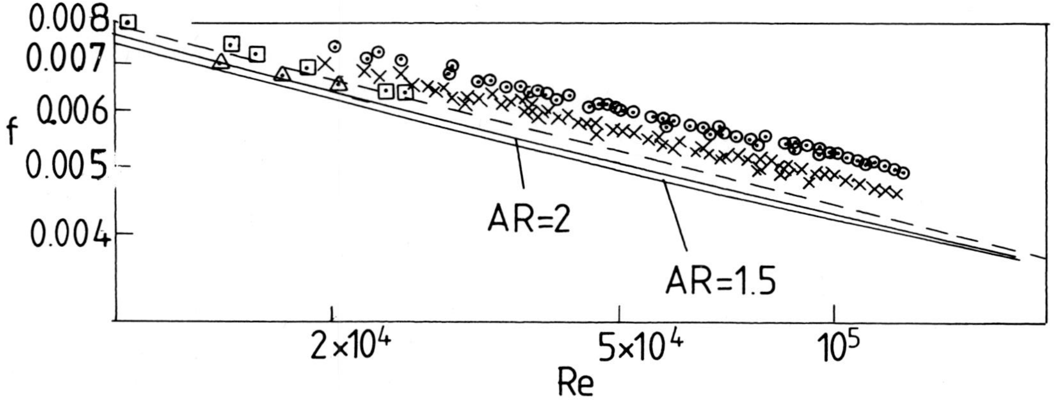

FIGURE 24 Friction factor characteristics for elliptical ducts

Experiment:

Aspect

× 1.5 } Cain & Duffy (1971)
⊙ 2.0

△ 2.41 } Barrow & Roberts (1970)
⊡ 3.16

Prediction:

——— Gosman & Rapley

– – – – Blasius equation

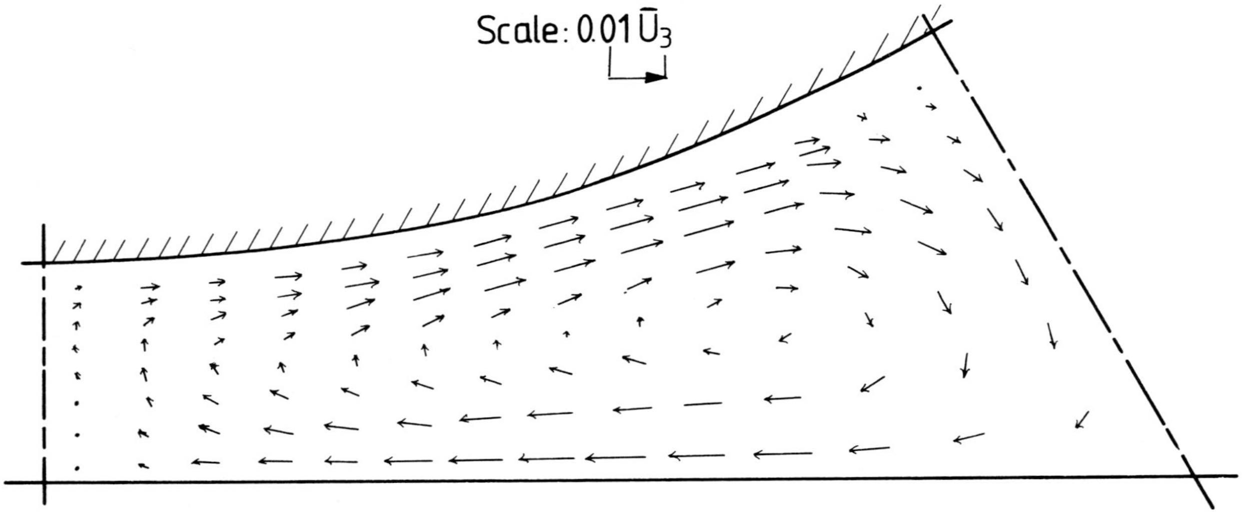

FIGURE 25 Secondary flow pattern in a triangular tube assembly of 1.123 pitch to diameter ratio, Re = 2.7×10^4

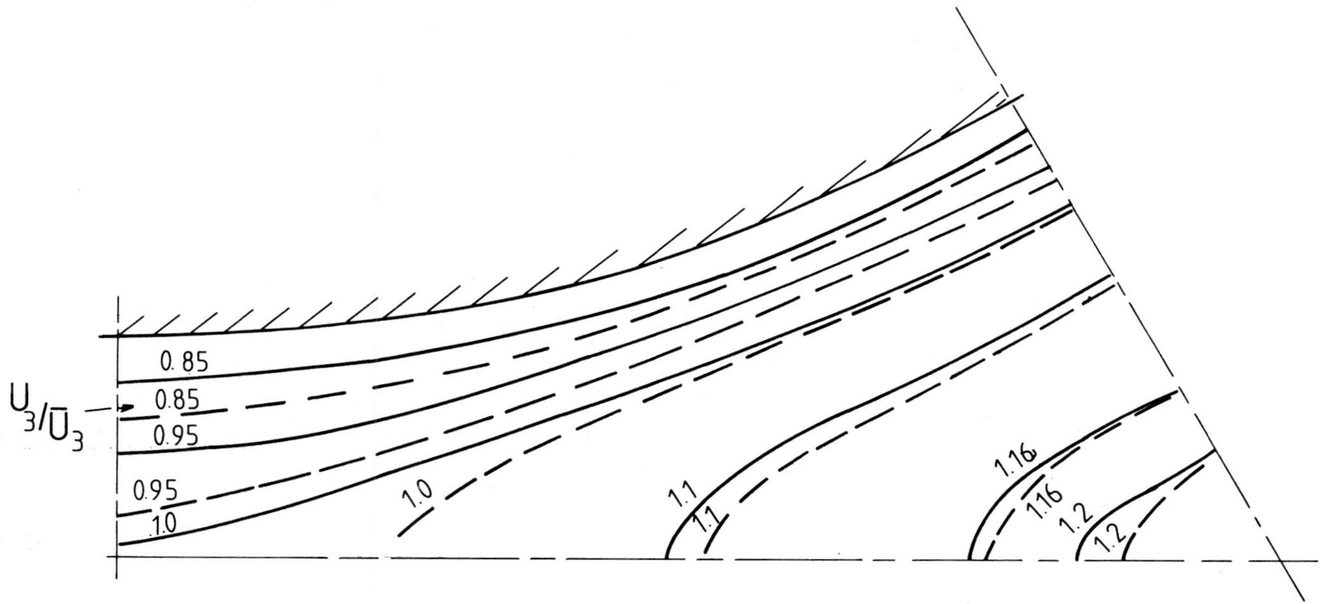

FIGURE 26 Axial velocity contours in a triangular tube-assembly of 1.123 pitch to diameter ratio. Re = 2.7 x 10^4

——— Experiment, Carajilescov and Todreas (1975)

– – – Prediction, Gosman and Rapley

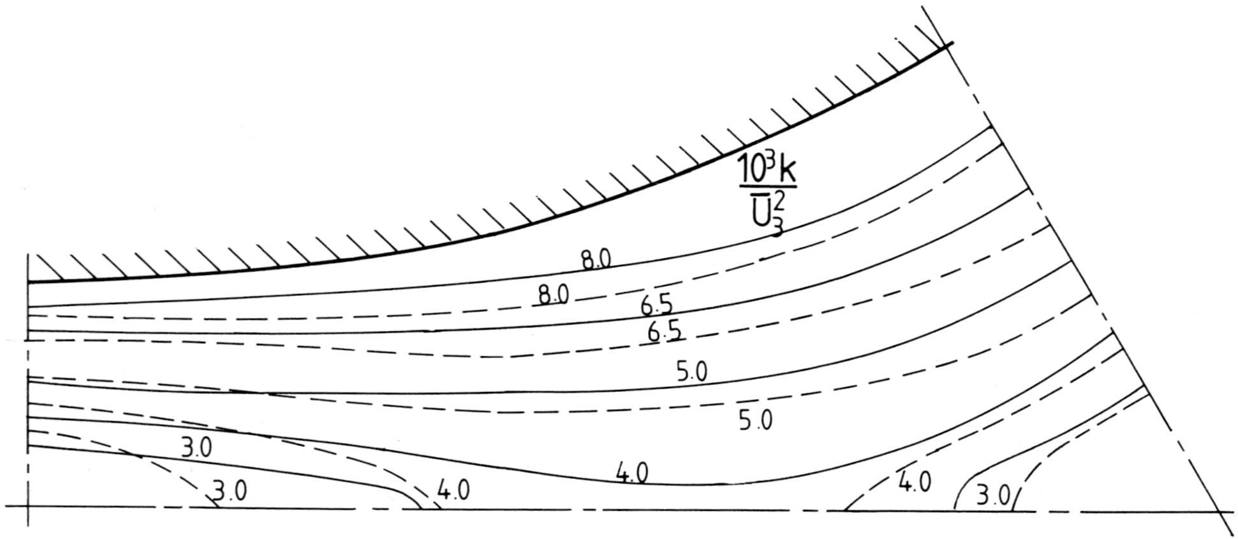

FIGURE 27 Turbulence energy contours in a triangular tube-assembly of 1.123 pitch to diameter ratio, Re = 2.7 x 10^4

——— Experiment, Carajilescov and Todreas (1975)

– – – Prediction, Gosman & Rapley

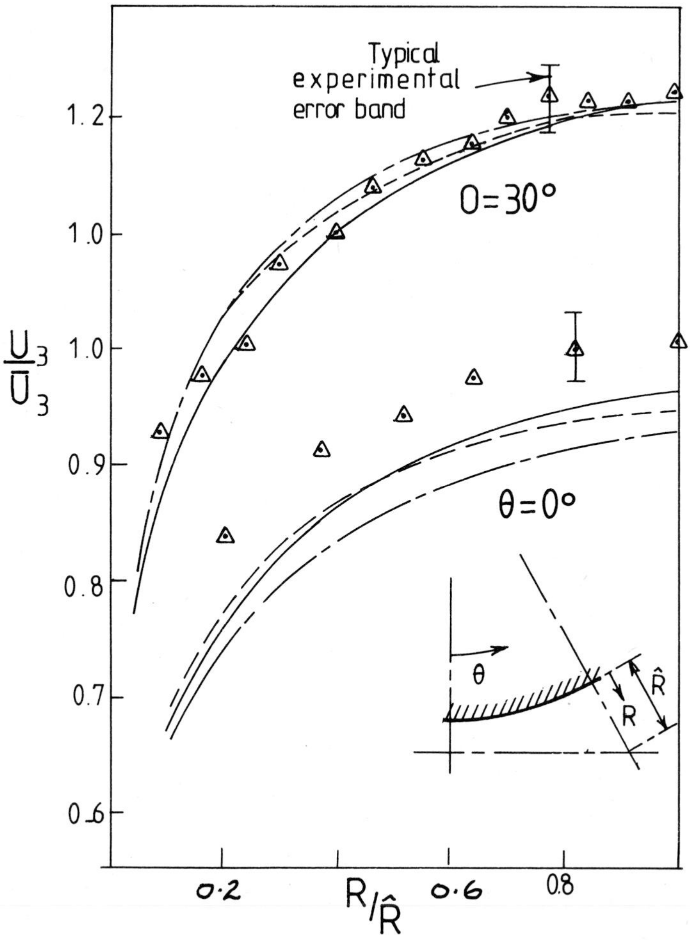

FIGURE 28 Axial velocity profiles in a triangular tube-assembly of 1.123 pitch to diameter ratio, Re = 2.7 x 10^4

△ experiment, Carajilescov and Todreas (1975)

– – – – – prediction, Carajilescov and Todreas (1975)

—— —— prediction, Trupp and Aly (1978)

———— prediction, Gosman & Rapley

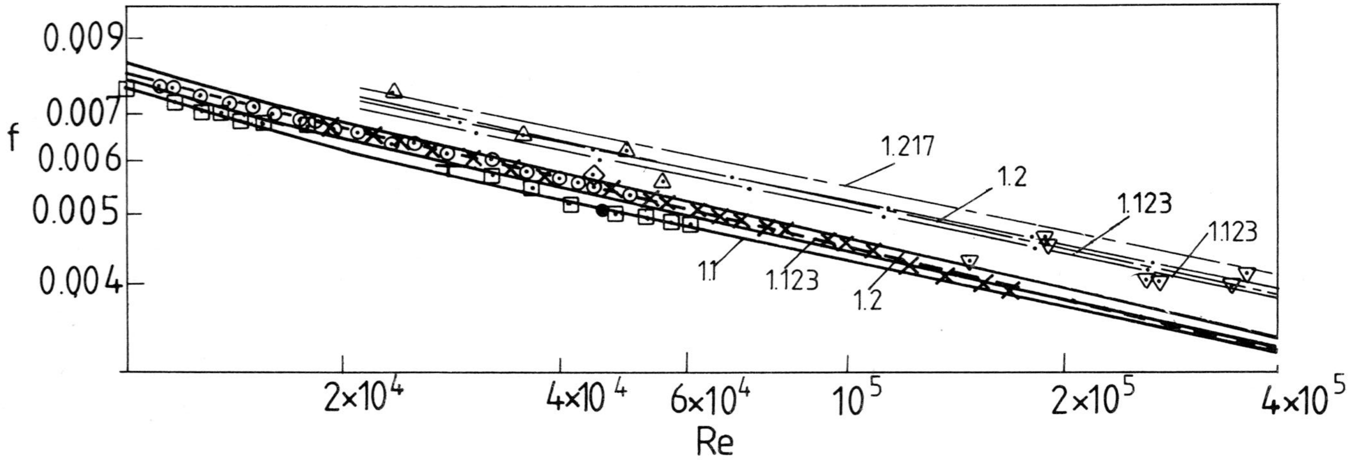

FIGURE 29 Friction factor characteristics for triangular tube-assemblies.

Experiment

	P/D ratio	Ref.
⊡	1.084	Rehme (1972)
⊙	1.125	Rehme (1972)
×	1.232	Rehme (1972)
●	1.1	Subbotin et al. (1971)
◇	1.2	Subbotin et al. (1971)
▽	1.217	Kjellstrom (1971)
△	1.2	Trupp & Azad (1975)
+	1.123	Carajilescov & Todreas (1975)
— — —		Blasius equation

Prediction:

— — — Carajilescov & Todreas (1975)

— · — Trupp & Aly (1978)

——— Gosman & Rapley

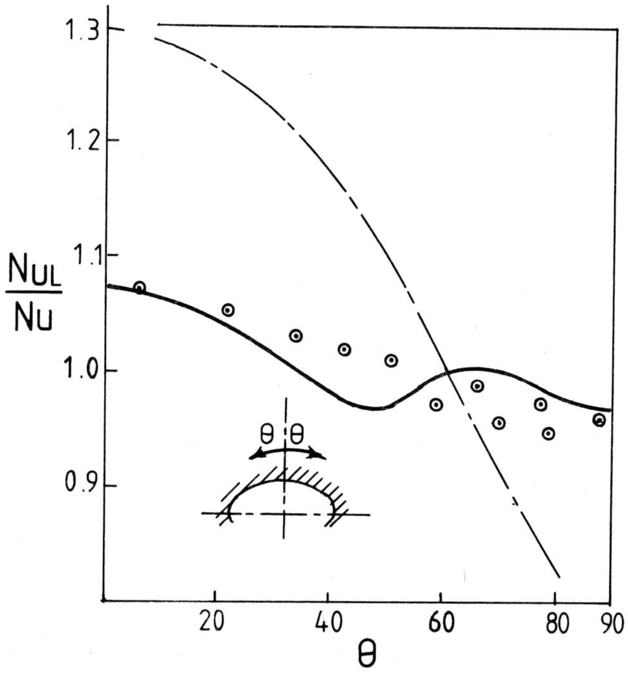

FIGURE 30 Local Nusselt number variation for air flow in an elliptical duct of aspect ratio 2, Re = 6.7 x 10^4

Pr = 0.7

⊙ experiment, Cain et al. (1973)

——— prediction, Gosman & Rapley

—·— prediction (no cross-flow), Gosman & Rapley

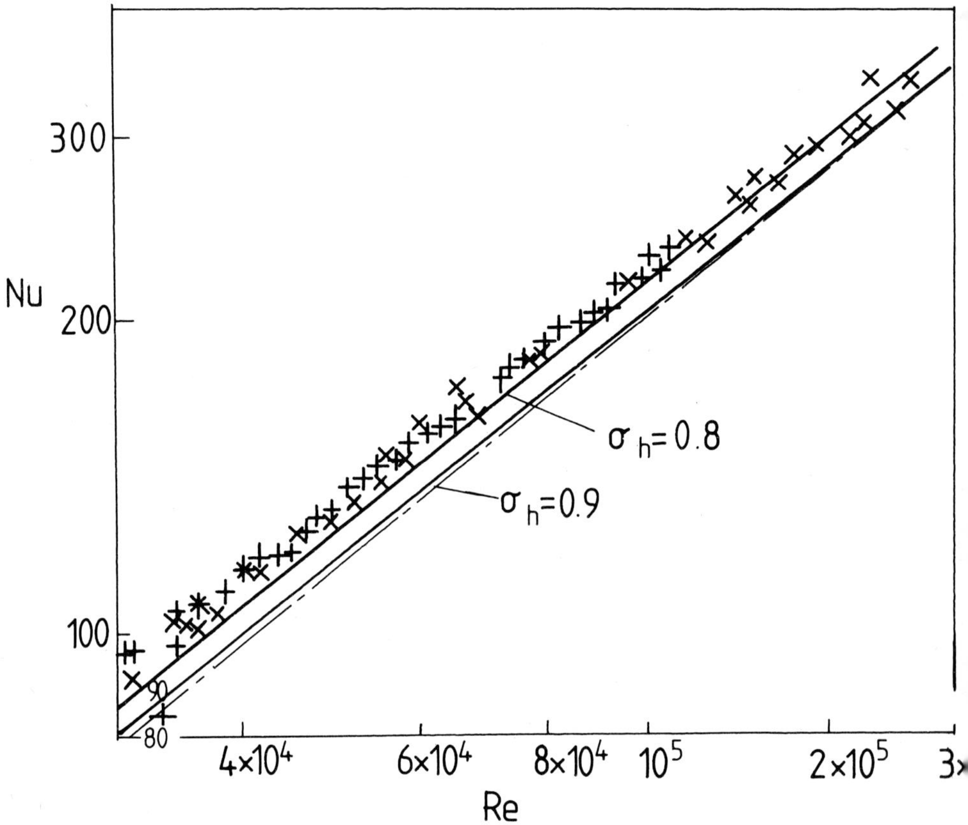

FIGURE 31 Average Nusselt number characteristics for air flow in elliptical ducts of aspect ratios 1.5 and 2

x Aspect 2 — Experiment, Cain et al. (1973)
+ Aspect 1.5

—— • —— Kays (1966) circular tube correlation ($0.022\ Re^{.8}\ Pr^{.6}$)

———— Prediction, Gosman and Rapley

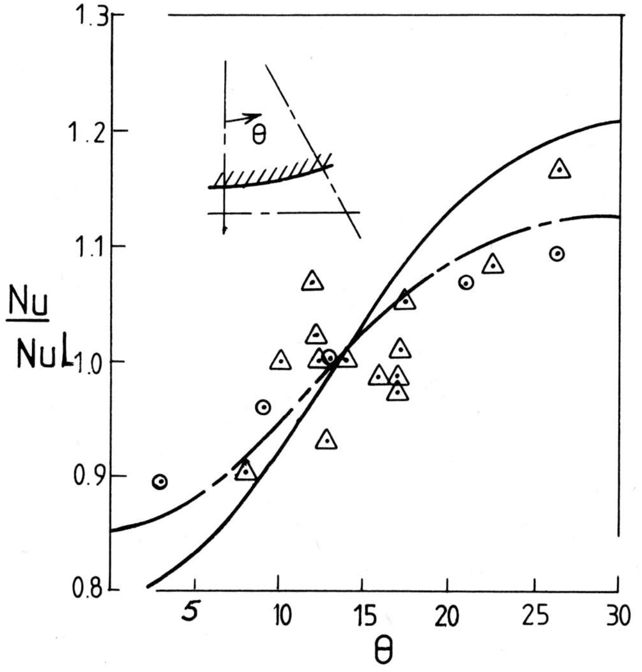

FIGURE 32

Local Nusselt number variations for water and air in triangular tube assemblies of 1.1 and 1.12 pitch to diameter ratio.

Experiment:

⊙ pitch/dia. = 1.1, air flow, Redman et al. (1966) $Re=2.2x10^5$

△ pitch/dia. = 1.12,water flow, Dingee et al.(1955) $Re=5x10^4$

Prediction:

——— pitch/dia. = 1.1, Pr = .7, $Re=2.2x10^5$, Gosman & Rapley

—·— pitch/dia. = 1.12, Pr = 1.75, $Re=5x10^4$, Gosman & Rapley

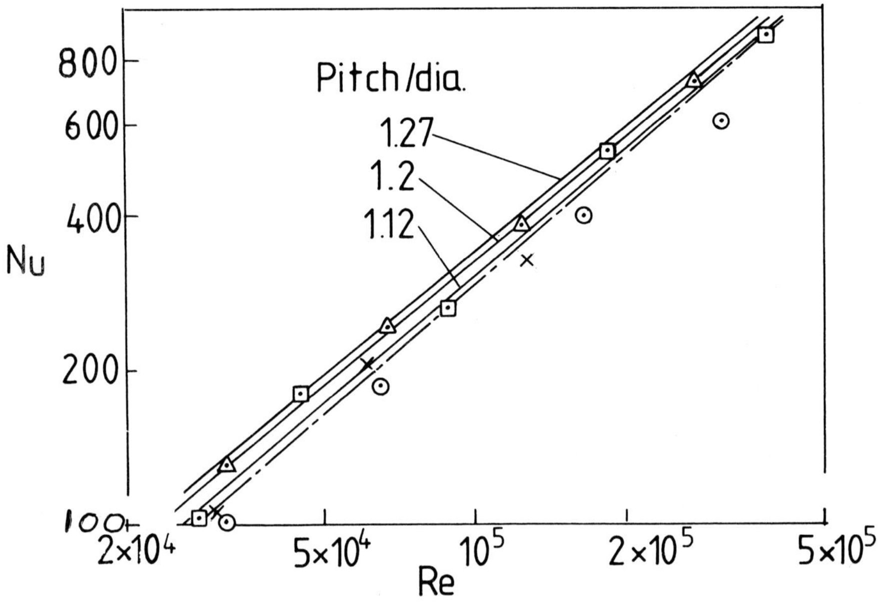

FIGURE 33

Average Nusselt number characteristics for water flow in triangular tube arrays

	pitch/dia	
x	1.12	experiment, Dingee et al. (1955)
⊙	1.2	
△	1.27	
⊡	1.2 (re-test)	

—— • —— Kays(1966) circular tube correlation ($0.0155\ Re^{.83}\ Pr^{.5}$)

———— Prediction, Gosman & Rapley

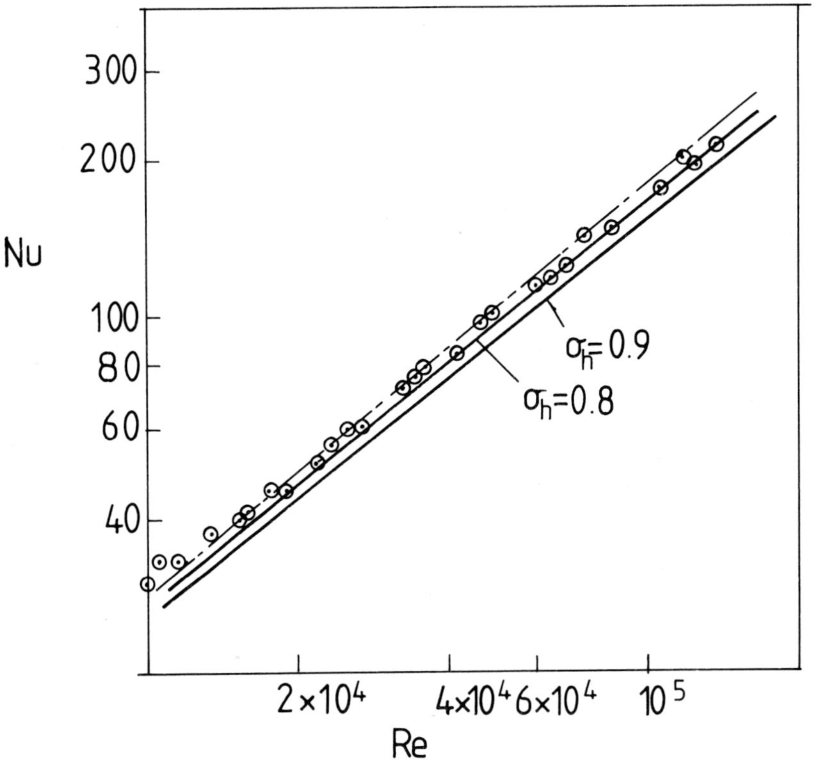

FIGURE 34

Average Nusselt number characteristics for air flow in an equilateral triangular duct.

⊙ experiment, Lowdermilk et al. (1954)

— · — Kays (1966) circular tube correlation ($0.022\ Re^{.8}\ Pr^{.6}$)

——— Prediction, Gosman & Rapley

SUBJECT INDEX